Contesting Megaprojects

Gerardo del Cerro Santamaría

Contesting Megaprojects

Complex Impacts, Urban Disruption and the Quest for Sustainability

Springer

Gerardo del Cerro Santamaría
United States Fulbright Fellow
New York Academy of Sciences
New York City, NY, USA

ISBN 978-3-031-98381-8 ISBN 978-3-031-98382-5 (eBook)
https://doi.org/10.1007/978-3-031-98382-5

Cover Source: iStock.com/cmart7327

This Springer imprint is published by the registered company Springer Nature Switzerland AG
The registered company address is: Gewerbestrasse 11, 6330 Cham, Switzerland

Advance Praise for *Contesting Megaprojects*

In a field that is overly focused on the cost and schedule overruns of big civil engineering projects, Gerardo del Cerro Santamaría's new book, *Contesting Megaprojects: Complex Impacts, Urban Disruption and the Quest for Sustainability*, provides a much-needed corrective by exploring the societal, environmental, and political-economic implications of very-large-scale urban development projects. Del Cerro Santamaría's use of a grounded theory approach to explore the relationships of megaprojects to their place-based contexts is especially welcome, as is his choice of example projects in cities as different as Istanbul, Shanghai, Sydney, and Toronto. For scholars interested in how urban megaprojects fit into contemporary city-making practices, *Contesting Megaprojects* may be the most essential read since Altshuler and Luberoff's 2003 book detailing the politics behind Boston's Big Dig project.

—John Landis, Professor Emeritus of City and Regional Planning, Weitzman School of Design, University of Pennsylvania, and author of *Megaprojects for Megacities*

Contesting Megaprojects is a deeply researched and analytically thoughtful assessment of the disruptions caused by urban megaprojects and what can be done in response and, as such, a much-needed corrective to the global fascination with large construction projects. This book is another impressive and thoughtful contribution from one of the most knowledgeable critics of urban megaprojects.

—Robert Beauregard, Emeritus Professor, Columbia University

With the masterful insight, deeply researched material on a range of widely discussed megaprojects in almost every region of the world, and a clear sense of complexity in both theory and practice, Gerardo del Cerro has sealed his reputation as a leading scholar of megaprojects. Beyond demonstrating historical trends in their funding, design, and implementation, this must-read book situates its focus on the disruptive impacts of megaproject development in the context of tensions between economic growth and environmental sustainability. With an eye to the ways that aspirations of economic competitiveness lead local and national authorities to build projects that harm both ecologies and societies, del Cerro sounds a warning bell that rings loud and clear. This book offers crucial first steps by revealing what went wrong in the past and why, thus laying the groundwork for how we might avoid destructive impacts in the future.

—Professor Diane Davis, Harvard University

Based on a remarkable amount of research in three continents, Gerardo del Cerro identifies cities as the "heat islands" that are mostly responsible for our sustainability crisis and indicts megaprojects as the main culprits. The book shows how the diverse and negative impacts of megaprojects are not compatible with any approach to urban sustainability that prioritizes socioeconomic equity and environmental preservation. This book is a must read.

—Mauro Guillén, best-selling author of *The Perennials* and Professor, Wharton School of the University of Pennsylvania

Contesting Megaprojects is an important contribution to the literature that demonstrates how large building projects aimed at increasing a city's competitiveness produce negative effects. Most important and original is its analysis of projects advertised as contributing to sustainability. Del Cerro shows how "green capitalism" is primarily symbolic with limited benefits to the larger population that continues to suffer from environmental degradation and widening inequality.

—Susan Fainstein, Senior Research Fellow, Harvard Graduate School of Design, and author of *The Just City*

Gerardo del Cerro's *Contesting Megaprojects* offers a refreshing and radical critique of megaprojects, which have become a controversial key feature of neoliberal urbanism. Megaprojects have routinely been criticized because of cost overruns, their spatial encroachment, or temporal uncertainties. Both their planning and implementation have raised more general concerns about the lack of transparency, accountability, or even some kind of democratic deficit. Del Cerro proposes that we should focus primarily on the disruptive impact of megaprojects as essentially unsustainable in nature. They are disruptive because of their sheer scale, and this *disruptive complexity* tears into the very texture that makes and sustains cities. If we adopt del Cerro's expanded understanding of urban sustainability, we may end up with fewer megaprojects but a far better and more livable urban environment. *Contesting Megaprojects* is a mandatory reading for urban and development professionals who want to promote a more sustainable and just future, as well as all citizens who are actively committed to participating in urban politics and want to have a say in the large-scale reconfiguration of urban space today.

—Professor Judit Bodnar, Central European University, Vienna

Although the global proliferation of urban megaprojects was initially driven by logics of economic competitiveness and city branding, they have increasingly been vaunted as opportunities for green growth. This book systematically dismantles the convenient fiction that megaprojects are advancing sustainable development as well as economic growth. Gerardo del Cerro Santamaría documents how often they lead to neither. His international examination of their multiscalar and long-term impacts reveals that even well-intentioned and faithfully implemented megaprojects are prone to gross underappreciation of their ecological disruptiveness. *Contesting Megaprojects* provides compelling and convincing explanation of why this approach to development is failing in both liberal democratic and authoritarian contexts and lays foundations for alternative ways forward.

—Professor Tim Bunnell, National University of Singapore

The question posed by this book is whether the impacts of megaprojects are compatible with the broader objectives of sustainability and socioeconomic equity. In this timely study, Gerardo del Cerro Santamaría examines the different political, social, and geographical contexts of megaprojects and their broader long-term impacts. It is based on a rigorous empirical study of projects from across the world that vary from the planned Istanbul Canal to the new World Trade Center in New York to city transport and greening projects and innovation districts. Megaprojects are deeply embedded in the thinking of the global capitalist system, and we are unlikely to be weaned off them soon. This book sounds an important note of caution to all policy planners. Megaprojects can bring benefits but only where carefully planned and well thought through. The problem highlighted in this well-argued and readable book is that this is often not the case. Even if these projects were to offer solutions, they are invariably socially disruptive and environmentally insensitive and have unforeseen side effects. In effect, the *engineering* becomes an end in itself, with the rights and voices of the citizen ignored. This book should be essential reading for economists, engineers and policymakers who believe that the simple solutions can be applied to complex problems.

—Professor Peter Bishop, Bartlett School of Architecture, University College London

In *Contesting Megaprojects*, del Cerro Santamaría critically discusses the disruptive and negative impacts of urban megaprojects on sustainability and a climate-adaptative built environment. The author argues that the negative impacts of urban megaprojects are an even larger problem than their notorious underperformance. Interestingly, he labels the cascading effects after the construction of megaprojects as *disruptive complexity*, which affects the life of a large number of citizens. I fully agree with del Cerro Santamaría's call for transdisciplinarity as the most robust analytical strategy for understanding the impact of urban megaprojects. He asks for *complex sustainability*, which is a socioeconomic, deep, non-anthropocentric, resilient, and transdisciplinary approach of such projects. This book is a critical and important contribution to the debate on the relation between urban megaprojects and city transformations and also to the debate on socioeconomic value creation through megaprojects.

—Prof.dr.ing. Alfons van Marrewijk, Delft University of Technology, The Netherlands

Contesting Megaprojects goes beyond mere rhetoric of mega-urbanisation to grapple with the messy complexities of unprecedented human agglomeration and envision what sustainability may look like under hyper-growth and varying scenarios of global dimension and consequence. From socio-politics to economics, technology and digitalization, to culture and environmental sustainability, this book enriches our understanding of megacities through a series of case studies across geographic regions. The book presents evidence and implications of how global economic flows and ambitious utopian development plans disrupt local contexts of everyday life in meaningful ways that portend both risk and opportunity for sustainable development. It examines both the historical context and current patterns of megacity development, and analyses how major challenges and opportunities may catalyse or inhibit the 'green' future of cites and indeed humankind. This book is an important resource for anyone interested in how globalisation, neoliberalism and global politics (including environmental politics) are impacting urban transformation at various scales.

—Professor Taibat Lawanson, University of Lagos, Nigeria

This book is dedicated to Professor Noam Chomsky, who has devoted his life to "simply helping people develop intellectual self-defense." With deep gratitude and admiration.

To the kind and considerate ones, for thinking about the other and for paying attention, which is the highest form of generosity.

To the city of New York and to New Yorkers anywhere.

The greatest thing above all things is to be a master of the metaphor. It's the only thing that one cannot learn from others; and it is a sign of genius and originality, because a good metaphor implies intuitive perception of similarities in dissimilar things.
—Aristotle, *Poetics* 1459a 8 (330 BCE)

One cannot suppress a certain indignation when one sees men's actions on the great world-stage and finds, beside the wisdom that appears here and there among individuals, everything in the large woven together from folly, childish vanity, even from childish malice and destructiveness. In the end, one does not know what to think of the human race, so conceited in its gifts.
—Immanuel Kant, *Idea for a Universal History from a Cosmopolitan Point of View* (1784)

My brain is only a receiver. In the Universe there is a core from which we obtain knowledge, strength and inspiration. I have not penetrated into the secrets of this core, but I know that it exists.
—Nikola Tesla, *My Inventions* (1919)

In the beginning was the Topos. Before—long before—the advent of the Logos, in the chiaroscuro realm of primitive life, lived experience already possessed its internal rationality; this experience was producing long before thought space, and spatial thought, began reproducing the projection, explosion, image and orientation of the body. [...] Nothing disappears completely ... In space, what came earlier continues to underpin what follows ... Pre-existing space underpins not only durable spatial arrangements, but also representational spaces and their attendant imagery and mythic narratives.
—Henri Lefebvre, *The Production of Space* (1974)

Entropy is the price of structure. [...] Whatever we call reality, it is revealed to us only through the active construction in which we participate. [...] The incompatibility between the ascetic beauty sought after by science, on the one hand, and the petty swirl of worldly experience so keenly felt by Einstein, on the other, is likely to be reinforced by another incompatibility, this one openly Manichean, between science and society, or, more precisely, between free human creativity and political power. In this case, it is not in an isolated community or in a temple that research would have to be carried out, but in a fortress, or else in a madhouse.
—Ilya Prigogine, *Order Out of Chaos: Man's New Dialogue with Nature* (1984)

The piece at the beginning is a vast ocean; by the end the ocean has receded to a single drop of living water. The piece ends exactly at the same point in space and time where it began. Since these are two-sided mirrors, for a total of four sides, the result is the reflection of the whole piece inside and an eternity from outside. It's infinity infinity. Put a different way, the sum total of this music is zero. And zero is eternity.
—Pianist Vladimir Feltsman on Johann Sebastian Bach's Goldberg Variationen—Aria mit Verschiedenen Veränderungen für Klavicembal mit Zwei Manualen (1991)

There is no tidy atom and no embracing world, only complex situations, long strings reptating as in a polymer goo, or in a mineral before it hardens.
—Harrison White, *Identity and Control* (1992)

El sabio quiere mantener la mente totalmente abierta, para aprehender la realidad tal como se presenta, asi, por si misma, e intentar captarla como un sonido emitido. No se trata por tanto de conocer definiendo los objetos, sino de tomar consciencia del fondo de inmanencia que dispensa lo evidente: aquello que, precisamente por tenerlo siempre ante los ojos, no vemos, no alcanzamos a ver.
—François Jullien, *Un sabio no tiene ideas* (2001)

Science is an alliance of free spirits in all cultures rebelling against the local tyranny that each culture imposes on its children.
—Freeman Dyson, *The Scientist as Rebel* (2006)

Atoms are musical instruments, and the light they emit makes their tones visible … Matter is built up from atoms that are, in a rich and precise sense, tiny musical instruments … From the heart of matter, music.
—Frank Wilczek, *A Beautiful Question* (2016)

Preface and Acknowledgments

My scholarly interest in urban megaprojects began many years ago with my research on the city of Bilbao. I presented my work at the 2006 ISA congress in South Africa, and I later published my book *Bilbao: Basque Pathways to Globalization*. One of the lessons of my research about Bilbao was that urban megaprojects (UMPs) cannot be understood in all their complexity from the perspective of a single discipline, not even from the broader angle of the social sciences. Another lesson learned during my research was that the ambitions of local elites to redevelop cities via urban megaprojects often meet severe resistance from civil society. These two ideas are among the drivers of the book you have in your hands.

I continued working on UMPs after my research on Bilbao ended. In 2008, I taught a graduate seminar at the Massachusetts Institute of Technology, on "Cities, Regional Restructuring and Globalization," where I devoted several sessions to discussing urban megaprojects and benefited from feedback by my MIT and Harvard students. In 2009, I participated in the Arrested Development conference at the Cooper Union in New York City and gained insights allowing me to better understand the fate of UMPs under constrained economic conditions. My graduate class in New York, "Urban Projects and the Architectures of Globalization," attempted to reconcile research on UMPs with social-scientific concerns about globalization, development discourses, and other large social forces. These new experiences and projects coalesced in an edited volume, *Urban Megaprojects: A Worldwide View*, published in 2013, with outstanding contributions from scholars from the United States, Europe, and Asia, which reflected an increasing interest worldwide in the megaproject phenomenon.

In the past ten years I have shifted my focus on urbanism and megaprojects to include concerns about sustainability, innovation, complexity, and the Earth system, which is reflected in some of my publications during this period. Some of these publications have informed the shaping of some sections in this book, and they are referenced in the relevant chapters. Chapter 2 uses material from my piece "Megaprojects, Development and Competitiveness: Building the Infrastructure for Globalization and Neoliberalism," published in the *Athens Journal of Social Sciences*. The chapter also includes some sections from the article "Complex

Resilience and Entropic Risks in Urban Megaprojects," published in the *Transdisciplinary Journal of Engineering & Science*. Some sections on Hong Kong in Chapter 3 were published in "Disruptive and Contentious Enterprises: Megaprojects in Bilbao, Istanbul and Hong Kong," in the proceedings of the XIII International Conference on Virtual City and Territory that took place in Barcelona in 2019. The first part of Chapter 4 is an updated version of my article "The Fading Away of the Bilbao Effect," published in the *Athens Journal of Architecture*. In Chapter 5, the section on innovation districts is an updated version of my article "Innodistricts, Complex Sustainability and Urban Redevelopment: Evaluating Multiple Success Factors," in the *International Journal of Economics, Business and Management Research*. Some material in Chapter 6 on New York, Paris, London, and Tokyo comes from my book chapter "Megaprojects, Ecology and the Environment: Dimensions of Sustainability in Global Cities," in F. Trapani et al.'s (eds.) *Advanced Studies in Efficient Environmental Design and City Planning*. Lastly, some sections in Chapter 7 have been adapted from two of my recent publications: "Complex Resilience and Entropic Risks in Urban Megaprojects" (mentioned above) and "Possibilities and Challenges for Intercultural Research in Global Urbanism," a book chapter in *Examining Cultural Perspectives in a Globalized World*, edited by Richard Brunet-Thornton. In spite of these adaptations from published material, the book's new material represents about 85 percent of its content.

Some professional experiences during the past ten years have had an influence on the crafting of the project and the final shape of the book. I was a *Visiting Scholar* at Columbia University in 2014–2015 and devoted my time at Columbia's Graduate School of Architecture, Planning and Preservation (GSAPP) to studying sustainability issues and debates in the context of built environment research. I applied some of my learning in 2017, when I was a Guest Speaker at a class taught by Professor Lily Song on "Urban Infrastructure, Environment, and Sustainability" at Harvard Graduate School of Design. Regarding megaproject impacts, a key notion in this book, I co-taught a class at the Cooper Union in New York, "Urban Megaprojects and Environmental Impacts," with Professor Constantine Yapijakis, an environmental engineer. The class helped me draft and test some initial ideas about impacts that are reflected in this book.

My research stay at LSE Cities in Spring 2023 allowed me to do fieldwork in London and analyze the relationships between innovation districts and sustainability (in King's Cross, Tech City, and Here East), focusing on the governance of sustainable innovation. My engagement with CIRET (Centre International de Recherches et études Transdisciplinaires, in Paris), my learning from the Santa Fe Institute, and my publications in the *Transdisciplinary Journal of Engineering & Science* have contributed a great deal to flesh out my approach to complexity and transdisciplinarity. My fieldwork and observations in 11 of the 16 cities (12 of the 18 megaprojects) discussed—Sydney, Bilbao, Denver, Helsinki, Barcelona, Boston, London, Toronto, Tokyo, Paris, and New York City—constitute a key element in this book's foundations. I am very grateful to my informants in these cities.

The manuscript has also been enriched thanks to my interactions with several colleagues. Evinç Dogan and Tuna Kuyucu (Boğaziçi University), as well as

Dikman Bezmez (Koç University), provided input on Istanbul that I used in Chapter 3. Besides my visit to Sydney in 2016, Mike Harris (University of New South Wales) offered suggestions on key literature about the city, and Agnes S M Ku (Hong Kong University of Science and Technology) gave helpful insights on Hong Kong, both in Chapter 3. Visits to both Helsinki and Denver helped shape Chapter 4. In Denver, my reference has been the work by both Georgia Lindsay and Penny Victoria Snyder. My conversations with Juulia Kauste, the former Director of the Finnish Museum of Architecture, helped me understand the urban context in Helsinki and some of the tacit complexities of the failed Guggenheim project there.

My research on Bilbao, a city I first visited as a tween with my father, has been enriched over the years by the work of Sara González (University of Leeds), Joseba Zulaika (University of Nevada, Reno), as well as Marisol Esteban, Arantxa Rodríguez, and Beatriz Plaza (University of the Basque Country), among others. Sharon Zukin from CUNY read an early version of Chapter 5 and provided very helpful suggestions. Regarding Chapter 5, I am also grateful to Eva Kassens-Noor and the anonymous reviewers of the *Journal of Mega Infrastructure & Sustainable Development*.

Regarding Chapter 6, I am indebted to some publications by Yasser Elsheshtawy (Columbia University) and Xianming Chen (Trinity College), which contributed to my understanding of the context of Saudi Arabia and China, respectively. I learned about Tokyo from my field observations and, particularly, from some Japanese colleagues, including my friend Hiro and his wife Setsuko, during my visit in 2014 to participate in the ISA Congress in Yokohama. Hiro and Setsuko hosted me in Osaka with grace and generosity. During my 2023 research stay in London, I held very helpful conversations about London's development with Harry Dimitriou, Claudio de Magalhaes, Max Nathan, and Peter Bishop (University College London), as well as Ricky Burdett, Ben Rogers, and Nuno Ferreira da Cruz from LSE Cities. While in London, I also had the fortune to hold an exchange with Xuemei Bai (Australian National University) on complexity and cities that helped me clarify some ideas. I learned about Paris from the work by Patrick Le Galès and his colleagues at Sciences Po, which aided me to place my field observations in context. In New York City, my hometown for the past decades, my sources of learning among colleagues are too many to list here. I did several rounds of fieldwork at the New World Trade Center starting in 2012 and up to 2021.

I am particularly grateful and indebted to Diane Davis (Harvard University) and Robert Beauregard (Columbia University). Both provided feedback about the project and also about some sections of the manuscript. Most importantly, both have been supportive of my career over the years. I would also like to mention two widely respected scholars who had an influence on my thinking during my early years as a scholar: the late Janet Abu-Lughod and Manuel Castells. Janet was my dissertation advisor at the New School for Social Research, and I took classes with Manuel in Madrid before I moved to New York City. I still try to learn from their scholarship today.

Thanks to the work by the anonymous reviewers of both the book proposal and the manuscript, this book is in better form than it would have been without their participation. Katy Harris and Ava Shopland from Foster + Partners provided the image of Masdar City included in the book. Joana Lazarova from Moreau Kusunoki Architectes provided the image of the projected Guggenheim Helsinki. Sarah Banks and Maryam Umar took care of editing and formatting matters with utmost professionalism, and they deserve to be mentioned here.

The staff at Springer Nature, New York, deserves gratitude for their support of my project and recognition for their professionalism, high-quality standards, and diligence. My publishing editor, Andrea Body-Barre, has been instrumental in keeping the project on track over the many months. She knows that I acknowledge and very much appreciate her kindness, world-class professionalism, and scholarly mindset. I'd also like to acknowledge the design and production teams, led by Tina Silva, and the copy editors at Springer Nature, New York, who are responsible for the fine quality of this book's published version.

As always, my family has provided love, support, and understanding.

New York City, NY,
United States of America
May 2025

Gerardo del Cerro Santamaría

Competing Interests The author has no competing interests to declare that are relevant to the content of this manuscript.

About the Book

Megaprojects are not merely transformative, but rather, they are inherently disruptive. They are incompatible with any approach to urban sustainability that prioritizes the well-being of urbanites and the boundaries of the Earth system. In fact, megaprojects undermine efforts to tackle the global impacts associated with both the destructive creation of capitalism and the new climate regime.

The reason for this state of affairs is not cost overruns and poor megaproject management, but the disruptive impacts that megaprojects have on urban areas. This book emphasizes the impacts imperative in megaproject planning: the importance of identifying disruptive impacts as a prerequisite for critically assessing whether megaprojects should be built.

Drawing on 18 megaprojects in 16 cities around the world (inter alia, Istanbul, Hong Kong, Abu Dhabi, Barcelona, Boston, New York and Shanghai), *Contesting Megaprojects* illustrates urban-regional trajectories over the past fifty years. These trajectories show the coexistence and entanglement of property development, culture, and iconic architecture, together with a more recent interest in science and technology, innovation, and sustainability as strategies to foster urban competitiveness.

Megaprojects cause massive disruptions everywhere, and the book offers a typology of megaproject impacts: socioeconomic, environmental, spatial, political and geopolitical, financial, cultural and systemic. As a response to this emerging complex web of interrelated megaproject impacts (*disruptive complexity*), the book develops the concept of *complex sustainability*, a transdisciplinary approach to urban sustainability, through socioeconomic sustainability and deep sustainability.

The critical outlook of this book contributes to fundamental debates around capitalist development: trickle-down urbanism, spectacularization, high-tech innovation and AI, planetary boundaries, global risks, resilience, the right to the city, geopolitics, democracy, prosperity, and power. *Contesting Megaprojects* can help urban scholars and analysts, practitioners, policymakers, and residents of cities better understand the complex impacts of urban megaprojects, their magnitude and consequences, and identify pathways of analysis and action toward a more sustainable and just future.

Contents

About the Author

Gerardo del Cerro Santamaría is a senior research scholar, professor, author, mentor, and international advisor in New York. His experience in research, teaching, mentoring, advising, executive leadership, and board service spans almost thirty years and includes prestigious institutions in the United States and abroad. He has taught at Massachusetts Institute of Technology, Skidmore College, and the Cooper Union for the Advancement of Science and Art in Manhattan, where he was Founding Director and Senior Executive Director of Strategic Planning and Innovation and Research Professor of Planning and Megaprojects for twenty-three years. He has also been a Visiting Scholar at Columbia University, London School of Economics, University College London, University of Illinois Urbana-Champaign, Centre for Policy Research (New Delhi, India), and the Tata Institute of Social Sciences (Mumbai, India) and a Guest Speaker at ETH Zürich and Harvard GSD. Del Cerro is a U.S. Fulbright fellow, a former Program Evaluation Director (Gateway Engineering Coalition) at the U.S. National Science Foundation, and a European Union Expert in Regional Policy (Regio Program). At the NSF, he was part of the effort to restructure STEM education in the United States. He serves as Senior Research Scholar and Mentor at Future Earth Urban Knowledge-Action Network, Earth System Governance Project, Global Land Program, and the New York Academy of Sciences. Gerardo has authored a large number of scientific publications on various aspects of urbanism and global capitalist development, megaprojects, iconic architecture, technology, innovation, complex sustainability, STEM education, neoliberalism in higher education, robotics, and mathematics of music. Some of his book-length publications, as author or co-author, are *Bilbao: Basque Pathways to Globalization* (2007); *Urban Megaprojects: A Worldwide View* (2013); *Technological Futures*, in Technology in Society (2015), *The Oxford Handbook of Megaproject Management* (2017); *Advanced Studies in Efficient Environmental Design and City Planning* (Springer Series on Advances in Science, Technology & Innovation, 2020); and *From Complex Systems to Transdisciplinarity: Exploring Resilience, Sustainability, and Knowledge Integration* (2024). He has received several professional recognitions, and his work has been featured in the media, in both the United States and Europe. He is a member of the American

Association for the Advancement of Science, the International Network for Engineering Studies, the Academy of International Business, the New York Chapter of the U.S. Fulbright Association, and the Centre International de Recherches et Études Transdisciplinaires (CIRET) in Paris. Del Cerro's background is in science, mathematical logic and philosophy of science, social sciences (political economy), and music. He holds a Ph.D. from the UAM in Madrid and a Ph.D. from the New School for Social Research in Manhattan, where he received a full-tuition scholarship and graduated with Honors. A music lover and classical piano player, Gerardo holds the degree of *Professor of Piano* from the Joaquín Turina Conservatory and actively participates in New York's musical life.

Abbreviations

ADB	Asian Development Bank
AI	Artificial Intelligence
AKP	Adalet ve Kalkinma Partisi (Justice and Development Party)
ANT	Actor-Network Theory
BCG	Boston Consulting Group
BDA	Barangaroo Delivery Authority
BDAC	Big Data & Analytics Center of Excellence
BDN	Blue Dot Network
BIM	Building Information Modeling
BRI	Belt and Road Initiative
CA	California
CACFC	Core Arts and Cultural Facilities Committee
CC	Consultative Committee
CCP	Chinese Communist Party
CCS	Carbon Capture Storage
CE	Common Era, same as AD (Anno Domini)
CG/LA Inc.	Creating Great Long-Term Assets, an infrastructure company
CIRET	Centre International de Recherches et études Transdisciplinaires, Paris
CKDs	Central Knowledge Districts
CNC	Computer Numerical Control
CNN	Cable News Network
CO	Colorado
COP 27	The 2022 Conference of the Parties to the United Nations Framework Convention on Climate Change
Covid-19	Coronavirus Disease
CT	Connecticut
CUNY	City University of New York
DAM	Denver Art Museum
DC	District of Columbia
DNA	Deoxyribonucleic Acid

eHighway	Electric Highway
ELISA	Elektrifizierter, Innovativer Schwerverkehr auf Autobahnen (Electrified, Innovative Road Freight Transport on Motorways)
EPAD	Établissement public pour l'aménagement de la région de la Défense (Public Establishment for Installation of La Défense, in Paris)
ESBs	Earth System Boundaries
ESCO	Energy Service Company
FANUC Robotics	Fuji Automatic Numerical Control Robotics
FAO	Food and Agriculture Organization
FDI	Foreign Direct Investment
FSIN	Food Security Information Network
G7	The Group of Seven
G20	The Group of Twenty
GA	Georgia
GBA	Greater Bay Area
GCLs	Global Corporate Landlords
GDP	Gross Domestic Product
GHG	Greenhouse Gas
GNP	Gross National Product
GSAPP	Graduate School of Architecture, Planning and Preservation, Columbia University
HBR Press	Harvard Business Review Press
HK	Hong Kong
HKU	University of Hong Kong
IARI	Istituto Analisi Relazioni Internazionali (Institute for the Analysis of International Relations)
IAs	Intangible Assets
IBM	International Business Machines
ICTs	Information and Communication Technologies
IFTC	International Foreign Trade Center
IISS	International Institute for Strategic Studies
IJEBMR	International Journal of Economics, Business and Management Research
IMF	International Monetary Fund
IoT	Internet of Things
IPCC	Intergovernmental Panel on Climate Change
IT	Information Technology
ITU	International Telecommunication Union
IUMPs	Iconic Urban Megaprojects
KAUST	King Abdullah University of Science and Technology
KPMG	Klynveld Peat Marwick Goerdeler, a services firms provider
KXC	King's Cross Central
LED	Light-Emitting Diode
LLP	Limited Liability Partnership

LSE	London School of Economics
MA	Massachusetts
MAD	MA Design or MAD Architects, a Beijing architecture firm
MIT	Massachusetts Institute of Technology
MO	Missouri
MoMA	Museum of Modern Art
NATO	North Atlantic Treaty Organization
NC	North Carolina
NEOM	Neo Mostaqbal (New Future)
NSW	New South Wales
22@	22@ Barcelona Innovation Megaproject
NY	New York
NYU	New York University
OCA	Olympic Coordination Authority
ODA	Olympic Development Authority
OH	Ohio
OL infrastructure	Overhead Line Infrastructure
OMA	Office for Metropolitan Architecture
PA	Pennsylvania
PBS	Project-Based Securities
PFAS	Per- and Polyfluoroalkyl Substances
PPPs	Public–Private Partnerships
PPWK	People's Panel on West Kowloon
PRCMEE	Popular Republic of China Ministry of Ecology and Environment
PRTS	Personal Rapid Transit System
R+D	Research and Development
RCF	Reference Class Forecasting
REIT	Real Estate Investment Trust
RPBW	Renzo Piano Building Workshop
RWTH	Rheinisch-Westfälische Technische Hochschule (Technical University of Rhenish-Westphalian, in Aachen, Germany)
SBWID	South Boston Waterfront Innovation District
SCIP	Shanghai Chemical Industry Park
SDGs	Sustainable Development Goals
SEZ	Special Economic Zone
SHKP	Sun Hung Kai Properties
SSEZ	Shenzhen Special Economic Zone
STEM	Science, Technology, Engineering, and Mathematics
STS	Science and Technology Studies
TAMI	Technology, Advertising, Media, and Information
TNH	The Next Helsinki
TOD	Transit-Oriented Development
TSMC	Taiwan Semiconductor Manufacturing Company
TX	Texas

UA&P	Urban Affairs and Planning
UAE	United Arab Emirates
UAM	Universidad Autónoma de Madrid
UCDRN	University of California Disaster Resilience Network
UCL	University College London
UHI	Urban Heat Island
U.K.	United Kingdom
UMPs	Urban Megaprojects
UN	United Nations
UNCTAD	United Nations Conference on Trade and Development
U.S.	United States
USD	United States Dollars
UTS	University of Technology Sydney
WA	Washington State
WBCSD	World Business Council for Sustainable Development
WEF	World Economic Forum
WFP	World Food Programme
WHO	World Health Organization
WI	Wisconsin
WKCD	West Kowloon Cultural District
WTC	World Trade Center
YLE	Yleisradio Oy (Finnish Broadcasting Company)

List of Figures

List of Tables

Part I
Introduction

Chapter 1
An Age of Megaprojects

Megaprojects are inherently tied to the logic of growth, development, qualitative urban transformation, wealth creation, competitiveness, and prosperity. Large construction projects have been a fixture of development processes throughout the 19th and 20th centuries. However, globalization and the neoliberal agenda from the 1980s to the present day have provided an additional impetus for the rise of many emblematic projects in cities and regions. Particularly from the 2010s, many high-tech and green public-private megaprojects have been developed worldwide, and governments have returned to using large infrastructure for development. The conquest of political economy throughout socio-ecological space has evolved, networked, and manifested as a process of building the infrastructure for globalizing economies during the neoliberal era. This process uses both human and nonhuman resources, demanding global and energy-intensive supply lines, equivalent types of wages and coerced labor.

Even in the face of ecological and climate catastrophes, this developmental trajectory continues unabated. Increasingly larger projects continue to be built even faster worldwide. Global partners made two significant decisions to encourage large-scale projects in the aftermath of the 2008 recession. In 2015, G20 countries agreed to increase their commitment to restoring economic growth by pledging tens of trillions in infrastructure spending. In December of the same year, at the UN climate summit in Paris, 195 nations committed around USD 10 trillion to large projects related to investments in low-carbon technology, energy efficiency, and climate change mitigation. Furthermore, the developing world was projected in 2015 to invest USD 9 trillion in infrastructure over the next 10 years to keep global warming within the UN limit.[1] More recent projections for the period 2013–2030 estimate worldwide spending to bridge the infrastructure gap to average $3.4 trillion annually or almost 4% of the total global GDP. Large-scale projects account for most of this spending. In emerging economies, the annual cost of infrastructure from 2009 to 2018 was USD 2.2 trillion. Global spending on large projects for the decade ending in 2030 could be around 8% of the world GDP.[2]

G. del Cerro Santamaría, *Contesting Megaprojects*,
https://doi.org/10.1007/978-3-031-98382-5_1

A Highly Distressing Scenario

Scientists at the Stockholm Resilience Center have expressed concerns about global warming and other planetary constraints that must be respected to achieve successful global economic expansion. A 2023 report published in *Nature* identifies safe and just Earth system boundaries (ESBs) at global and subglobal scales for climate, biosphere, water and nutrient cycles, and aerosols.[3] Seven of the eight globally quantified ESBs, and at least two regional ESBs, have been exceeded in over half of the global land area. Engineers and scientists are focused on decarbonizing cement, steel, and plastics; shipping, trucking, and aviation; and agriculture, energy, and construction.

A growing number of experts believe that market-based solutions are unlikely to be adequate or effective to reduce carbon emissions. Markets frequently experience overshoots. If price or quantity overshooting exceeds a threshold that causes nonlinear change, such as that triggered by carbon emissions, biodiversity, or other similar challenges, market systems will not function. An alternative to market-based strategies is to enable robust global government cooperation, most likely through the creation of supranational institutions. The planetary boundaries school, and more broadly, the notion that natural science should be more heavily relied upon as a crucial input into sustainable development, urges for far more state planning and action than would be consistent with still-dominant global capitalist narratives and practices.

However, to date, governments have been unable to implement the Green New Deal worldwide.[4] The covid-19 pandemic has revealed the flaws of laissez-faire global capitalism by exposing the irresponsibility of decades of commercialization, underfunding, and austerity in the health and social care systems of rich economies. Further, the "economy" is still conceptualized narrowly, ignoring the wider social and socio-natural linkages that constitute our complex and linked reality, despite current concerns about ecology and sustainability. Economics is solely seen in terms of the production, distribution, and exchange of goods, obfuscating the contestation of each of these social ills as well as exploitation, alienation, extraction, various forms of exclusion, and processes of ecological damage.

The investment commitments agreed during the Paris Summit were insufficient to halt increasing global warming. However, the Summit served as a starting point. All nations agreed that the only way to prevent a global climate catastrophe was for the public and private sectors to invest heavily in coordinated big projects to reduce carbon emissions. The scale and design of these projects would safeguard cities and farmlands from the potential for climate damage (break-it-fix-it model).

Green capitalism, therefore, has proposed massive investments in megaprojects as a solution for both growth and sustainability, which is a remarkably convenient equation. In fact, several drivers contribute to the current age of megaprojects: geopolitical competition; the pursuit of economic growth; interfirm competition and the quest for new markets; the search for natural resources; population increases (expected surge in the demand for food, energy, and water); reengineered development finance (using public money to offset risk); recovery from the pandemic recession via infrastructure investments; the rise and expansion of megacities and urban population; and interurban race for competitiveness. Remarkably, all these factors contribute to increased nonlinear, and likely irreversible, transformation of built environments and the planet.

Megaprojects contribute to urbanization and, consequently, to the resultant sustainability challenges, both socioeconomic and environmental. A considerable concentration of human resources, economic activity, resource consumption, and the generation of emissions occur in cities because of urbanization, which is a crucial feature of economic development. Cities account for approximately 75% of the world's resource usage despite making up approximately only 2% of the planet's landmass. Nearly 50% of annual worldwide CO_2 emissions are caused by the built environment. Building operations account for 27% of the total annual emissions, while building materials and construction (often referred to as embodied carbon) account for another 20%.[5]

Generation of emissions through urbanization will become more pronounced in the coming decades in Africa and, more significantly, Asia. There are many repercussions from urbanization and the subsequent generation of emissions. The rising population density and demands in urban areas worsen poor air and water quality, lack of water supply, waste disposal issues, and high-energy use. Rapid urbanization can exacerbate poverty and socioeconomic inequalities because local governments cannot adequately serve everyone.

Environmental risks such as flash flooding may be increased by urban expansion. Toxic substances, automobiles, and the destruction of habitats and food supplies adversely affect animal populations. Notably, the "sand crisis" (a potential shortage of sand) directly and increasingly challenges the very process of creating built environments, let alone large construction projects.[6] Megacity concentrations in low-income countries are most likely to be affected by climate change. Most megacities are situated extremely close to the coast, which raises the risk of flooding as a parallel challenge. All the repercussions listed above make the threat that megacities face from climate change highly relevant. Newly built projects must be integrated into existing (and occasionally failing) infrastructure, but they must also be robust and sufficiently stable to accommodate significant population increases. Projects also need to address the socioeconomic disparity that is an existing and increasing issue in large cities, megacities, and city regions, as well as the high demand for housing brought about by rapid urbanization.

Given these increasing difficulties and risks, further development of robust sustainable approaches for urban areas worldwide has become a matter of urgent strategic necessity. Megaprojects, even when successful from a planning, management, and construction perspective, are hazardous and have disruptive effects. We know that, unless they are subject to strict and independent public monitoring, megaprojects, like the neoliberal ideology they epitomize, undo the demos and do away with democratic procedures[7]; as John Keane put it, "they resemble forms of emergency rule in the heartlands of democracy."[8] We know comparatively little about the factors that may enable or constrain sustainable megaproject construction in urban areas.

Furthermore, little is known about the impact of sustainability on urban, regional, and national competitiveness. Research results on these topics may have important consequences for future investment in megaprojects as a primary mode of regional development, given the fragilities of contemporary growth and development processes, in which the rising competition between major infrastructure investment plans can contribute to worsening outcomes.

Although urban development through megaproject implementation can lead to better living conditions, it remains an open question whether this strategy contributes to urban and national prosperity, particularly when megaproject impacts are factored into the equation. A key question, then, is whether the diverse and negative impacts of megaprojects are compatible with any approach to urban sustainability that prioritizes socioeconomic equity and environmental preservation. Many countries embrace sustainability as the main paradigm of urban policy and development and have implemented sustainable practices in their architecture, engineering, and construction.

Research on megaprojects, in connection with sustainability, is increasingly attracting the attention of scholars and policymakers in various fields (urban studies, globalization, political science, new economic thinking, management, anthropology, international development, etc.). After the urban impacts of the covid-19 pandemic, a "G7-G20 pact for sustainable urbanization" has been proposed.[9] Several multilateral organizations have undertaken initiatives related to urban "resilience" to foster a green and fair economic recovery under the assumption that the absence of integrated investment policies and public responses leads to widening disparities between high- and low-income countries, cities, and regions, as well as lost opportunities for sustainable growth. While the focus is on addressing global issues that require massive investments, such as social infrastructure and a net zero economy, the question of sustainability associated with megaprojects is customarily framed within the usual growth and development objectives, and megaproject impacts are largely overlooked or treated as unavoidable externalities.

Aims of This Book

In this book, megaprojects and their impacts are empirically examined, conceptually framed, and positioned vis-à-vis urban sustainability. We show the resemblances among the projects we analyze, despite the very different political, social, geographic, and developmental contexts in which they bloomed within the past 40 years. We link the development and impact of megaprojects to key causal forces, relevant governance systems, and related developmental discourses. We argue that these increasingly important and deeply disruptive projects need to be situated (embedded) within the multiscalar political, economic, and socio-cultural manifestations of global capitalism. Crucially, we emphasize that large projects have highly complex impacts that tend to damage or inhibit long-term socio-ecological sustainability. Megaprojects are insensitive to their environment, are inherently disruptive, exacerbate inequalities, and destroy the connectivity that makes cities generate creativity and community. Following an assessment of the complex impacts of megaprojects in several cities (rather than the more conventional account focused on planning, implementation, and stakeholder agency), we champion a more complex and robust understanding of urban sustainability, via "socioeconomic sustainability" and "deep sustainability." Our analytical lens of *disruptive complexity* provides insights that most critiques of megaprojects do not include.[10] Thus, we show that cost overruns and poor management, although both are common, are not the only or the worst effects of megaprojects; rather the primary reason to criticize megaprojects is their disruptive impacts on the cities where they are built—socioeconomic, environmental, spatial, political and geopolitical, financial, cultural and systemic impacts. Thus, we emphasize the megaproject impacts imperative: the importance of identifying impacts as a prerequisite for critically assessing whether megaprojects should be built. We aim to help urban analysts, practitioners (including policymakers), and residents of cities better understand how to make sense of the complex impacts of urban megaprojects across the globe.

Urban megaprojects (UMPs) present themselves in many different packages and resist easy definitions; however, a close link between megaprojects and development is evident. We previously conceptualized UMPs as *large-scale urban development projects that sometimes have an iconic design component, that usually aim at transforming or have the potential to transform a city's or parts of a city's image, and are often perceived and promoted by the urban elite as crucial catalysts for growth and even as linkages to the larger world economy.*[11] In this book, we enhance this characterization by conceptually linking megaprojects to disruptive impacts and sustainability, using a complexity perspective. A megaproject is initially defined as sustainable if planned and executed to account for the capacity, fitness, resilience, diversity, and balance of its urban ecosystem; this definition of sustainability evolves as we conduct our research. We consider sustainability as an organic process, including the environment, economy, and community. Sustainability includes form and efficiency (environmental factors in design, architecture, engineering, and construction) and policy (urban plans and practices that explicitly maintain and can sometimes improve the social and economic well-being of citizens). We use this concept of sustainability when reflecting on the data collected on the effects of megaprojects and propose a series of conclusions centered around a more robust and ambitious notion of sustainability—that of *complex sustainability*.

Notably, in the context of rapidly expanding urbanization, virtually all megaprojects are urban in nature and location or are built to have a direct effect on cities and the urbanization process. They are also built to have a direct effect on urban development and competitiveness, because megaproject construction has been a major response to adapting to neoliberalism and globalization in the urban realm. Over the past few decades, we have witnessed an increasingly active and worldwide process of *megaproject urbanization* that shows no signs of slowing in the age of megaprojects. Hirschman calls megaprojects "privileged particles of the development process" and points out that often they are "trait making," that is, they are designed to ambitiously change the structure of society, as opposed to smaller and more conventional projects that are "trait taking," that is, they fit into preexisting structures and do not attempt to modify them.[12] Importantly, megaprojects do not just "transform" cities; rather, their Hirschmanian "trait-making" is disruptive, drastically altering urban structures and livelihoods by causing disturbances and problems even in optimal planning, construction, and management scenarios where the best combination of strategies for rational decision-making has been deployed. This indicates that megaproject disruptive impacts, which take shape through the embeddedness of projects in the polity, economy, and society, must be part of any meaningful plan intended to address the question of urban sustainability.

Some approaches to studying megaprojects have underplayed the many types of risks and negative outcomes they produce. Such research has tended to model megaprojects as closed systems and has focused on providing insights to contribute to megaproject management and performance improvement. From this viewpoint, "external risks" and "wider impacts" are factored in as a source of threats for projects, but these categories are not placed at the center of analysis. Instead, megaprojects have been portrayed as necessary features in the process of development, where megaproject performance can be improved by maximizing the processes of rational decision-making.[13]

To be sure, some megaprojects are not always the failures they have been claimed to be—at least if judged by criteria beyond "iron triangle" project delivery concerns. Some projects have contributed to improving or solving specific urban problems, as was the case with Boston's Big Dig project, which, although way over budget upon its completion, knitted Central Boston back together by burying its Central Artery highway in a tunnel. Or take the case of the Guggenheim Museum in Bilbao, which was a positive addition to the city (although not the key element for the positive urban transformation that took place there). However, we are not primarily concerned with assessing success or failure in megaproject management or delivery. Rather, we put megaproject ***impacts*** at the forefront of our research, under the premise that they are a constitutive element of urban projects and a key to assessing urban sustainability. We thus focus on megaproject impacts and show that megaprojects are inherently unsustainable endeavors, and their construction compromises any attempt (even modest) to achieve urban sustainability.

Analytical Structure

To address the key questions about impacts and sustainability, we approach megaprojects as fundamentally embedded in and disruptive to wider networks of production, consumption, power, and meaning. The key question of embeddedness also requires that we approach the web of connections and flows between megaprojects and urban society using a complexity perspective that analyzes megaprojects as complex adaptive systems. Thus, we contend that megaproject embeddedness begets complexity—disruptive complexity—as the central organizing and explanatory principle to describe the many and variegated impacts of megaprojects on cities, regions, and the wider world economy. In turn, a focus on disruptive complexity leads us to (1) assess a possible response scenario (which we conceptualize as complex sustainability) to such disruptions caused by megaprojects and (2) use transdisciplinarity as the most robust and nonreductionist analytical strategy for understanding megaprojects in the world economy, their impacts, and prospects for urban sustainability.[14]

Succinctly, the embeddedness of megaprojects in the world economy begets complexity and, in turn, complexity begets disruption, which leads us to the concept of complex sustainability as a response to disruptive complexity. The sequence is also valid in the reverse way: asking about sustainability requires examining megaproject impacts (disruptive complexity), and integrating impacts into the analysis implies an embeddedness approach to megaprojects that begets complexity.

In addition to interrelations and interdependencies, we focus on trajectories and transitions. This refers to two distinct and interdependent sets of processes: (1) the larger causal context of neoliberalism, globalization, and competitiveness, megaregionalization, strategic connectivity, and increased geopolitical competition that explains the rise and development of megaprojects and their linkages with socioeconomic development worldwide, and (2) the coexistence and entanglement of urban political economies in the past 40 years utilizing property development, culture, and iconic architecture, together with a more recent interest in science and technology, innovation, and sustainability as strategies to foster urban competitiveness. We highlight the strengths, weaknesses, contradictions, and challenges of megaproject development at different periods during the past 40 years. This shows that the notion of disruptive impacts is not just a new or recent feature of megaprojects but has been present in their development at any given moment in the past four decades. We believe that this continuing trajectory of megaproject disruptive impacts adds relevance to our understanding of megaprojects in the world economy. In our discussion of the empirical evidence, we have focused on showing this trajectory rather than providing the latest developments in each megaproject, unless these latest developments add to, or qualify, our main argument.

Focusing on disruptive impacts, Chapters 3, 4, 5 and 6 utilize and interpret empirical evidence from 18 megaprojects in 16 cities around the world (Table 1.1). Chapter 3 illustrates our main substantive theme, the disruptive force of megaprojects, in three separate cases worldwide (Istanbul, Sydney and Hong Kong). In Chapter 4, we show evidence of the fading away of both the Bilbao Effect and iconic architecture as paramount strategies for urban redevelopment and revitalization in recent times (Bilbao, Denver, Helsinki, and Abu Dhabi). In Chapters 5 (Barcelona, Boston, London, Abu Dhabi, and Toronto) and 6 (Tokyo, London, Paris, New York, Shanghai, and Riyadh), we focus on the simultaneous shift toward new strategies based on science and innovation megaprojects under the promise of increased urban sustainability.

The recent proliferation of sustainable and innovative megaprojects is a reflection of the evolution of the world economy, characterized by (1) the widespread use of technological innovations, particularly in the Global North, to secure a new round of global capital accumulation; and (2) the global quest for sustainability as a tool

Table 1.1 List of research megaprojects

Chapter 3	Istanbul Kanal (Istanbul, Türkiye) Barangaroo (Sydney, Australia) West Kowloon Cultural Center (Hong Kong, China)
Chapter 4	Abandoibarra/Guggenheim Museum (Bilbao, Spain) Denver Art Museum (Denver, United States) Guggenheim Museum Project (Helsinki, Finland) Guggenheim Museum (Abu Dhabi, UAE)
Chapter 5	22@ District (Barcelona, Spain) South Boston Waterfront Innovation District (Boston, United States) King's Cross (London, U.K.) Masdar City (Abu Dhabi, UAE) Quayside (Toronto, Canada)
Chapter 6	Tokyo Bay (Tokyo, Japan) 2012 Olympic Park (London, U.K.) La Défense (Paris, France) New World Trade Center (New York City, United States) Pudong's Waterfronts (Shanghai, China) Quality of Life Megaprojects (Riyadh, Saudi Arabia)

for resilience in the face of complex, systemic, and planetary risks, the crisis tendencies of capitalism, and the disruptive character of recent and current global socioeconomic and technological developments. These newer megaproject developments create distinct challenges and have significant drawbacks, including the usual negative impacts associated with more traditional megaprojects, as we will show in Chapter 7.

Given that, by and large, the direct benefits of megaprojects accrue to a small portion of the surrounding urban population, most residents see the need to develop adaptation and resilience strategies because of disruptions and disorders caused by large urban projects. However, should urban sustainability be considered as just a strategy to pursue the goal of adapting to change? Can transformative, proactive changes in urban settings leading toward sustainable urban futures be shaped mainly by resilience strategies? In Chapter 8, we develop the idea of "complex sustainability" as a twofold response (socioeconomic and deep sustainability) to megaproject disruptive impacts.

Method

We use grounded theory inductively as an analytical strategy for producing conceptual categories and theory building. The research methodology aims at both rigor and innovation through the use of heuristics and sensitizing concepts that maximize the discovery and structural variation of perspectives in data collection. Since grounded theory is used as an analytical approach, this research was organized as a "proof-of-concept" scenario. This scenario was built on archival research, arrived at preliminary conjectures and conclusions, identified a set of probability statements about the relationship between concepts, and developed an integrated set of conceptual hypotheses from empirical data. Evidence was gathered from direct on-site observations at different times over the past ten years in 11 of the 16 cities.

In addition, empirical evidence was gathered from archival research and secondary sources. Conceptual research was conducted by analyzing the relevant literature to select the key, driving ideas, and developing a new interpretative framework. The logical structure of this methodological approach is as follows: (1) Study selection topic, (2) collecting relevant evidence, (3) identifying emerging variables, and (4) generating an interpretive framework.

We promote a comprehensive urban analytical project by analyzing evidence from many different urban contexts. To make urban experiences comparable across a much wider range of situations, we concentrate on particular features or processes (impacts and sustainability) in different cities and countries. This technique is assumed to be suitable for the intricate spatiality of urban forms and considers megaproject diffusion processes and urban policy mobility. In our view, this complex scenario undermines comparative approaches that treat "cases" as discrete units. Comparators—the elements and processes being analyzed—need to be selected as relevant to a diversity of urban contexts, rather than the conventional approach of analyzing megaprojects in relatively similar cities. By doing so, we can push theoretical notions to the breaking point necessary for the advancement of urban studies for global analysis, rather than merely reinforcing preexisting notions. The megaprojects chosen are not "cases" in the analytical framework of this book because we compare processes across cities and nations rather than megaprojects.[15] This enables the construction of transdisciplinary theory in connection with both the larger empirical processes influencing specific outcomes and the conceptualizations that are a key focus of this work.

The selection of megaprojects that constitute the empirical evidence in this book obeys a methodology that focuses on critical, exemplary, or paradigmatic evidence in diverse geographic contexts. This allows medium-range generalizations to be built based on evidence, comprehensiveness in analysis, and the likelihood of producing meaningful and valuable insights. Sets of examples cluster in each of the four empirical chapters. The examples in each cluster were interconnected due to similar attributes. A strategic selection of examples aids the development of new concepts, variables, and theories that can account for phenomena previously thought to be outliers as well as the understanding of the limitations of existing theories. Critical examples also address the issue that it is often more crucial to identify the root causes of an issue and its effects than to describe its symptoms and how frequently they occur. As they involve more individuals and fundamental mechanisms in the issue being investigated, critical examples—paradigmatic evidence—may disclose more decisive information about the research problem at hand. Scholars often utilize their expertise and intuition to determine whether they think a certain instance is intriguing in a paradigmatic context and whether they can articulate reasons for selecting generally accepted evidence. We present substantial empirical evidence from around the world about the many disruptive impacts caused by the large urban projects under consideration. Other similar large urban projects would be expected to be as problematic, regardless of geographic location. This is primarily because large scale and complexity, irrespective of cost, define megaprojects, and constitute preconditions for highly disruptive and negative impact scenarios. Our approach also means that we do not focus on "North–South" differences or any other geographic cleavage. While acknowledging that context and difference are key notions in urbanism, our analytical strategy fully concurs with Randolph and Storper when they state the following:

> Our reading of the evidence indicates that parts of the Global South may be urbanizing along historically and geographically specific trajectories; however, we argue that these differences are best understood through a unified set of global urban theories. Rather than flattening or silencing differences, theories that seek generalization across time and space sharpen the identification and appreciation of key differences in urbanization processes. Analyzing how the fundamental dynamics of urbanization recombine and interact with one another in different contexts offers insight into policy challenges that cut across cities, both within and between the Global South and North, as well as context- specific policy issues that arise through the interaction of global urbanization forces and local specificities.[16]

The previous analytical and methodological considerations mean that the paradigmatic empirical evidence in this book has been carefully selected according to the themes in each of the four empirical Chapters 3, 4, 5 and 6. Accordingly, the claims presented for each theme in each empirical chapter can be expected to hold for other projects that fit each theme. This means that the critique of megaprojects developed in this book is tied to, and becomes meaningful in, the specific developmental contexts and urban growth strategies analyzed in each of the four empirical chapters.

Our paradigmatic evidence is not treated in the book as "cases" in the usual sense. We do not focus on describing each megaproject as a whole, and we deliberately overlook matters of design, planning, and construction in most instances—these aspects have been widely discussed elsewhere. Because of this, complex decision-making around the analyzed projects and the agency of leaders and stakeholders is not brought to the forefront. Our aim is to focus on description and analysis of megaproject ***impacts*** on the cities where they are built.

Therefore, this book is ***not*** a multicase study of megaprojects. We show evidence of disruptive impacts in projects that are varied in scope, timing, and intention. Further, we aim to show such processes at various points in time rather than prioritizing a focus on the latest developments. In sum, we do not show or compare megaprojects but rather processes—impacts and sustainability—associated with megaprojects across cities and nations in the past decades.

Chapter Summaries

The book unfolds logically from an *Introduction* (Chapters 1 and 2) to a set of empirical chapters on *Repertoires of Disruption* (Chapters 3, 4, 5 and 6), followed by two chapters on *Inferences and Conjectures* (Chapters 7 and 8), and a chapter on

Conclusions (Chapter 9). This sequence is intended to build a cohesive framework developing in inductive form, where the book's foundation is the four empirical chapters providing the material for elaboration in the subsequent analytical chapters.

Chapter 2, Trajectories and Transitions, addresses the planning, design, and project management aspects of megaprojects (locally executed institutional arrangements and financing at various spatial scales), as well as the larger causal context (trajectories and transitions) of neoliberalism, globalization and competitiveness, megaregionalization, strategic connectivity, and increased geopolitical competition, which explain the rise and development of megaprojects and its linkages with socioeconomic development worldwide. The title of the chapter also refers to the trajectories and transitions in urban political economies toward the coexistence of the focus on property development, culture, and iconic architecture over the past 30 years with a new interest in science and technology, innovation, and sustainability as strategies to foster global competitiveness (Chapters 4, 5 and 6). This chapter analyzes megaprojects in a dual, interconnected way: as spatial settings and infrastructure units that obey specific planning, design, and management priorities, as well as reflections or expressions—as consequences or crystallization—of larger socioeconomic forces. This chapter introduces the reader to the analytical themes that frame Chapters 3, 4, 5, 6, 7 and 8: complexly disruptive impacts and sustainability.

Chapter 3, Landscapes of Disruption, discusses megaprojects in Istanbul, Sydney, and Hong Kong. We offer a panoramic view of the various controversies, difficulties, obstacles, negative impacts, and civic opposition associated with constructing these megaprojects. We show that the nature and shaping of the projects owes to the socioeconomic, developmental, institutional, and geographic contexts in which they emerge. However, we also see that these urban megaprojects—regardless of context—constitute disorderly and contentious enterprises and have intrinsic potential (often realized) to elicit substantial controversy and criticism that fundamentally questions the parameters of the projects as envisioned and publicly presented by their promoters. The megaprojects are disorderly enterprises, in that they substantially modify the physical appearance of cities and their urban fabric, often triggering socioeconomic imbalances and realignments in urban power arrangements in growth machines and civil society. Megaprojects are contentious enterprises because of the complex composition of stakeholders with conflicting interests in their planning, construction, management, and governance. This complexity often results in major obstacles to megaproject implementation, strategic misrepresentation of costs and benefits, optimism bias about megaproject impacts among planners and promoters, and a widespread perception among urbanites that these structures are negative for their cities.

Chapter 4, Emblems of Capitalism, uses several cases of attempted urban regeneration via cultural megaprojects to show the drawbacks and failures of the Bilbao Effect, a model of urban revitalization based on iconic architecture that became commonplace among urban managers worldwide after 1997 with the alleged success of the Guggenheim Museum Bilbao. This Chapter discusses the case of Bilbao together with other similar efforts in urban reimaging via iconic cultural projects: Denver, Helsinki, and Abu Dhabi. Such iconic projects can be questioned in three main ways: (1) iconic architecture is no longer a hegemonic visual discourse in urban revitalization; (2) the franchise model imposed by Guggenheim means that local officials have no autonomy to make major decisions on matters ranging from exhibition calendars to budgets and investments; and (3) local cultural identities are usually neglected in a foreign global art model. The second part of the chapter, Architecture After Spectacle, suggests that architectural discourse in the West has been moving away from spectacular, iconic architecture and toward embracing new paradigms and ideas. Among these new paradigms, we see renewed interest in linking aesthetics to topography using the concepts of tectonics and rebuilding resistance over time and history, as well as through the practice of places as axes of architectural production. We can also see a fuller embrace in architectural theory and practice of the complexity paradigm and the crossing of disciplinary boundaries, particularly when dealing with issues of urban ecology and sustainability.

Chapter 5, High-Tech Megaprojects, analyzes megaprojects characterized by processes of high-tech innovation in urban settings, focusing on urban technopoles, manufacturing megaprojects (factories of the future), innovation districts (innodistricts), the pervasive expansion of the smart-city paradigm, and its gradual evolution into a nascent AI urban regime. Such an AI regime is characteristically disruptive, and it begins to show some distinct features that impose valuation and devaluation strategies over entire neighborhoods and urban areas, thus reinforcing marked inequalities and a dual pattern of benefits and high risks in already "quartered" cities. By focusing on high-tech development and utilization, particularly new kinds of automation capabilities and AI, this chapter offers some elements for a critical analysis of the "smart-city" paradigm and the promises and discontents of urban sustainable infrastructure centered on these new technologies. The paradox of artificially intelligent urban environments lies in the promise of attaining a certain type of sustainability at the risk of further jeopardizing urbanity, equality, and livability. AI alone does not necessarily foster urban sustainability, even if specific AI innovations can. Furthermore, in the case of innovations that foster sustainability, one needs to factor in both the benefits and risks of the innovation strategies. The key question is how to craft and build a framework for sustainability that can discriminate and identify which innovations contribute to the desired sustainability outcomes. This chapter discusses the empirical evidence from five megaprojects in Barcelona, Boston, London, Abu Dhabi, and Toronto.

Chapter 6, Greening Megaprojects, focuses on the impacts of megaprojects on the surrounding environment in six different cities (Tokyo, London, Paris, and New York, Shanghai, and Riyadh). The chapter also analyzes some of the strategies used to develop allegedly "sustainable" megaprojects to counter environmental damage. The "greening" of megaprojects has become a widespread strategy for urban development worldwide. However, we contend that allegedly "sustainable" megaprojects follow the logic of strategically disinvesting in issues that appear secondary to the competitiveness of cities, while investing in issues that are crucial for forming urban regions as strategic centers for capital accumulation. Why are urban environmental policies frequently motivated by the goal of improving cities' competitiveness or limiting elements (pollution, traffic, urban sprawl, etc.) that are likely to impair competitiveness? These policies are characterized by their thematic, social, and spatial selectivity by opting to pursue goals seen as win-win from an environmental and economic standpoint, and created to meet the aspirations of the middle and upper classes. Although the approach to sustainability used in "sustainable" megaprojects of "green capitalism" is a feasible strategy, it does not adequately address the main challenges posed by megaproject construction. We propose some essential elements to begin understanding megaproject sustainability as a complex, multidimensional concept.

Chapter 7, Disruptive Impacts, develops a set of inferences from the empirical chapters. This chapter sheds light on the notion of disruptive complexity, which aims to express the variegated and interdependent web of impacts that megaprojects have on urban areas: cost overruns, negative environmental impacts, gentrification risks, accentuation of socioeconomic inequality, drawbacks of top-down cultural engineering, neglect of local cultural identities, uncertain economic success, population displacements, spoiling of cities' visual and structural uniqueness, rise of congestion and overpopulation, political gridlock, threats to ecosystems, environmental risks of many kinds, and others. The chapter offers a typology of disruptive impacts: socioeconomic; environmental; spatial/territorial/ visual; political (including geopolitical); financial; cultural; and systemic. Impacts can generally be defined as probable consequences or materializations of risks, exposure, and vulnerability. Risk can then be defined as the probability of an impact. Some materialized risks (impacts) are never identified in the original risk assessment and are not included in any sustainability planning, leading to negative megaproject impacts. Thus, uncertainty—the reason for conceiving a risk—is also present when risks materialize into impacts. This is expected because the urban impacts of megaprojects are a function of megaproject embeddedness in the polity, economy, and society. Therefore, impacts can sometimes, but not always, be predicted by examining the risks. In other words, impacts are *emergent properties of megaprojects* that cannot always be predicted in advance in any meaningful or comprehensive manner.

Chapter 8, Complex Sustainability, develops a set of conjectures to flesh out a robust concept of sustainability, coalescing around the notion of *complex sustainability*. We present complex sustainability as an assemblage of two main components: socioeconomic sustainability and deep sustainability. The question about socioeconomic sustainability revolves around four axes: the need to go beyond green capitalism (or decoupling sustainability from profitability), sustainable innovation in the new AI urban regime (or framing innovation around sustainability goals), cohesion versus competitiveness (or finding ways to design equity policies with a local focus), and public value (or developing bottom-up market-shaping and value-co-creating eco-strategies). These four elements represent fundamental vectors or directions for urban sustainability that dispute existing strategic public policies and private priorities. We define and articulate deep sustainability around the following discussions: urban ecology and nature (as self-organizing complex systems), low entropy (a desired sustainability state would be one of minimum entropy production), materialities (contributing to the nonanthropocentric thrust of complex sustainability), urban experience (on the role of agency in shaping human understanding and inquiry through civic frictions), and transdisciplinary sustainability (the integrative analytical approach for complex sustainability). In summary, this chapter fleshes out the notion of complex sustainability, which includes two different yet complementary and interrelated responses (socioeconomic sustainability and deep sustainability) to the disruptive complexity generated by megaprojects.

Chapter 9, Contesting Megaprojects, can be viewed as a coda for Chapter 8 and the entire book. The chapter begins by summarizing our findings and then presents a set of concluding remarks and questions for further research.

Chapter 2
Trajectories and Transitions

Introduction

For the past few decades, megaprojects have been a fundamental part of the infrastructure of globalization and neoliberalism and, as such, have actively contributed to rapid global urbanization. Megaregionalization, the drive for strategic connectivity, and increased geopolitical competition have become key drivers for megaproject development. To meet their stated goals of contributing to urban and national growth and development, megaprojects are usually shaped as public-private institutional arrangements by elites and pro-growth coalitions, which are able to gather the necessary political support, funding, and expert knowledge to build and manage these complex structures. Local, regional, and national elites as well national and sometimes transnational capital play prominent roles in reconfiguring spaces for urban regeneration schemes, transport and energy infrastructure, industrial corridors, city clusters, new towns, innovation districts, science and technology parks, sports infrastructure, and other megaprojects. Their design and implementation require bringing together and aligning several scales of power for two reasons: (1) increased urban competitiveness and global visibility are perceived as essential outcomes of the development of these projects and (2) in diverse sociopolitical contexts, political power is expressed through various relationships among local, regional, national, and global domains of social action.[1]

The neoliberal era is marked by the shift in urban governance from managerialism to entrepreneurialism, in which cities are thought of as nodes in a global network of relationships. In this context, urban elites often perceive linkages to the global economy as fundamental to ensuring sustained local economic development.[2] Thus, attaining or regaining global visibility is an essential economic strategy[3]; it also serves the purpose of enhancing positive symbolic or representational transformation in cities or regions. Both objectives—the material and the symbolic—shape many cities' recent attempts to regain or maintain their status as

G. del Cerro Santamaría, *Contesting Megaprojects*,
https://doi.org/10.1007/978-3-031-98382-5_2

globalizing metropolises by engaging in urban revitalization and developing megaprojects.[4] Revitalization itself is a political strategy that questions developmental approaches that emphasize exclusively the economic and financial characteristics of globalization.

This chapter addresses the planning, design, and project management aspects of megaprojects—for example, their institutional arrangements and financing at various spatial scales that are executed locally—and the larger causal context of neoliberalism, globalization, megaregionalization, strategic connectivity, and increased geopolitical competition that explains the rise and development of megaprojects and their linkages with socioeconomic development worldwide. As an umbrella concept, competitiveness includes innovation capacity, regional endowment factors, macroeconomic approaches, and microeconomic strategies that link productivity to prosperity. Development encompasses both the infrastructural improvements epitomized by megaprojects and the evolving dynamics of globalization (fostered by developmental states turning into entrepreneurial states), geopolitics, geoeconomics, and geotechnology.[5]

The title of the chapter refers to the trajectories and transitions from a focus on property development, culture, and iconic architecture over the past thirty years to the recently heightened interest in science and technology, innovation, and sustainability as strategies to foster global competitiveness (see Chapters 3, 4, 5 and 6). This chapter analyzes megaprojects both as (1) spatial settings and infrastructure units that are shaped by specific planning, design, and management priorities and as (2) reflections or expressions of larger socioeconomic forces. The chapter also introduces the reader to the book's analytical approach that is centered on complexity, impacts, and sustainability and that frames Chapters 3, 4, 5, 6, 7 and 8.

Development and Competitiveness

To understand the links between megaprojects, development, and competitiveness, we must briefly discuss the processes of globalization and neoliberalism as causal contexts. Megaprojects are shaped by a complex set of global geographic, economic, and political restructuring processes that first arose in the late 1970s and the early 1980s and are still dominant.[6] Globalization and neoliberalism became ascendant, manifesting in urban areas as megaprojects[6] through four global processes: (1) the rise of city-based international competition, (2) the mobility and growth of knowledge economies, (3) the redirection of global investment from physical to human capital, and (4) the dominance of market- rule ideology and politics.[7]

Megaprojects are products of political and economic changes and rhetorically minimal state involvement. However, these projects are "clearly, and almost with no exception, led by the state and often financed by the state."[8] This gap between framing and reality regarding the degree of state participation symbolizes the tension between the ideology and practice of neoliberalism, as crudely revealed in megaprojects.

Neoliberalism has generated a complex reorganization of the relationship between the state and economy, in which government actively enables and promotes market-based regulatory agreements that favor the private sector—often in the form of public-private partnerships. Urban development today depends on this peculiar and perverse form of mobilization of state power to a greater degree than before 1970.[9]

The neoliberal, "entrepreneurial state" has replaced the "developmental state," the paradigmatic example of which is China, as controlled by the powerful Chinese Communist Party. One of the first expressions of the Chinese entrepreneurial state was the construction of the Shenzhen Special Economic Zone (SSEZ) by teams of the Corps of Engineers of the People's Army; these teams took on this project after they finished rebuilding Tangshan, which had been leveled by an earthquake in 1976. The first skyscraper in the SEEZ, the International Foreign Trade Center (IFTC), opened its doors in 1985 and was, at the time, the tallest building in China. The IFTC was inspired by the Hopewell Center of Gordon Wu in Hong Kong and soon became a widely replicated type of building throughout China. From this skyscraper, Deng Xiaoping issued a historic call for China to join the free market in January 1992.[10]

The increase in foreign direct investment (FDI) during the past few decades of globalization is an indicator of the impact of financial and human capital mobility on urban spaces. These increases have occurred in the context of a massive economic transition from the dominance of manufacturing to the rise of the service, information, and knowledge sectors beginning in the second half of the 20th century. Global FDI in services and IT has continued to grow steadily, whereas it has declined in terms of primary products and raw materials. Megaprojects have been developed to capture the share of the world's mobile wealth. A narrative of international competitiveness to ensure economic survival frames most megaprojects.[11]

The ideology behind megaprojects offers the promise of growth, development, competitiveness, wealth generation, and prosperity directly for the actors and sectors planning and building them and indirectly for the public. However, there is often no mechanism to guarantee that these benefits will materialize, particularly for urban residents. Urbanites' perception of urban prosperity is influenced by their quality of life, cost of living, conveniently located housing, and the availability of recreation, programs, and services. And according to Harris, the presence of these factors depends, to a large extent, on the degree of accessibility in the urban space:

> This requires that certain patterns of land use and transportation be coordinated to achieve broad-scale benefits. The premises may be, in principle, correct, but it has been shown that the relationship between urban growth and prosperity under urban policies guided by neoliberalism favors certain groups and places while harming others, accumulating benefits that are geographically and, in terms of socio- economic status, very unequal.[12]

Another important aspect of megadevelopments is that they have similar objectives and results, despite being planned and built in very different geographical, political, and social urban contexts. All follow, to varying degrees, a clear model of competitive urban entrepreneurship, which seems to be copied from one place to another without making the necessary adjustments to specific contexts.[13] Most urban megaprojects, particularly mixed-use projects, can be categorized as global models of urban development that are lucrative for the capital markets that benefit from public financing, even though public-private partnerships are driven by the desire to reduce public spending.[14] Capital markets, a major driver of megaprojects, are part of powerful coalitions promoting urban growth that defend and benefit from their development and performance.

Megaproject protagonists typically embrace a globalization discourse in which international economic competitiveness is seen as of paramount importance for the prosperity of the city and state. This rhetoric dominates public discourse, frames objectives, and guides decision-making processes, despite being rarely operationalized into project management processes. The structural change that these projects aim to bring about, who precisely stands to benefit, and more importantly, what alternatives might be available remain shrouded in a generic "glossy globalization" discourse that glorifies potential investment and growth while obscuring real urban displacement and socio- spatial polarization.[15]

Growth Machines

In 1987, John Logan and Harvey Molotch published *Urban Fortunes: The Political Economy of Place*, a book that fundamentally altered urban theory. The authors laid a new groundwork for urban studies by applying the Marxian framework of use value and exchange value to the urban fabric, their central idea of the "growth machine," and an analysis of opportunities to thwart unwelcome expansion and development.[16] This framework of use and exchange value, together with the concept of adaptive reuse inspired by Jane Jacobs, has replaced Robert Moses' and Justin Herman's destructive urban redevelopment and growth ideology. Elite gentrification has slowed in favor of the "suburbanization of poverty" and predatory gentrification. As visitors and immigrants alike look for "genuine" and "gritty" districts, Boston's Faneuil Hall's fake urbanity has not held up well, eventually becoming gentrified. Fostering the revitalization of urban life as opposed to "value-free development" has emerged as a growth strategy in and of itself.[17] Cities and communities that are safeguarded against rapid growth and that are frequently strongholds of a generally progressive citizenry, divert development away from less-favored areas. By using real estate investment trusts (REITs), cutting-edge financial products, and speculative hedge funds, speculators have discovered new ways to extract the value inherent in the built environment. The alliances that support expansion have, in some circumstances, become far more "democratic" than in earlier decades, or at least they are formally open to participants beyond the economic elite, such as working-class community organizations.[18]

We can debate whether megaprojects only are legitimate growth engines and serve commercial interests or whether they reflect the key role of government actors and agencies in urban restructuring. To be sure, city and state boosterism to reposition cities in the global economy frequently takes the shape of machine coalitions. Megaprojects may be more significant as emblems or accelerators of economic improvement and prospective wealth than for their specific economic and urban impact, while being a pro-growth tactic favored by political and economic elites.[19]

Cities and towns in the United States have historically embraced capital investment, with little regard to its social or financial repercussions. The "New Federalism" of Ronald Reagan is an example of national politics and policy that benefit capital investment by incentivizing the development of favorable urban sites.[20] However, with the tensions and conflicts brought about by the increasing globalization of capital, communities are finding that the prices of such investments have increased. The property-oriented "rentier" elite, as well as the middle and working classes, occasionally attempt to replace the typical pro-growth agenda in places that reject it.

With the advent of neoliberalism, this growth model has expanded worldwide: pro-growth coalitions in many cities now strive to become more visible globally so that they can draw in flows of capital and labor, thereby enabling growth. However, experience has shown that the growth model will only be successful if it identifies the precise local outcomes produced by the interaction of local, regional, national, and global forces as they evolve over time. Accounting for many scales of social and political action enriches the model rather than turning it into a metanarrative, especially in cities and regions that have tense political connections with their nation-states. To understand how growth and development agendas form and change because of new socioeconomic frameworks and situations, investigations into path dependence in redevelopment schemes and institutional alliances are helpful. Connections between regional arrangements for development and global political and economic activities must be also strengthened.[21]

What has changed is the growth machine concept, which reflects shifting urban economy–environment connections. The growth machine is a collection of methods used to control and reconcile altering and conflicting demands—economic, political, and sociocultural—at different spatial scales.[22]

Modern applications of the growth machine concept are deeply ingrained in the changeable geometry, particular paths, and urban impacts of global forces, as well as in the mediation of developmental states. As shown elsewhere, these pathways include (1) regionally developed economic and political strategies to create flows and transnational connections with the global economy (such as state-led export-oriented growth strategies), (2) local developments to deal with global changes (such as industrial restructuring), and (3) structural and territorial adjustments to place the city in the context of global cities (such as megaproject development). Coalitions in favor of growth typically support the objectives of developing nations, particularly (but not exclusively) in emerging markets. In a comparative paradigm that offers local elites and actors external models to imitate, local development machines may be portrayed as pursuing national or regional competitiveness on a global scale. Given that growth and productivity were national priorities for many countries and regions even before the current period of globalization, modern growth machine applications do not fundamentally change the direction and goals of state policy. In fact, during the 16th century when modern capitalism first developed, local growth engines and commercial–state coalitions aimed at preserving local trade privileges could be found in numerous European regions. This emphasis on expansion and development occasionally influenced local legal systems.[23]

To offset the negative effects of comparative disadvantage, late industrializers need a strong state that places a high priority on growth. Given the prominence of the issue of effectiveness, local growth machines may be seen as addressing (and benefiting from) official priorities for industrialization and development in this context. The maximizing of benefits from international commerce and export-led development and the strategic use of domestic resources of capital, information, and labor are shared objectives of both state and pro-growth coalitions. Yet, aligned state and pro-growth actors may have additional shared strategic objectives. International recognition may be gained by using local economic influence to better protect regional political autonomy and identity. Local elites in reterritorialized national states and local and regional governments implement measures to preserve their autonomy and identity. [24] Regional governments in multinational states may prioritize the corporate aims of international trade and investment to improve their capacity for political maneuvering in relation to other nation-states. The specific methods and strategies used by the local growth machine are directly affected by this process.[25]

Financing and Management

The ability of cities to stimulate growth through urban megaprojects and share their costs and benefits fairly depends largely on the distribution of public resources.[26] Yet, all too often, state actors involved in planning megaprojects ignore regional planning frameworks, conventional democratic channels of participation, and accountability.[27] Because of this, civil society perceives these large developments as the result of secret, backroom deals, which contributes to strengthening opposition initiatives and movements.[28] In addition, urban megaprojects are generally developed as public-private partnerships, which frequently privilege growth, competition, and profitability over socially progressive objectives, typically leading to little dedication to policies for public goods or social justice.[29] Thus, urban megaprojects are the paradigm of the entrepreneurial city and present most of the problems of this mode of urban governance.

The financing mechanism for such projects is of great importance. Initial costs to decontaminate the soil, eliminate physical barriers, and create the necessary infrastructure are extremely high. They require large-scale investment that may have low expectations of returns and considerable risks, which makes private sector involvement unlikely in the preconstruction phase. Despite the antistatist and promarket rhetoric of neoliberalism, state investment and involvement are required for megaprojects to get off the ground. However, given the limited financing capacity and budgets of most municipalities and metropolitan regions, finding the needed resources to develop large-scale operations is very difficult. When the state does provide public financing, the impact of that investment on municipal budgets is typically high, yielding an obvious opportunity cost. That is why many urban megaprojects are perceived as the centerpiece or catalyst of revitalization plans in their respective cities. Seen as "connecting" cities and regions to the global economy, local politicians often overpromote the potential of the project.[30]

The global rhetoric of urban planners and leaders and the image of renewal aimed at generating investment in cities and regions contrast with the limited capacity of cities to attract and secure foreign capital investments. Many supposedly "global" megaprojects have only a local or regional scope, an indication of the many contradictions inherent in most urban regeneration plans based on requalifications and new land uses, as Fainstein has pointed out.[31] The multiple risks of megaprojects and the uncertain financial viability of flagship projects severely limit their prospects for success and, at times, force developers to redesign initially ambitious projects.

Too often, megaprojects are financially unsound. They are motivated by a desire to show the city's supposed economic well-being and, as a result, establish unproductive businesses in urban areas that are frequently unable to support them. Rent-seeking conduct and optimistic bias in projecting outcomes occur frequently because of the high stakes involved, as Bent Flyvbjerg argues:[32]

> The subliminal rapture experienced by engineers and technologists in building large and innovative projects is reinforced by the attraction to politicians of large ribbon-cutting events attended by the media, whereas financiers and trade unions expect to gain profits and jobs. Aesthetes who appreciate good design and enjoy viewing iconically beautiful structures such as the Bilbao Guggenheim, the Golden Gate Bridge or the Sydney Opera House also influence the funding of ever more monumentally grand projects.

Yet, despite all their complexities, challenges, and shortcomings,

> Megaprojects are not only large and growing constantly larger, they are also being built in ever greater numbers at ever greater value. The McKinsey Global Institute (2013) estimated global infrastructure spending at USD 3.4 trillion per year from 2013 to 2030, or approximately four percent of the total global gross domestic product, mainly delivered as large-scale projects. *The Economist* (June 7, 2008) similarly estimated infrastructure spending in emerging economies at USD 2.2 trillion annually for the period 2009–2018. Spending on infrastructure to accompany these megaprojects also increased tremendously in the early 21st century: In the 5 years from 2004 to 2008, China spent more on infrastructure in real terms than in the entire 20th century. In other words, an increase in the spending rate by a factor of 20. Similarly, from 2005 to 2008, China built as many kilometers of high-speed rail as Europe in two decades, and Europe was extraordinarily busy building this type of infrastructure during this period. Not at any time in the history of mankind has infrastructure spending been this high measured as a share of world GDP, according to *The Economist*, who calls it "the biggest investment boom in history.[33]

If we include not only infrastructure, but also the many fields where megaprojects are the main delivery model—oil and gas, mining, aerospace, defense, ICT, supply chains, mega events, regeneration schemes, transport and energy infrastructure, industrial corridors, city clusters, new towns, innovation districts, science and technology parks, and sports complexes—a reasonable estimate for the global market for megaprojects is USD 6–9 trillion annually, or approximately 8% of the global GDP. To put this into perspective, this amounts to spending amounting to five to eight times the total United States debt owed to China annually.

Flyvbjerg posits that the vested interests of stakeholders' coalitions are driving the global boom in megaprojects. While cost overruns are the "iron law" of megaprojects, Flyvbjerg also postulates a paradox. Although there is a growing demand for both public and commercial megaprojects, their management in terms of sticking to budgets and schedules and producing the expected benefits has not improved throughout the century. This situation may be ameliorated by the increasing role over the past twenty years of private financing of megaprojects. This indicates that banks, pension funds, and capital funds all have growing management influence. To be clear, private finance is not a cure-all for problems with megaproject management; in some instances, it can worsen matters. However, because private investors are putting their own money at risk, they have added incentive to ensure it is not being misspent. It is a positive step that banks frequently hire their own advisers to do independent risk evaluations, due diligence, and forecasting.[34]

Approximately 50 years ago, the American economist Albert Hirschman proposed the principle of the *hiding hand*: projects are viable because unexpected events or setbacks that cause cost overruns also spur project managers to be more creative in overcoming adversity and even increasing benefits.[35] In addition, project managers and promoters are increasingly aware that different stakeholders have widely varying predictions and that these forecasts are as influenced by power and negotiation as by data and mathematical modeling.[36]

As a result, confidence in the inaccurate estimates made by project promoters has been eroded. Ideally, *reference class forecasting* (RCF) is used to make accurate projections based on actual performance in a reference class of comparable actions; its use circumvents both optimism bias and strategic misrepresentation.[37] Recent advances in megaproject management research have greatly improved both our knowledge of the factors underlying the numerous delivery failures in megaprojects and our ability to prevent them. For instance, we now recognize that optimism bias and strategic misrepresentation have a much greater influence than earlier hypotheses, such as Hirschman's "hiding hand" and Sawyers' "creative error," in explaining megaproject outcomes.[38]

Managing megaprojects to achieve sustainability not only in terms of the environment but also economically, politically, and institutionally is a significant challenge. The San Francisco–Oakland Bay Bridge, which was damaged during the 1989 earthquake and reopened in 2013, was $5 billion over budget and took ten years longer to construct than originally projected. However, the bridge was built to last 150 years and to withstand earthquakes and seismic activity of the highest magnitude, which ultimately resulted in substantial savings. Thus, construction costs and schedule are not the only ways to judge the Bay Bridge's cost effectiveness. There needs to be a planning framework to help governments and the public understand the larger benefits of a project and to include the impact of economic and social development in the final analysis. This does not mean that residents will be less irritated by the daily disruptions of construction projects that never seem to end.

However, by looking at the bigger picture, they may consider it worth the temporary annoyances.[39]

The report "*Mega-Projects: Lessons for Decision-Makers*," prepared by the Omega Center at the Bartlett School of Planning at University College London and led by Professor Harry Dimitriou, discusses thirty $1 billion-plus megaprojects around the world. Most were close to being on time and on budget. However, the team looked well beyond the "iron triangle" of fulfilled schedule, budget, and specifications—and indeed, that is their major takeaway: big projects must be judged for how they meet their objectives over time amid shifting societal, political, and environmental values. Measuring the success of a megaproject is nonlinear. There are twists and turns not only in terms of engineering and the emergence of new technologies, for example, but also in the moving targets of public expectations. New problems always crop up that such projects are expected to solve, long after the first blueprints were approved. The long timeframe of megaprojects ensures that these changes will occur. This is particularly true of the largest megaprojects, which always tend to be proposed by political leaders who are almost always out of power by the time of completion.[40]

After the global financial crisis of 2008, new types of housing financialization appeared. A global environment of excess liquidity, housing foreclosure, and nonperforming loans spurred the channeling of financial capital flows onto residential and mixed-used real estate, often in the form of newly constructed megaprojects. Large international residential property portfolios were amassed by multinational investment funds such as those developed by Blackstone. They have undertaken aggressive property management tactics as global corporate landlords (GCLs) to increase returns on their investments.

Tenants in their properties frequently experience lease terminations or unexpected rent increases, generating increased opposition to the actions of GCLs in these places. What are the sociospatial effects of asset restructuring? How do state policies encourage this type of transnational real-estate investment? What types of tenant resistance and strategies are GCLs facing, and how effective are those strategies? These are some questions that should be addressed in research on the urban impacts of these processes.[41]

Planning and Partnerships

In *Mega-Projects: The Changing Politics of Urban Public Investment*, Alan Altshuler and David Luberoff drew attention to the United States federal government's role in infrastructure provision, which has increased significantly since the 1960s; they also highlighted that the megaprojects receiving congressional funding were those that had built winning political alliances. Consumer autonomy and cost effectiveness seemed to be irrelevant factors in these allocations, which, after all, were funded by nonlocal taxpayers. The federal role in funding has expanded with the help of pro-growth coalitions.[42]

Public-private partnerships connecting public officials and private investors continue to be the principal vehicles for urban megaprojects across the globe. These megaprojects can no longer be built by either the private or public sector alone but require the pooling of resources and risk sharing, resulting in projects that are jointly owned even after they are completed and are operational.

Public-private partnerships (PPPs) began to be adopted by city officials in the United States in the 1970s for downtown redevelopment projects, setting a policy shift in motion that has since had a profound effect on the practice of urban planning at home and abroad.[43] Although its origins can be traced back to the patronage partnerships of the late 19th century in the United States, its modern roots lie in the previously discussed changes in global relations that transformed urban economics and increased interurban competition.[44] Since the earlier widely publicized flagship projects such as Baltimore's Inner Harbor and Boston's Faneuil Hall Marketplace, public-private partnerships have increasingly become the vehicle of choice for public agencies seeking to engage with the private sector in redeveloping urban sites in United States and Western European cities.[45]

Earlier studies by political scientists focused on the imbalances of power within the new joint ventures and the favorable treatment given to developers to the detriment of communities' needs and other socially necessary uses.[46] Commenting on the United States and British experiences, particularly the new waterfront and innerharbor development in Baltimore. Harvey argued that such public-private partnerships were inherently speculative in execution and design, because the risk was to a large extent borne by the public sector. Thus, they amounted to little more than a subsidy for affluent consumers and corporations at the expense of local collective consumption for the working poor.[47] Such criticisms, however, betrayed an ideological aversion to market-based solutions and lacked systematic analyses of riskreturn or cost-benefit relationships or whether the projects actually performed as their initial financial projections predicted.

Susan Fainstein's detailed study of public-private partnerships in London and New York City in 1980–2000 provides a more balanced analysis of the role of the public and private sectors and the usefulness of such structures under certain conditions. Nevertheless, she sides with those arguing that public-private partnerships were unequal and tended to over-rely on property development as an economic growth strategy while leaving unpursued other strategies that would develop worker skills and directly spur job creation and placement.[48] Fainstein argues that public redevelopment programs and assistance to the private sector should form part of the comprehensive development effort, guided by economic planning and supporting revitalization without glutting the market; this requires the control of price levels and participation in development profits.[49]

Focusing on the redevelopment of Times Square district in New York City during the 1980s and the 1990s, Sagalyn points out that the essence of public-private development is an inherent asymmetry in the degrees of freedom that private investors and public officials have to bear in partnerships. Private developers benefit from greater maneuverability because they face fewer political risks. However, for city and state officials, the combined economic and political costs of severing ties are often unpredictably high; therefore, new compromises must be found in the form of revised deals. The political imperative thus forces a solution, and the challenge is to find forms of engagement that afford some protection to the taxpayer while taking on enough risk to allow the government's political goals to be achieved.[50]

One of the most comprehensive studies done within the European context is Moulaert and coeditors' book analyzing 13 large-scale urban development projects in 12 European Union countries. Its focus is on how globalization and liberalization articulate with the emergence of new forms of governance and on the relationship between large-scale urban development and political, social, and economic power relations in cities. Among their findings, the authors argue that, in contrast to their market-led and entrepreneurial activities and their predominantly privatized management structures, urban development projects are almost without exception state-led and often state-financed. Additionally, the configurations of project-based institutions created to implement projects result in an extraordinary degree of selectivity. This, in turn, produces a significant deficit in accountability, representation, and formal rules of inclusion or participation. Most importantly, participation is rarely statutorily mandated and instead tends to happen through ad hoc co-optation and invitation, usually extended by key power brokers within institutions. In a general evaluation, the authors decry a lack of democracy and social policy in the new urban development policies and the poor integration of large urban projects into wider urban processes and planning systems.[51]

In a more recent study, Fainstein argues that public-private partnerships can be a vehicle for the provision of public benefits, including job development, cultural facilities, and affordable housing. However, such projects are risky for both public and private participants, must be primarily oriented toward profitability, and typically produce a landscape dominated by bulky buildings that does not encourage urbanity. Comparing recent megaprojects in New York, London, and Amsterdam, the author concludes that they represent a convergence between United States and European approaches to government intervention, as embodied in private sector involvement and market orientation.[52]

In *The Oxford Handbook of Megaproject Management* (2017), edited by Flyvbjerg, the chapter authors do not propose alternatives for megaproject development but argue that megaprojects should be the last development option, rather than the first one to be considered. They recommend the use of improved cost-estimation tools and note that learning from completed projects would enable developers to use successful techniques and avoid unsuccessful ones. In addition, Graham Winch states that a project stakeholder is any group or individual who can affect or is affected by the development of the project. The key issue that he identifies is the absence of stockholders with a fiduciary claim on the project or organization, although he ends up recommending more research on megaprojects' political and social aspects. Rhodante Ahlers and coauthors noted that dam financiers typically have no obligation to share information with the public; most dams, and most megaprojects, are built by public-private partnerships (PPPs). Graeme Hodge and Carsten Greve ask what problems are solved by PPP structures use and how well. Because PPPs are effective megacredit cards for the government, they can address budget constraints by drawing on private finance, although the U.K. government was forced to rescue projects financed by bank loans during the 2008 global financial crisis. A common rationale for megaproject PPPs is that private sector expertise and discipline can ensure efficiency, timeliness, and expenditure accuracy; however, Hodge and Greve note that there has been little rigorous and independent empirical work in the area to verify this claim.[53]

The study of urban megaprojects should be positioned within the context of urban institutional change. Processes of decentralization and recentralization in governance produce a continuous rescaling of interrelationships in the metropolitan arena. Large urban projects can no longer be considered as local projects because of their scope and their embeddedness in the frames of multiactor and multilevel governance. Many projects have been conceived from the get-go as national and global projects.[54] Such conditions call for multilevel frames of analysis that can grasp the complex relations between new projects and the multilayered networks in which they are embedded.

Global Territorial Development

In the introduction, I noted that the title of this chapter, trajectories and transitions, refers to two distinct but interdependent sets of processes: (1) the larger causal context of neoliberalism, globalization and competitiveness, megaregionalization, strategic connectivity, and increased geopolitical competition that explains the rise and development of megaprojects and their linkages with socioeconomic development worldwide; and (2) the coexisting focus in urban political economies of the past 30 years on property development, culture, and iconic architecture—together with a more recent interest in science and technology, innovation, and sustainability as strategies to foster urban competitiveness. These megaproject trajectories and transitions introduce a disorderly, disruptive, and contentious complexity into economic planning that is best managed from a critical transdisciplinary perspective focused on complexity, impacts, and sustainability. I also contend that any meaningful typology of megaprojects (see Table 2.1) that classifies them according to location, size, form, function, or purpose can be interpreted in the dual explanatory framework of trajectories and transitions. This chapter focuses on the drivers of the larger context of capitalist evolution: megaregionalization, strategic connectivity, and geopolitical competition in a transitional phase of globalization that requires economic policies that go beyond the declining neoliberal paradigm.[55]

Table 2.1 Megaproject urbanization. A typology of megaprojects

Megaprojects
URBAN-REGIONAL MEGAPROJECTS: downtown, waterfront, and neighborhood regeneration/redevelopment; convention centers; sports and culture megaprojects (stadia, museums, etc.), casinos, Olympic sites, urban technopoles, innovation districts, high-tech manufacturing megaprojects, factories of the future, water and electricity projects, energy megaprojects, IT infrastructure, economic and industrial corridors, logistics centers, SEZs, transportation megaprojects, port facilities, ship canal projects, greenfield megaprojects (new towns, new capital cities in Egypt, Indonesia, etc.).
EXURBAN AND RURAL MEGAPROJECTS: megaprojects in vulnerable rural areas, extractive megaprojects (fracking, mining, etc.), waste sites, incinerators, wind turbines, solar power stations, forest cleanup, disaster area cleanup, sea exploration, Arctic and Antarctic exploration, containerized shipping, ship canal projects, geological projects, cattle movement management projects, golf course construction, aquatic parks, smart theme parks, transfrontier parks, manufactured nature projects, forestry plantations, megahydraulic engineering projects, moving rivers, water transfer schemes, and reshored manufacturing plants.
VIRTUAL MEGAPROJECTS: public policies, electoral campaigns, high-tech infrastructure, science projects, virtual infrastructure processes (contracting, purchasing, design, labor regimes, regulatory processes), property ownership (global corporate landlords), retail (amazon.com), aerospace projects, virtual reality designs (Metaverse), project bookkeeping automation
BLUE OCEAN MEGAPROJECTS: crop biotech projects, ocean fisheries projects, deep-sea mining, climate geoengineering, industrialized harvesting, global motor vehicle assembly, geospatial modeling of supply shocks, new transnational smart highways (Amazonia, Trans-Saharan, Mongolia, etc.), permafrost-related projects, Bosphorus tube crossing project, intermodal and smart logistics, hyperloop projects, border policy megaprojects, retirement community projects, and terraforming projects

Own elaboration; several different sources

Megaregions and Strategic Connectivity

As the urbanization process develops relentlessly around the world, and global city regions become economic units of development and competitiveness in their own right, the role of megaprojects in providing infrastructure for development expands. They represent a globally embedded approach to city making, development, and competitiveness that spans cultural and geographical contexts.

These megaprojects and the megaregions where they are built are manifestations of agglomeration economies and obey the simultaneous logic of the dispersion–concentration of economic activity. Just as the conceptual phrases "cities in the world economy" and "regions in the world economy" are accepted terms in the literature, [56] it is now meaningful to use the expression "megaprojects in the world economy." The combined impact of technology and trade favors the global dispersion of economic activity. At the same time, the proximity of innovation activities and decision-making centers promotes economic concentration in large megaregions, exemplifying the intensive functional relationships of global cities in large megaregional spaces. Recently, forty megaregions were identified worldwide, representing 18% of the world's population, two-thirds of the world's economic activity, and 86% of patented innovation.[57] Thus, megaregions are a new urban form of globalization that is increasingly shaped by megaproject urbanization.[58]

In the United States, there are several differentiated megaregions, defined by economics and demography: the northeast corridor, from Boston to Washington, DC (Bos-Wa); Northern California, around San Francisco; Southern California, around Los Angeles; the Great Lakes area, with Chicago as its epicenter; the Arizona Sun Corridor, from Phoenix to Tucson; the Front Range Corridor, from the city of Salt Lake to Denver and Albuquerque (New Mexico); the Cascadia Belt, from Vancouver to Seattle; the Piedmont Atlantic group, from Atlanta, Georgia, to Charlotte, North Carolina; the Gulf Coast area, between Houston, Texas, and New Orleans; the Texas Triangle area, with Houston, Dallas, Austin, and San Antonio; and Florida, which includes Miami, Orlando, and Tampa. The aggregate population of these megaregions is estimated to reach 277 million people in 2025, and 360 million people in 2050, equivalent to 80% of the projected United States population. The gross product of two of these regions (Bos-Wa and Southern California) is equivalent to one-third of the United States GNP.[59] Although the 50 states continue to be the basic organizing principle politically, the United States is increasingly reorganized around regional infrastructure lines and metropolitan clusters that ignore state and even national borders.[60] These city regions are more economically relevant than most United States states.

Federal policy can help these nascent archipelagos thrive and help others emerge in places like Minneapolis and Memphis by funding infrastructure megaprojects that will create a grid efficiently connecting them through better roads, railroads, and fiber optic cables. In the coming decades, more than half of the population growth of the United States and almost two-thirds of the economic growth measured in terms of output will occur in United States megaregions. These demographic and economic concentrations are becoming increasingly connected by economies, population patterns, land use, infrastructure systems, topography, environmental systems, and a common culture and history. As they consolidate, they experience significant governance and decision-making challenges that cannot be resolved at the urban or metropolitan level.

Bruce Katz, of the Brookings Institution, pointed out that, of the 350 major metropolitan areas in the United States, cities with more than three million residents recovered much better from the 2008 financial crisis than did smaller cities.[61] Smaller cities and disconnected towns across the country have lost economic power since then. Yet, the 50-state model means that government resources are concentrated in a state capital, which is often a small isolated city, and funding is allocated with little sense of the whole.

The United States Congress was once the world leader in regional planning. The Louisiana Purchase, Pacific Railroad Act (which authorized the use of government bonds to fund railroad expansion from Iowa to San Francisco), and the interstate highway system are examples of the federal government's interventions into economic development on a continental scale. The Tennessee Valley Authority was a catalyst for renewal after the Depression by creating infrastructure, jobs, and industrial modernization across six states. Even though there are entities, such as CG/LA Inc., that promote public-private investment in a new national infrastructure project, Congress and the federal government still act in terms of individual states.[62] What we need today is a return to national and regional planning characterized by flexibility and guided by a shared vision, alternative energy, digital technology, and other advanced sectors.

Fortunately, on the state level, efforts are underway to coordinate regional planning and investment. Regional cooperation and planning are priorities for the National Association of Governors, as it seeks to facilitate operating across borders and replacing the logic of competition with one of coordination, planning, and suprastate cooperation. For example, the Western High Speed Rail Alliance, a quasi-governmental entity, aims to link Phoenix, Denver, and Salt Lake City with next-generation trains to promote next-generation industrial development.

Megaregionalization is proceeding throughout the world. Despite the millenary history of its cultural and linguistic provinces, China is transcending its traditional internal borders to become an empire of 19 megacity groups, each with populations of up to 100 million inhabitants. The three main Chinese megalopolises—centered around the Pearl River Delta, the Yangtze River, and Beijing (Jing Jin Ji)—stand out for their large scale (each has more than double the population of Tokyo, the largest megaregion today), their massive development made possible by numerous high-speed rail lines, and their top-down planning in an authoritarian political context. Despite the possible fragility of the Chinese model of massive investments in infrastructure, these groups of cities will be, over time, the nuclei around which the central government allocates subsidies, designs supply chains, and builds connections with the rest of the world.[63]

Western countries are following China's example. As of 2015, the most important political actors in Italy were no longer individual provinces but instead were 14 "metropolitan cities," such as Rome, Turin, Milan, and Florence: each of these cities has merged economically with surrounding municipalities to form economically viable subregions. This Italian megaregion is the third most important in Europe in terms of population and the seventh most important in the world. Britain is also undergoing an internal reorganization, with the government steering investment toward a new corridor extending from Leeds to Liverpool, known as the "Northern Powerhouse," which can become a third economic anchor after London and Scotland. Together with the London region, it is the second-largest European megaregion, trailing the huge economic and population conglomerate (60 million inhabitants) that includes Amsterdam and Rotterdam, the Ruhr area, Cologne, Brussels, Antwerp, and the Lille region and has an economic output higher than that of Canada.[64]

The connectivity that underpins the growth of megaregions is not solely a product of building more megaprojects; it derives not only from infrastructure but also from strategic development. It is not just about building more roads, railways, telecommunications, manufacturing plants, and data centers but also about carefully locating them to maximize public investment free of the constraints of regional or state borders. At the same time, for a megaregion to be successful, its development must reflect the dynamic relationships among its components and respond to the specific elements of its geographical and sociopolitical contexts.

Efficient connectivity is likely to drive the evolution of activities and economic processes. It is the strategic connection between cities and regions that will enable the United States, China, and other countries to win the battle of world trade, investment flows, and supply chains. It is possible that these economic efforts will determine which country stands as the world's first superpower in the 21st century.[65] Megaprojects play a fundamental role in maintaining the global competitiveness of nations in the future. To some extent, this competition has already begun with the ambitious Chinese Belt and Road Initiative.

Infrastructure and Geopolitics

In my earlier work, I examined megaproject development as a process of "building the infrastructure" of neoliberalism and globalization.[66] To be sure, megaprojects serve as indicators of modernization, and infrastructure megaprojects provide a framework for the global exchange of commodities, services, and resources, as well as connecting individuals, societies, and locations. Infrastructures are both physical (roads, channels, ports, airports, electricity and broadband cables, pipelines, satellite networks, energy systems, and ICTs) and virtual (contracting, purchasing, design, labor regimes, regulatory processes, and property ownership). To the extent that such megaprojects contribute to global urbanization processes, they can be considered as urban megaprojects.

Megaproject construction firms—typically multinational businesses (MNEs)—must face the differing political realities of the numerous nations they operate in, given the unpredictability of the environment, populism, and politics in both home and host countries. [67] Neoliberalism began to lose its luster after the 2008 financial crisis, and its decline has accelerated in recent years with the global shift toward a new multipolar world order.[68] Global supply networks have broken down because of the gradual weakening of international standards governing trade and the movement of money. Although geopolitics incorporates sovereign states, it is now understood to be much more complex than country—state politics.[69] It encompasses nested dimensions (such as nations within supranational organizations) and overlapping dimensions (such as ethnic groups that cross national boundaries), as well as meta-organizations or groups of organizations such as the European Union, which are made up of other groups of organizations. [70]Today businesses operate in an even more complex environment: they must deal with the effects of geopolitics and geopolitical risk on FDI and firm capabilities, the differences between developed and emerging economies, and the relative importance of political, economic, and social factors in various geopolitical environments, among other issues.

The development of increasingly sophisticated infrastructures, both physical and virtual, has enabled capitalism's globalization, molded power structures, and supported forms of dominance throughout the global urban system.[71] Consequently, they frequently are contentious. Megaprojects made possible by those infrastructures are also influenced by geopolitical and geo-economic rivalries, are frequently beset by international security tensions, or are even created with the intention of fostering political control and enhancing security. National security concerns often override commercial considerations. This has become evident in the technological infrastructure sector and the race for AI innovation. Political pressure is pushing the technological sectors of China and the United States away from each other to the point where collaboration between Chinese and American researchers and corporations may end. In the event of a new Cold War, the United States would lose a significant portion of the resources it currently depends on for innovation, while China would lose access to the markets necessary for its expansion. The mutual suspicion between the United States and China and the growing belief that artificial intelligence is a winner-take-all game are pushing the world toward a potential rupture.

Another competitive front is the Belt and Road Initiative (BRI), the trillion-dollar transcontinental plan to reshape the global economic order with Chinese characteristics. It is the most ambitious of all the current infrastructure development projects in the world. The BRI was designed to address several interconnected domestic, economic, and strategic priorities, including (1) fostering economic development in China's landlocked and underdeveloped central and western regions; (2) increasing export opportunities for Chinese manufacturers; (3) supporting national champions and modernizing China's industrial policy; and (4) guaranteeing Chinese access to strategic assets and resources throughout the Indo-Pacific region.[72] It has the potential to remake Eurasia (from China to Western Europe) into a vast, interconnected development area. Chen explored the implications of a China–Europe freight train as part of an analysis of the BRI.[73] Thees conducted a thorough analysis of the literature and concluded that the BRI prioritizes economic expansion at the expense of environmental concerns, particularly sustainability at the local level.[74] The BRI's enormous size and scope have made it vulnerable to a variety of problems, including claims of widespread corruption, lack of transparency, and "debt trap" diplomacy. [75] The United States administration and important allies have moved forward with a variety of bilateral, multilateral, and minilateral counterinitiatives to curb China's alleged predatory economic diplomacy.

Despite China's massive BRI, the ongoing race for infrastructure worldwide is more competitive than it appears.[76] The International North–South Transport Corridor and the East–West Corridor were established by India, the second-largest economy in Asia, in the early 2000s to connect nations in South Asia, Central Asia, and West Asia to a new global trade and investment network. In an attempt to provide a credible alternative to the BRI, leaders of India, the United States, Saudi Arabia, the UAE, Europe, and Japan announced in 2023 a mega joint infrastructure deal to connect Asia with Europe via ports and railroad. Since 2017, South Korea has similarly established its own "New Northern" and "New Southern" policies that are designed to strengthen its commercial ties with the resource-rich Eurasian landmass and the expanding economies of Southeast Asia. Japan, a treaty ally of the United States, launched its Connectivity Initiative and Partnership for Quality Infrastructure in 2016, which gained the EU as a partner in 2019.

Increased urbanization, connectivity, and investment in urban infrastructure in Asia are having profound effects on geopolitics and territorial dynamics. There is usually some resistance to creating a supply- chain empire, and we are seeing this worldwide, but especially in Pakistan, Southeast Asia, and Africa. The economic powerhouses' militarization of their supply networks, at times triggered by anti-Chinese unrest, is a response to this resistance. These cities, in which resistance is centered, serve as power bases for their nations or regions; many are located where the new urban infrastructure of Eurasia intersects. The enhanced communication between China, West Asia, Gulf States, and Europe will also have a significant territorial influence on Central Asian nations.

These rapid changes in geopolitical power on the Eurasian continent and elsewhere are related both to the complicated history of Western outsourcing and globalization politics in the 1950s and the 1960s and to the growth of Asian megacities as global manufacturing hubs. Simultaneous processes of urbanization and globalization have created Asian megacities as the foundations of geopolitical power.[77] They play key roles in internal and global economic connections and access to global supply chains (via the creation of special economic zones [SEZs]), illustrating the clear relationship between geopolitics, urbanization, and megaprojects. [78]

Western countries are also developing connectivity strategies. According to the Paris Club criteria for debt and environmental sustainability, G7 nations donated more than $113 billion in development assistance for international infrastructure projects between 2016 and 2019. The Biden administration has undertaken several bilateral, minilateral, and multilateral projects. A $4.5 billion joint high-tech infrastructure project that the United States and Japan inked in 2023 is designed to increase cooperation in the development of cutting-edge industries, from 5G networks to artificial intelligence and renewable energy. A trilateral United States–Australia–Japan initiative, the Blue Dot Network (BDN), also recently held its first stakeholders' meetings in Paris. Top experts, international investors, and managers of sovereign wealth funds from around the world gathered at the event to talk about new platforms for increasing high-quality infrastructure investments.

Shortly after the BDN's first gathering, the Biden administration unveiled the Build Back Better World (B3W) program as a "values-driven, high-standard, and transparent" rival to China's BRI. Its mission is to "collectively catalyze" and consolidate the strategic overseas commitments of G7 powers and to secure trillions of dollars held by Western investors and financial institutions, including sovereign wealth, pension, and mutual funds, as well as insurance company holdings, toward this end. The BDN's and B3W's mission is also to strengthen international standards for infrastructure investments.[79]

Thus, megaprojects are indicators of the interrelationships between connectivity and geopolitics.[80] They are not only drivers of economic development but also serve as key elements of geopolitical and international relations strategies.[81] The intersection between international relations and infrastructure development via megaprojects occurs at several levels—from global geopolitical competition to regional competition and in the daily lives of those participating in and affected by infrastructural developments. The relationships between infrastructure and security interactions in a global world order may help untangle how logistics and infrastructure relate to both geopolitical power and commercial interests. It may also explain how the intertwining of military and commercial interests led to the spatial materialization of unequal power relations, some of which generate extreme violence.

Megaprojects that rely on interconnected global supply chains, complicated financing, and the power of authoritarian leaders are subject to increased geopolitical risk and tend to generate massively disruptive effects. We examine one such project, Istanbul's Kanal, in Chapter 3. It is evident that, without an understanding of urbanization and today's cities (increasingly driven by megaproject development), it is barely possible to comprehend geopolitical struggles in the 21st century. The reverse also holds: we should know the fundamental dynamics of global and regional geopolitical development in recent and current global megaproject development as the so-called knowledge-based economy unfolds.[82]

Authoritarian Developmentalism

Despite the appeal of the Chinese infrastructure-based authoritarian development approach, critics have highlighted its weaknesses. Ren points out that the large investment in and development of megaprojects was made possible by deregulatory measures in the real estate and housing sectors, as well as the establishment of local investment corporations. Before 2008, holding megaevents such as the Olympics frequently served as a rationale for the development of megaprojects. After 2008, the economic downturn provided a new justification: the Chinese government put in place a significant stimulus program that increased investment in infrastructure megaprojects. Many local governments (including those in Shanghai, Beijing, and Guangzhou) found themselves in serious debt because of overinvesting in infrastructure megaprojects over the previous twenty years, which caused the Chinese economy to slow down in the middle of the 2010s.[83]

In a persuasive study published in 2016, Ansar, Flyvbjerg, Budzier, and Lunn show that infrastructure investment fails to generate a positive risk- adjusted return and thus is far from being a driver of economic growth. [84] A boom occurs by investing in unproductive projects while construction is ongoing, followed by a bust when projected benefits are not realized and the projects end up being a drain on the economy. When investments are debt-financed, excessive investment in unproductive projects leads to debt accumulation, monetary growth, financial market instability, and economic fragility, as observed in China at the time of this study. The authors remind readers that major projects that have a positive effect on the economy are less common than is typically believed. They conclude that China's emerging economic and financial problems are mostly explained by poorly managed infrastructure investments. If China does not reduce spending on high-quality infrastructure, it will face a financial and economic crisis driven by its infrastructure that is likely to extend to the global economy.[85] Events over the years since this study partly confirm this prediction.

Many high-profile projects, particularly megaprojects such as new skyscrapers, rapid transit lines, airport expansions, sports and entertainment complexes, and future cities, are being planned in the Middle East and are either in the design phase or under development. Rich Arab oil-producing countries have been able to finance a large portion of the costs of building sparkling offices, hotels, resorts, and entertainment facilities. However, these ostentatious projects must navigate a minefield of dangers, including fluctuating revenue from oil prices, the possibility of a worldwide recession, increased material costs, and labor shortages. Because of the magnitude and scope of these new megaprojects, as well as variable (and perhaps reduced) earnings from dropping oil prices, Arab countries are being forced to look outside their borders for funding. [86]

These projects also require substantial personnel and stringent quality controls, both of which may be in short supply due to their sheer magnitude and complexity. The manufacturing and distribution of building supplies have been slowed by trade disputes, covid-19, and slowdowns in China's production, potentially delaying the completion of these massive projects.[87] Given geopolitical unrest in the area, it is uncertain whether new businesses or tourists want to come to the area.

Arsel and coauthors argue that authoritarian developmentalism, with its emphasis on infrastructure investment, is the latest stage of neoliberalism.[88]There are varieties of authoritarian developmentalism in Morocco, Egypt, and other countries that differ from that in China. Chapter 3 discusses the case of Istanbul, [89] as well as the Barangaroo megaproject in Sydney, an instance of neoliberal urbanism ("financialized municipal entrepreneurialism") in a Western-style democracy. The differences between liberal democracies and authoritarian regimes regarding sustainability strategies and outcomes are not large, especially given the shortcomings and deficiencies of megaprojects that are common in both contexts.

Embeddedness Begets Complexity

We adopt a transdisciplinary approach to the analysis of megaprojects that includes insights from complexity theory and practice, urban and regional political economy, urban design, architectural theory, impact assessment, urban studies and sustainability studies. "Complexity" here refers to assemblages in which inseparability, inter- retroactivity, interactivity, and interdependence prevail between the elements that form it and the subject of knowledge and its context. French *savant* Edgar Morin's description of complexity informs this book:

> Pertinent knowledge must confront complexity. *Complexus* means that, which are woven together. In fact, there is complexity whenever the various elements (economic, political, sociological, psychological, emotional, mythological, etc.) that compose a whole are inseparable, and there is inter-retroactive, interactive, interdependent tissue between the subject of knowledge and its context, the parts and the whole, the whole and the parts, the parts amongst themselves. Therefore, complexity is the bond between unity and multiplicity. Developments proper to our planetary era confront us more frequently, ineluctably with the challenge of complexity.[90]

The Latin word *complexus* means "intertwined" or "twisted." It can be defined as a union of two or more elements that constitute a unit comprising different elements. Here, we find the basic duality between parts that are simultaneously different and connected: to be complex, a unit requires two or more components linked in such a way that it is difficult to separate them. We cannot analyze these units by breaking their elements into independent modules because those components cannot be separated without destroying them. Thus, analyzing complex entities, using models for prediction or control, and solving complicated issues will be challenging: they are wicked problems. In addition, complexity simultaneously includes order (the connection between components) and disorder (i.e., variety and heterogeneity). Therefore, units are permanently in an unstable equilibrium, "even to the edge of chaos."[91]

Urban complexity can be said to emerge from decentralized and self- organizing webs, assemblages, and networks of interactions and transactions among many different individuals, agencies, and stakeholders; these interactions generally occur at several scales in dynamic, hazy, changing, and uncertain metropolitan settings. These cooperative and competitive transactions and exchanges, influenced by chance and serendipity, reflect the actors' perceptions, decisions, and preferences.[92] Agents, actors, and stakeholders can be individuals, a community, a city, or a region, and social, economic, and political institutions. Their mutual interactions produce feedback loops that allow the adaptation of individual and group actors and the emergence of phenomena, patterns, and outcomes—physical, behavioral, social, economic, ecological, and environmental—that cannot be predicted by analyzing particular webs, assemblages, networks, and their constituents and components.[93]

Insofar as they are influenced by a variety of interrelated causes, social and behavioral phenomena are inherently complex. Thus, it is necessary to examine how several elements interact or combine to produce certain results. Both megaprojects and cities are complex systems situated (embedded) within multiscalar political, economic, and sociocultural systems and contexts. Megaproject impacts are also complex, disruptive, and emergent (as we will see in Chapter 7), producing the concept of *complex sustainability* proposed in Chapter 8.

Project complexity can be approached from the perspectives of project type, technology, organization, and environment; actors involved in megaprojects can be understood from the perspectives of leadership, stakeholder behavior, resource capabilities, and human behavior. In addition, biases, emotions, and emotional intelligence can be examined as sources of decision-making complexity. Uncertainty, an inherent feature of complex systems, can be analyzed through effective risk management, project planning, and proactive project monitoring and control systems.

"Embeddedness" refers both to the structural and relational interdependencies of projects to the urban systems in which they are located.[94] Cities are viewed as complex adaptive innovation ecosystems and large-scale matter systems far from equilibrium that consume energy, process information, and adapt to the environment. Their subsystems are interconnected through multiple relationships and interdependencies. Complex cities self-organize in ways that show unpredictable nonlinear dynamics and can modify themselves to fit new environments.[95] Sets of relational and structural interrelationships and interdependencies develop between the project and urban system. In both, internal and external environmental conditions affect organizations, people, systems, and processes. In summary, embeddedness begets complexity, which manifests as *disruptive complexity*: complex webs of interrelated project impacts on urban systems.

Complexity perspectives in the social sciences are not new: a distinguished tradition of social-scientific scholarship has embraced complexity science in various ways and extents. [96] Here, I briefly mention the work of Immanuel Wallerstein and Manuel Castells, both of whom decisively contributed to the advancement of empirical and theoretical knowledge in the social sciences since the mid-20th century. Wallerstein tells the story of his relationship with Ilya Prigogine, a pioneer in complexity science and a Nobel awardee in chemistry, in *Uncertain Worlds*.[97] Prigogine famously proposed the idea that equilibrium processes are a special and unusual case of physical reality, which can be demonstrated in a core element of classical physics: dynamic systems. This idea translates into the fundamental indeterminacy of all reality—physical and social—which exists in a mode of "deterministic chaos": order exists for a while, but then it undoes itself when it reaches a point of bifurcation.

Wallerstein points out that this is not a matter of incomplete knowledge but of the impossibility of foreknowledge. In turn, it calls for an "unexcluded middle"—made up of determined order and inexplicable chaos—that, according to Wallerstein, contributes to our understanding of Braudel's rejection of both particularisms and universals; however, there are orders, such as structural time, that undo themselves and come to an end. The application of Prigogine's ideas to world-systems analysis led to abandoning methodologies based on scientific, Newtonian certainties and mechanical worldviews and to differentiating between periods of "normal" development and moments of structural crisis. Thus, systems have lives, moving from beginnings, to normal development, and to terminal crises. Further, the idea that there is no time reversibility (the arrow of time) implies that social systems are historical systems, and the building of the discipline of historical social science is a way of overcoming the distinction between economic, political, and sociocultural realms.[98]

The concept of complexity highlights how systems balance order and chaos, that a system does not always advance toward equilibrium, and that events can have unforeseen outcomes and irreversible impacts. Consequently, certain occurrences can have unintended consequences that are frequently out of the ordinary in terms of time and place. Complexity suggests that global emergent forms arise from the intricate interactions between networks and flows. Any global theory based on well-bounded systems or clear-cut impacts cannot adequately explain megaprojects. Complex situations lack clear boundaries: what happens "outside" continuously restructures what is "inside," and various processes interact to produce change. Transdisciplinarity is a result of complexity.[99]

Manuel Castells entered the field of urbanism with his 1972 book, *La question urbaine*[100] and later embraced a complexity perspective in his acclaimed theory on the information age in *The Information Age: The Rise of the Network Society*. In that book, written in 2000, he explores the patterns and dynamics of the network society in its cultural and institutional diversity.[101] Castells defines the network society in his next book as "the social structure that results from the interaction between social organization, social change, and a technological paradigm constituted around digital information and communication technologies."[102] He analyzes the technological, cultural, and institutional transformation of societies around the world through the lens of electronic communication networks in business, everyday life, public services, social interaction, and politics. Using a relational, complexity perspective, Castells is interested in variations in degree and complexity visible in the penetration of the network society around the world and the many different manifestations it adopts in response to different societal and cultural preconditions.[103] Thus, understanding complexity requires observing the complex, multifaceted, and uneasy balance between a general theory and the contextual production of societal outcomes.

In keeping with the complexity approach, we do not curtail the fuzziness of empirical openness and complexity so as to produce a clear-cut set of theoretical propositions. The common thread among all the chapters is their emphasis on megaprojects as an embedded organizational form and a boundary object: this explains why the focus of the book is not just on these projects but rather on the articulations of megaprojects as embedded in the world economy. To avoid the "presentism fallacy," I suggest considering a time perspective, in a limited sense. This is consistent with a complexity perspective and is especially important in our current transitional era. In this book we treat megaprojects as complex, disruptive structures; they are also fragile as "dissipative structures" far from equilibrium.[104] Complexity analysis has both a topological, descriptive dimension (of parts that are interrelated in networks) and a dynamic dimension that accounts for contradictions, obstacles, and disruptions.

We posit that megaproject impacts are "entropic impacts" generating disorder and disruption. Chapter 7 explores the idea of "disruptive complexity" (any given web of interrelated megaproject impacts), which the book's empirical analysis shows to be a common feature of megaprojects regardless of type and location. The embeddedness of megaprojects in the world economy begets complexity, and in turn, complexity begets disruption. The question of sustainability needs to be pursued within this analytical framework, which the book does mainly in Chapter 8: it explores the idea of "complex sustainability" as a multifaceted response to megaproject disruptive complexity.

Disruptive Complexity: Risks and Impacts

Megaprojects and Risk

An important part of the literature on megaprojects focuses on the analysis of risks affecting them in their planning, construction, and management phases. In *Megaprojects and Risk*, Flyvbjerg and his colleagues identify a "megaprojects paradox": an increasing number of these projects continue to implemented, often with substantial cost overruns and market shortfalls, despite their dismal performance records. According to the authors, this poor performance occurs because many participants in the process have incentives to underestimate costs, overestimate revenues, undervalue environmental impacts, and overvalue economic development effects. These incentives exist because of inappropriate risk sharing and a lack of accountability, which can be addressed by reforming the institutional arrangements of decision-making and by instituting accountability at the project development and evaluation stages.[105]

Flyvbjerg and colleagues also argue that innate human biases, such as "uniqueness bias,“ undermine the standard assessment tools and processes utilized to evaluate megaprojects, such as Monte Carlo simulations. As an alternative, “reference class forecasting,” based on the work by Kahneman and Tversky, proceeds by collecting data on past megaprojects, so that future megaprojects can learn from real-world case studies.[106] Black swan events also affect megaprojects with devastating effects because megaprojects take such a long time to complete.[107] Furthermore, the scale and complexity of tasks involved in megaproject planning and construction mean that risks may stem from or be magnified by a lack of effective communication among stakeholders.

The literature offers many different classifications of risk in megaprojects.[108] The following typology is a basic classification:

- **Task-dependency risks**. This occurs when Task P (predecessor) must be completed before Task S (successor) can begin. In these dependent situations, a delay in one task is likely to affect all future tasks, thereby disrupting the entire project. Modularity has been proposed as a possible solution to isolate and control task-dependency risks.
- **Communication-related risks**. The two typical features of megaproject construction are (1) a multilayered organizational structure and (2) the existence of a significant number of dependencies. Communication risks come not only from the usual human cognitive biases but also from the complex picture created by these two typical features. Communication failures can be due to the so-called Rashomon effect, in which the same event, process, task, or need is described—and understood—in slightly or significantly different ways by all the people who were involved with it.
- **Regulatory risks**. This refers to threats to disrupt the progress of a project derived from regulatory requirements (including building size, health and safety regulations, or sanctions for noncompliance). Project leaders should devise a comprehensive strategy to ensure compliance by paying close attention to the regulatory framework and policies throughout the megaproject life cycle.
- **Environmental risks**. Environmental risks are becoming increasingly important and range widely: civil society protests in preconstruction, issues of contamination or pipe discovery during construction, problems related to maintenance and operations, weather impacts, ground conditions not conducive to construction, and bureaucratic problems related to licenses.
- **External risks**. These risks exist outside the project team and organization and so are often more difficult to predict and control. They include many kinds of

events derived from political and socioeconomic conflict situations, as well as processes of policy making and other routine workings of socioeconomic and political systems. External threats must be identified and categorized before they are considered within a risk breakdown structure, with special attention to their impact on the project plan.

From Risks to Impacts

To better understand the significance of the idea of risk and the practice of risk management as fundamental components of contemporary society and the analysis of megaprojects, we need to briefly discuss the work of Ulrich Beck. In his work on risk, Beck reflects on how the forces of globalization, individualization, the gender revolution, underemployment, and global hazards are interwoven and manifested in events such as ecological disasters and the collapse of global financial markets. Beck examined the concept of "risk calculus," the sociology of risk—how some interest groups benefit from "manufactured uncertainty"—and the use of guesswork in response to crises. In his *Risk Society: Towards a New Modernity* (1992), Beck states that "risk may be defined as a systematic way of dealing with hazards and insecurities induced and introduced by modernization itself."[109] Later, in the book, he writes, "In contrast to all earlier epochs (including industrial society), the risk society is characterized essentially by a lack: the impossibility of an external attribution of hazards. In other words, risks depend on decisions; they are industrially produced and politically reflexive."[110]

Beck rejected the conventional idea that risk management ought to be practiced as merely a technocratic and bureaucratic exercise in which views from civil society and citizens are treated with contempt.[111] His analysis shows some overlap with the debate on "the limits to growth" that the Club of Rome undertook in the 1970s. The mathematical calculations and probabilistic scenario of World-3—the computer model with which the Club of Rome analyzed the multiple global interactions among population growth, industrial production, food production, and ecosystem limits—were used to define current and tolerable standards of living. The perceived threat then spawned an ad hoc global risk community and study and discussion groups around the world that analyzed the Club of Rome report. World-3's simulation of the ecological trends challenged many taken-for-granted rules of everyday life, perhaps because the simulation offered an unprecedented dystopian projection of global decomposition. Beck, for his part, conducted a less mechanistic and a more reflexive and constructivist analysis adjusted to the parameters of complexity and uncertainties of sociopolitical reality and its ideological constructs.

Neither the notion of the "limits to growth" nor that of "global risk" can be understood without accounting for the capitalist project. Today, capital is presented as a Schumpeterian force that is quasi-revolutionary and capable of resolving the antinomies of the global risk society through its hegemony over the future. However, there has been a paradigm shift in futures studies from "forecasting" to critical analysis, because predictive and extrapolation tools have not been proven effective or accurate in a global risk society.

Beck defines risk analysis as the modern approach to anticipate and control the future consequences of human action, particularly those that are unintended results of radicalized and accelerated modernization. Systemic events—geopolitical and financial crises, the Exxon Valdez oil spill, the Challenger disaster, and the covid-19 pandemic, among many others—broadcast worldwide by the mass media have shown that modern societies are generating risks that they cannot control and that threaten their survival.

Growing awareness of the reality of risk has dramatically altered our perception of time. A future attack—something that is nonexistent by definition—has been displacing the past as an obsession that has a decisive influence on the present, producing distortions in the understanding of the problems of that present and their possible solutions. This change in perspective—from the historical perspective to the universe of projections—is reflected in international risk management, which was developed by financial and insurance systems. The emphasis in corporate planning models on the construction of scenarios is omnipresent. Michael Porter's "Five Forces" and McKinsey's "7S" models reflect these altered perceptions.

A risk society is a "dappled world," in Nancy Cartwright's formulation.[112] No single elegant theory can account for a world that is not completely ordered; some features are precisely ordered, others act with some regularity, and others behave in diverse ways. This patchwork makes sense when we realize that laws are special productions of nature that require special arrangements for their generation. In this context, risk perceptions and materializations contribute to disorder and yet warrant a transdisciplinary approach to the dappled world. In this paradigm, there is no longer any trace of the Newtonian universe of order or determinism in the Western modernist project.

The concepts of risk and resilience are best understood as part of a conceptual set. As Davis argues, the notion of risk permeates resilience strategies in a fundamental, unavoidable way because it is essential to design actions that mitigate or minimize adaptive vulnerabilities, instead of exacerbating them.[113] Risk is an essential and foundational element of human and nonhuman adaptive processes and, therefore, of complex societies, as Ilya Prigogine persuasively explained.[114] However, risk does not lie only in the realm of science and the "factual." Risk as a readable phenomenon or concern is informed by power and social issues, including who has the right or authority to define it, how it is distributed, and who pays for and who benefits from it.

> It is therefore important to avoid the "tyranny of risk" as a defining principle of action and understand that risk discourses can be abused to justify oppression, controls on citizenship, the appropriation of the right to territory, and other forms of exclusion. This challenges the principles of fairness and justice that should guide civic and professional behavior, both individually and collectively.[115]

Impacts can be defined as probable consequences or materializations of risks, exposure, and vulnerability. A risk is then the probability of an impact. Because some materialized risks (impacts) are never identified in the risk phase and therefore are not included in any sustainability planning, impacts can sometimes, but not always, be predicted by examining the risks. As stated earlier, megaproject impacts are *emergent properties of megaprojects* that cannot always be predicted in advance in a meaningful or comprehensive manner. The uncertainty of emergent impacts manifests in three main ways: (1) impacts cannot always be predicted, (2) impacts evolve over time, and (3) impacts are interrelated or interdependent. Chapter 7 in this book develops the notion of *disruptive complexity*, which refers to the complex web of interrelated and interdependent impacts that emerge after megaproject construction and affect the urban living conditions of large numbers of citizens, particularly those who are more exposed and vulnerable. Beck's view of risk, which portrays risk as fundamentally embedded in society, and Davis' conceptualization of risk as part of a conceptual set, together with the idea of resilience, are fruitful avenues for shifting the focus of analysis from risks to impacts. The embeddedness and complexity approach we follow in this book focuses on the types of risks and impacts produced by megaprojects because of their embeddedness in the polity, the economy, and society. Megaprojects constitute landscapes of disruption, and such disruptions are often the cause of systemic disorders or increased entropy. This shift from risk to impact represents a fundamental departure from many conventional analyses of megaprojects as organizational systems.

The analysis of megaproject impacts on urban systems cannot overlook responses to impacts, which consist of both adaptation and change. These two response types lead into a discussion of the concepts of resilience and sustainability as a logical consequence of an analysis of megaprojects that focuses on embeddedness, complexity, and impacts.

Resilience and Sustainability

In 1973, Holling introduced the concept of resilience to the ecological literature as a way to understand nonlinear dynamics and the processes through which ecosystems self-maintain and persist in the face of disturbance and change. According to Holling's definition, resilience highlights the conditions of a complex system far from equilibrium that can be transformed so much by instabilities that it then presents as a different behavior regime. Thus, resilience is measured by the magnitude of disturbances that can be absorbed by the system before it is reorganized into an entity with different variables and processes. In turn, sustainability is the ability of a complex system to maintain itself over time, despite environmental volatility fostered by learning, transformation, renewal, and evolution.[116]

Van Meerbeeck, Jucker, and Svenning argue that resilience increases the probability of avoiding unwanted changes to "stability domains" and also provides the flexibility and the opportunity required to develop sustainable systems. The idea of "stability domains" is a concept from the field of ecology. Ecological stability is defined as the overall ability of a system to remain in the same attraction domain and retain its function and structure in the presence of perturbations. As it relates to current socioglobal events, avoiding unwanted changes to "stability domains" will be one of the most important challenges in a world increasingly dominated by human beings who are in increasingly aggressive interactions with their environment. Another factor conferring resilience to ecosystems is the diversity of biological communities. This is because diverse communities have a higher chance of including one or more species with traits that can adapt to changing environments.[117] And resilience requires an ability to adapt to change.

Resilient systems can recover quickly from changes, disturbances, and perturbations, such as fires, flooding, windstorms, insect population explosions, and human activities: deforestation, fracking for gas and oil extraction, pesticide spraying in soil, and the introduction of exotic plant or animal species. If the ecosystem is affected by changes and disturbances to the point that a threshold is passed, the system changes qualitatively and a different regime of processes and structures predominates. This new regime may constitute a critical transition if it is associated with "bifurcation points."[118]

The concept of resilience has three defining elements: (1) the amount of change or transformations that a complex system can withstand while maintaining the same functional and structural properties; (2) the degree to which the system is capable of self-organization; and (3) the ability of a complex system to develop and increase the ability to learn, innovate, and adapt.[119] Gao and coauthors in 2016 developed methodologies to analyze multidimensional complex systems that enabled the separation of system dynamics, thereby enabling one to understand the behavior of each element of the system. They determined the point at which a net reaches its critical point of *resilience*, whether it is an ecosystem (the interrelationship between plants and animals) or a technological system (the cascading collapse of servers on the Internet). Their work implied that the resilience of the original network is predictable. Knowing the critical point of resilience strengthens the system's resilience (its networks, nodes, and flows), thereby preventing small changes from becoming tipping points leading to irreversible damage; for example, the total loss of biodiversity or significant population displacements caused by infrastructure construction.[120]

Urban resilience can be an emergent property of a city system—seen as a socio-ecological system—that arises only through the interactions of its components. Thus, a strategy to increase urban resilience cannot rely solely on the individual resilience of its components but must focus on their interactions. Increasing resilience requires new approaches to explain the dynamic relationship between shocks and stressors, as well as the outcomes of well-being. One approach involves complex adaptive systems that exhibit historical dependency characteristics, discontinuous changes, multiple balances, and nonlinearity.[121] Hopkins (2014) points out that increasing resiliency and strengthening the local economy require making more preparations now for a future without waste, with greater self-sufficiency; one strategy is to prioritize what is locally produced over what is imported.

To reorganize and recover from disturbances, a system should have three fundamental characteristics: diversity, modularity, and effective feedback.[122] Researchers from the Santa Fe Institute (especially Erica Jen) have studied "robustness" mechanisms in economic, social, and ecological systems (which are properly considered complex phenomena). They characterize "robustness" as the magnitude of volatility that can be compensated for by a complex system before the collapse of its main characteristics, processes, and functions. Their work aims to identify the dynamics common to these systems so that they can develop a theory to guide the complex systems of our world today to become increasingly sustainable.[123]

Going beyond robustness, Taleb (2014) proposes the notion of "anti- fragility" as a disposition whereby systems—and, presumably, people—learn from disorder and benefit from stress, volatility, and turmoil. What Taleb terms "antifragile" refers not only to systemic situations that benefit from chaos but also to those that need such chaos to survive and flourish. Avoiding disruption for fear of its consequences is an indicator of fragility, and the illusion of safety makes systems vulnerable to shocks. Robustness means standing up to shocks without compromising the essential features of the system.[124]

The concept of resilience has changed from being purely descriptive to including a normative agenda. Resilience sometimes focuses on individual entities and at other times on the system as a whole. This leads to the question of the "resilience of what, to what, and at what scale," to which geographers can contribute from their space–time perspective and society–environment systemic approach.[125] Governments and organizations must view resilience as a process, state, and quality. It ranges from the global level, focused on food security; the national, related to critical infrastructure (energy and water) and the economic sector; to the local, in terms of climate change. On the international level, the *Sendai Framework for Disaster Risk Reduction, 2015–2030*, focuses on improving urban resilience through the planning and development of quality infrastructure. This effort goes hand in hand with the adoption of ecosystem-based approaches and integrated data-driven policies for disaster risk reduction and management.

It is somewhat surprising that, despite decades of examining the concept of resilience in the social sciences, questions of power, governance, and social capital still do not play an important role in theoretical and practical approaches. Building resilience requires addresses elements related to historical and sociopolitical processes that create and maintain social vulnerability, as well as developing intervention projects that guide cities and societies toward less vulnerable paths.[126] This neglect is also surprising because the notion of "resilience" can easily be skewed toward the ideological—either because it can offer an excuse to leave citizens to fend for themselves while governments and markets help each other, or because resilience refers to a desire to return to a normality that has been and continues to be the cause of the planetary problems we face, according to Davis. It can be mistakenly conceived as "getting back to normal" after a disaster or as a means of restoring system balance after a shock. Focusing on resilience often means avoiding difficult questions about power, inequality, and the impact of limited resources on people.[127] Embracing resilience can translate into having faith that, with enough attention and adaptive effort, the future can be better, an expectation that needs to be contextualized or nuanced in some way.

In this book, I propose a complex conceptualization of sustainability in Chapter 8 as a response to megaproject impacts. The complex nature of sustainability contributes not only to the scope and variable shape of sustainable practices but also to the well-being of the overall socioeconomic context, characterized by crisis and uncertainty in which sustainability practices are implemented as *strategies to contain systemic risks*. Challenges facing sustainability include (1) unpredictability introduced by mechanisms of globalization that produce action at a distance and the increased inequalities and consequent transnational migration flows they have caused; (2) complexity in the global territorial organization, which reflects not only an incessant and rapid global urbanization drive but also the formidable challenges of the ecologies of towns and regions;[128] (3) the relative decline of the West and the tectonic shift in the center of gravity of the global economy to Asia, coupled with geopolitical multipolarity and the rise of geo-economics and geo-technology;[129] (4) the profound disruption of production and labor triggered by the informational and technological revolution of the last thirty years;[130] (5) the emergent understanding of the Earth system as a variable, responsive, adaptive, and self-regulating mechanism in the Anthropocene, which calls for recentering (or, better, decentering) within the universe of life, the human being, and its mechanisms for the production of knowledge and transformation of the environment.[131]

Much research has studied how systems can develop, learn, adapt, and persist over time, based on the fundamental concept of organizational resilience, which is directly associated with sustainability in any complex system. There has been a shift in the understanding of resilience from "bouncing back" to "bouncing forward." As a result, "resilience" can be understood as a part of a conceptual set together with the idea of sustainability and ought to be approached from a dynamic perspective depending on variations in time and space. Further, if a system begins to "lose" resilience, the "potential for change" increases; that is, the possibilities of moving to a different organizational state or configuration increase, even if the system had already been subject to small disturbances or disturbances that were previously insignificant or had not produced any adverse effect. From this angle, both resilience and sustainability should be understood as components or vectors in any systemic self-organizing process.

Resilience and sustainability are directly related to the long-term consequences of transformations and changes and their impact on the future profile of societies, economies, and the human system as a whole. Urban systems are complex adaptive innovation ecosystems with the potential for both resilience and sustainability. Complex system transformations are inevitable, and these changes do not follow a logical linear course: they can be continuous, inevitable, gradual, abrupt, local, global, required, promoted, or unexpected. This leads to the concept of "panarchy." Panarchy is a conceptual framework that accounts for the dual and seemingly contradictory characteristics of all complex systems: stability and change. It posits that socioecological systems operate at multiple geographic scales and that feedback operates both internally and externally.[132] Small systems can undergo changes in short periods because of the possibility that individual actors can exert great influence; those that operate on larger scales may experience change only after long periods of time, since a greater number of interactions between a large number of actors will be required.

Complexity theory suggests that properties of larger systems are generally shaped by interactions at lower levels.[133] In any complex system, individual nodes (individuals, businesses, communities, and even entire countries) are able to draw support and resources from elsewhere but should be also self-sufficient to meet their essential needs in an emergency. Yet, the key task of a living organization, whatever it may be, is not only that of "functioning" but also that of being able to face errors, uncertainties, and dangers: they need to have both strategic and evolutionary abilities.[134] The important thing is not only to adapt but also to learn, invent, and create.[135] History shows us that cities are less resilient to sociopolitical and socioeconomic events than to physical events, such as pandemics, droughts, and floods. However, when physical events trigger sociopolitical and socioeconomic problems, cities may not recover.[136] As discussed in Chapter 8, sustainability strategies as a response to megaproject impacts may need to focus on the socioeconomic dimension of sustainability, an element largely neglected in recent transitions to "green" and "high tech" forms of urban projects.

Conclusion

Megaprojects are deeply disruptive and not just "transformative." They drastically alter urban structures by causing disturbances or problems, even in optimal planning, construction, and management scenarios that deploy the best combination of strategies to optimize rational decision-making. For instance, when gentrifying effects and displacements occur after a megaproject occupies an urban neighborhood, the driver of such disruptions is not the megaproject performance quality level; rather, it is the mere fact that the megaproject was built and started to disrupt urban life in the neighborhood. This shows that disruptive megaproject impacts, which take shape through the embeddedness of projects in the polity, economy, and society, must be a part of any meaningful understanding of large-scale projects. Project embeddedness directs our attention to issues such as clustering and proximity, centrality of location, being part of a strategic initiative, iconic elements, involvement of urban elites, possible transnational connections, and the redefinition of urban public space.

Thus, rather than focusing on the risks affecting megaproject performance, we focus on the impacts of megaprojects on urban areas. Some projects become "megaprojects" because of their regional economic and environmental impacts and the resulting complexity of their external stakeholder management challenges.[137] Given the fact that megaproject impacts develop, materialize, and change over time, I refer to *megaprojectification* as the process by which some urban projects become megaprojects, from planning to impacts.[138]Megaprojects are, then, a *process*. Although some projects are planned, designed, and built as megaprojects, I do not use any mechanistic, quantitative, and universal definition of a megaproject based exclusively on cost, size, or scale and instead emphasize their embeddedness and impacts. It follows from this that a "sustainable megaproject" (to the extent that this phrase is not an oxymoron) can only be planned and built by factoring in its impacts on cities and regions. Accordingly, the empirical chapters in this book (Chapters 3, 4, 5 and 6) are largely devoted to exploring and assessing the impacts of megaprojects on their urban areas.

Part II
Repertoires of Disruption

Chapter 3
Landscapes of Disruption

Introduction

This chapter discusses urban megaprojects in Istanbul, Hong Kong, and Sydney, giving a panoramic view of the controversies, difficulties, obstacles, negative impacts, and civic opposition associated with their construction. Each project is shaped by the socioeconomic, developmental, institutional, and geographic contexts from which it emerged. However, these urban megaprojects, regardless of context, are contentious enterprises that often elicit substantial controversy and criticism that fundamentally question the parameters of the projects as envisioned and as publicly presented by their promoters.

One reason why megaprojects are contentious is that they require substantial financial investments, which may drain local budgets and substantially alter the priorities of local and even national governments. Megaprojects are also contentious enterprises because their stakeholders often have conflicting interests regarding planning, construction, management, and governance. These conflicts frequently produce major obstacles for megaproject implementation, strategic misrepresentation of costs and benefits, optimism bias among planners and promoters about megaproject impacts, and a widespread perception among urbanites that these structures are harmful for their cities.

Megaprojects' disruptive nature also manifests itself in that large projects substantially modify the physical appearance of cities and their urban fabric, often triggering socioeconomic imbalances and realignments in urban power arrangements in growth machines and civil society. Thus, urban megaprojects, regardless of their context, constitute *landscapes of disruption*. This chapter illustrates three cases in which disruption is a consequence of megaproject embeddedness and can be viewed as materialized megaproject complexity, as argued in Chapter 2.

Indeed, as we shift the focus from megaproject planning and management to their impacts, the disruptive character of megaprojects comes to the fore. Such

G. del Cerro Santamaría, *Contesting Megaprojects*,
https://doi.org/10.1007/978-3-031-98382-5_3

disruptive complexity (interrelated and interdependent sets of impacts) is shown in the *repertoires of disruption* discussed in this chapter, as well as in those discussed in relation to the themes of the next three chapters: megaprojects as emblems, smart destructive creation, and greening. The idea of disruptive complexity also centers the analytical discussion in Chapter 7.

In contrast to the iconography of order and control presented in the megaproject literature that analyzes megaprojects as closed systems, we see that they generate disorder, controversy, and many negative impacts in the cities where they are built. To the extent that disruptive impacts signal megaproject failure, we posit that such failure is not only a matter of "underperformance" that can be solved by maximizing rational decision-making: it is a result of the structural mismatch between the nature, goals, scope, and magnitude of (most) urban megaprojects and the well-being of urban citizens. Such structural mismatch ought to lead us to rethink development pathways focused on growth, international visibility, and competition in the global arena. Such pathways are still hegemonic despite recent transformations in the global political economy.

Kanal Istanbul: The *Crazy* Project

The complicated urban renewal processes in Istanbul reflect the national-level dialectics of political and economic power that have implications for Türkiye's international standing and prospects for economic growth. The Justice and Development Party (Adalet ve Kalkinma Partisi; AKP), which is still in power after many years, has used megaproject urbanization as a key tool for nation-building, globalization, and international visibility. In a city that produces approximately 40% of the national GDP, urban renewal through megaprojects has ignited local citizen resistance and contributed to political confrontations and the realignment of political options to foster a more environmentally friendly Istanbul under the comprehensive Vision 2023 plan, where megaprojects, including the Kanal, are one of the key components.

During the Erdoğan era, Istanbul has been experiencing an extraordinary construction boom that has clearly been engineered by the party in power. A key component of the political economy of the AKP era has been acquiring privately owned land for public purposes. As Yesilbag recently argued, the AKP has been able to develop strong ties with a new generation of firms in the construction arena, to strengthen its control of domestic capital, and to create new mechanisms to finance party politics through acquiring this land.[1]

The politics of growth became extremely attractive to the AKP as a means to maintaining political hegemony. It was seen as generating attract larger amounts of FDI and thus continuing a process that uses megaprojects as its main tool to acquire and commodify land. From 2011 onward, the AKP has also developed megaprojects as a means of holding onto and reenergizing its electoral base.[2] As argued by the Center for Strategic and International Studies,

> Erdoğan's national projects have furthered a foundation of political patrimony achieved via urban development primarily in the country's northwest, with megaprojects surpassing the conceptions of Ottoman sultans. By announcing Vision 2023 in 2011 to commemorate a hundred years of the modern Turkish Republic, Erdoğan is attempting to consolidate his authority by providing impressive physical symbols of progress, fitted with Türkiye's historical context.[3]

The government started several infrastructure projects in 2013 as part of Erdoğan's 10-year, USD 200 billion construction plan. Road and rail tunnels beneath the Bosphorus Strait, one of the biggest airports in the world, and two large suspension bridges in the world connecting Europe and Asia were completed. In Dardanelles, a second international bridge was recently completed. Several smaller projects were also built. President Erdoğan's reelection in 2023 neither changed the plans nor the timetable for the Kanal's completion, even if the project development seems to have recently slowed down, and completion is uncertain. In the case that the Kanal never materializes, it would be due in part to the widespread controversy generated among leaders and citizens in Istanbul in view of the massive negative impacts of this megaproject.

The Kanal's Disruptive Impacts

The Kanal Istanbul (Fig. 3.1), dubbed "a crazy project" by Erdoğan himself when he first proposed it in 2011, would likely transform Istanbul and Türkiye in very significant ways, if construction proceeds and is completed.[4] Its aim is to reduce current navigation and ship traffic problems in the Bosphorus Strait, easing the passage of oil tankers through it. Cargo traffic in the strait has increased greatly; in the last five years, an average of 42,000 ships passed through the strait annually. This is approximately three times the number of ships passing through the Suez Canal and four times the number through the Panama Canal. Simultaneously, the share of dangerous goods also increased from 25% ten years ago to 35% today.[5]

As a parallel seaway to the Bosphorus Strait, the Kanal would bisect the current European side of Istanbul and thus form an island between Asia and Europe. The Kanal would not only provide an alternate route for ships but would also integrate highway and railroad transit through the bridges across it that would connect the island to the mainland. From the outset, it was understood that it would generate the development of new towns and subcenters, producing changes in urban density and design.[6] This development was projected to benefit Türkiye's struggling economy, reduce unemployment, and attract FDI, particularly from Gulf and Arab countries.

Given that Türkiye's foreign currency reserves are decreasing and the Turkish lira has consistently been depreciating against the United States dollar and the euro,

Fig. 3.1 Location of the projected Kanal Istanbul and the Bosphorus and Dardanelles Straits. Source: Shutterstock.com

investment in the real estate industry has become the primary method of bringing in foreign money and strengthening the economy ahead of future elections. Real estate investments made up 57% of net FDI in 2020, compared to 37% in 2019 and 13% in 2007.[7] The funds generated by new zoning regulations and the approval of real estate projects on 36,453 hectares of land surrounding the Kanal will underwrite most of the project's estimated costs.

Urban Spatial Impacts

To build Kanal Istanbul, two new smart cities must be built along its banks. These cities would house at least 500,000 people and include marinas, container ports, logistics hubs, six bridge crossings, two deep subway crossings, and a traditional railroad crossing that would connect to the new airport in Istanbul.

Despite these promises of development and wealth, the Kanal has many vocal opponents. Haluk Levent, an economics professor at Istanbul Bilgi University, described the project as "the first stage of a Ponzi scheme" that would fuel inequality: "The growth model predominant in Türkiye for a long time is based on the monetization of rent, and, as a result, has led to a deepening of social problems such as unemployment and income inequality." According to the Turkish Green Party spokesperson Koray Dogan Urbarli, "The agricultural land that was acquired cheaply will suddenly be turned into residential areas, and we will watch as some of the lucky few get richer."[8] Ekrem Imamoglu, the mayor of Istanbul and a member of the main opposition party, describes the Kanal project as "one of the worst hazards this city faced in its entire history." He advocated that funding be used instead for social welfare programs and to improve education, the water supply, public transportation, and seismic safety: "The Canal Istanbul project is a major and risky surgery that no one would want to go through unless it was necessary. We are talking about a surgery that will chop off the city and fundamentally damage its vital systems."[9]

Strategic and Geopolitical Impacts

The Turkish Strait comprising the Bosphorus Strait and the Dardanelles links the Black Sea to the Mediterranean Sea: it is a major shipping chokepoint for the transport of oil and food. Four percent of the world's oil passes through the Turkish Strait; given Kazakhstan's increasing output, it is anticipated that oil traffic across the straits will increase. In addition, much of the oil destined for the EU comes from Central Asia and the Caucasus via this strait. The Turkish Strait is also part of the world's most important grain transit route, which originates in Ukraine and Russia: it carries one-fifth of the world's exports of wheat and one-sixth of its maize exports.[10]

One of the AKP's major goals is to make Türkiye a significant player on the geopolitical stage. According to Constantin Corneanu, president of the Gheorghe Bratianu European Association of Geopolitical and Strategic Studies, "The Kanal Istanbul will become an instrument in the Erdoğan regime's arsenal in his constant efforts to include Türkiye in regional geopolitical games."[11] Its establishment will strengthen Türkiye's position in discussions about the status of the Black Sea, enabling Ankara to hold direct talks with significant international players on global issues.

Although Turkish politicians frame their support for the Kanal in economic terms, Türkiye's main aims involve realist geopolitics. "Ankara raises its geopolitical leverage vis-à-vis the international community by having an exclusive authority over this geo-strategically pivotal strait," according to the NATO Defense Foundation. In creating the Kanal,

> Türkiye will have the capacity to strategically set regulations on maritime traffic at discretion and hence increase its own geopolitical influence in the area. The opening of a new waterway, subjected to Türkiye's sovereignty and disengagement from the Montreux Convention, will give Ankara the chance to bypass the limits imposed on its strategic control of the straits by the Convention.[12]

Concerns about Ankara's compliance with the Montreux Convention have also been expressed in relation to the proposed Kanal. The Convention, which went into effect in 1936, grants Türkiye authority over the Bosphorus and Dardanelles and ensures access for civilian vessels during peacetime, while restricting the passage of warships. It limits the number of ships, their tonnage, and their ability to carry weapons through the two Turkish straits. It distinguishes between the rights of Black Sea states and those of other nations, thereby undermining Türkiye's authority over the straits. Erdoğan has stated, however, that the Kanal has "nothing to do with Montreux" but would allow Türkiye to "completely establish our own independence, our own sovereignty." Emphasizing the geopolitical importance of the project, Erdoğan has said, "If Türkiye is going to be a global actor, the Kanal Istanbul Project is a necessity, rather than only a dream."[13] Because of its noncompliance with the Montreux Convention, a significant impediment to the construction of Kanal Istanbul is concern expressed by other nations about Turkish control of the project.

The major powers have also expressed a lack of support for the project. Russia is opposed to it because it would potentially decrease its strategic power by giving NATO ships a more direct path to the Black Sea. Nor does the United States seems to be supportive of the project, even if it has not yet publicly rejected the proposal. The development of the Kanal has both benefits for the European Union, by limiting Russian influence, and drawbacks—increasing Ankara's international role and influence.[14] There are internal political challenges to the project as well. In an April 2021 open letter to the Turkish people, more than 100 retired Turkish admirals condemned the plan to construct a waterway parallel to the Bosphorus. Ten of these signees were later taken into custody by the authorities on suspicion of plotting a coup.

The Turkish megaproject is thus a striking illustration of how geography can be used strategically and geopolitically. In times of increasing geopolitical and geoeconomic competition, the Kanal would elevate Türkiye's role from that of a member of an alliance that does not prioritize its interests to that of a principal able to govern its own affairs.[15] Its completion is projected to improve Türkiye's geopolitical position in the area, increase other nations' flexibility in dealing with it, and highlight Türkiye's significance as a significant player. It will serve Türkiye's own geopolitical objectives and can be used as a negotiating chip with significant international players.

Environmental Impacts

Urban development in Istanbul has reduced its air quality, the amount of green space, and the health of its flora and fauna while increasing waste. For example, the Third Bridge and its associated motorway pass through the northern border of Belgrad Forest (adjacent to Istanbul) and the Bosphorus Biodiversity Area on the European side, posing serious threats to a wide range of local habitats and exacerbating the heat-island effect. The airport, which opened in March 2019 and is located on bird migration routes, has increased noise and air pollution. Higher levels of air pollution decrease the quality of life in the city and lead to a higher incidence of respiratory diseases and deaths.

Gulcin Ozkan, professor of finance at King's College London, argues that the cost-benefit analysis of such projects does not take sufficient account of environmental harms: "Most megaprojects in Türkiye entail significant environmental costs, as consistently put forward by experts, environmentalists, and professional bodies such as the Union of Chambers of Turkish Engineers and Architects, and consistently disregarded by the government."[16] In highlighting the risks of deforestation, anthropogenic climate change, and other ecosystem problems, Dogan and Stupar note that, for the Kanal as for most megaprojects, "The key issue is the danger of losing green areas for business development. The forests in the north contain rich flora and fauna, water basins, and natural resources, representing protected areas."[17] The residents of Istanbul are generally unaware of the environmental threats posed by the Kanal nor that many "will lose their homes as lands are emptied for construction."[18]

According to environmental experts, large construction projects (Kanal and the recently constructed third airport) could jeopardize the city's water supply.[19] The Kanal is being dug into a wooded area, which is home to freshwater sources that have been crucial parts of the city's water supply system since the 17th century. The projected route would damage marine and freshwater ecosystems, as well as lakes and rivers supplying Istanbul's water, Lake Durusu (Durugöl) among them. One of the reasons why Mayor Imamoglu opposes the project is that it will be built on the European side of the economic capital, where 40% of the city's water originates.[20]

The effects on the ecosystem will likely be made even more harmful. The new Kanal is anticipated to upset the natural balance of currents and countercurrents between the Black Sea and the Sea of Marmara. Some scientists assert that the human-made canal will work as a siphon, drawing contaminated Black Sea waters into the Mediterranean Sea. [21] Much worse, development of the Kanal is likely to result in irreversible environmental harm, triggered by the expected inversion of the hydrological balance between the cold and fresh waters of the Black Sea and the warm and salty waters of the Mediterranean Sea.

The Sea of Marmara has recently experienced an alarming increase in white foam production. White foam is known popularly as "sea snot"—a clear portrayal not only of its harm but also its unpleasant nature. It is produced when industrial and other pollutants are combined with organisms at sufficiently high temperatures to form a goop that covers the water and can harm marine wildlife. The construction of an artificial passage such as the Kanal, according to experts, could further pollute the Sea of Marmara. Additional disastrous consequences affecting water supply have been reported in the case that the Kanal project proceeded. [22]

Cemal Saydam, an environmental engineering professor at Hacettepe University in Ankara who has studied the area for many years, claims that if the Kanal is built, it will create problems with waste treatment. The current system makes use of Bosphorus crosscurrents, which have a denser flow moving north at the bottom than the surface flow from the south. Wastewater and leftover waste are currently being discharged at the confluence of the Bosphorus with the Marmara Sea, where it is carried north by the undercurrents in the Black Sea. In the 1990s, a total of 67 waste treatment facilities were built that are claimed to take care of 97% of Istanbul's waste and garbage. The city is also adding several more advanced biological treatment plants.[23]

Saydam and other scientists have expressed concern that the Kanal Istanbul project would upend the delicate balance of aquatic life and disrupt the city's distinctive waste system, which was designed after years of modeling and scientific research. The city might begin to smell of hydrogen sulfide because of the shift in salinity, which might also result in an anoxic state in the water. Saydam stated,

> We are warning about this project, saying it's not feasible due to oceanographic reasons. If you decide to join the two seas, you cannot think of the next five or ten years, the next election, or the anniversary of the Turkish republic—you have to think in terms of geological time, because once you do this there will be no way to turn back.[24]

Financial Risks and Impacts

The megaproject's financing mechanism is based on a guarantee for its corporate investors of a certain number of ships passing through the Kanal, so the accuracy of estimations of the growth in shipping traffic has significant financial implications. If that guaranteed number is not achieved, the government will have to pay companies an amount established in their contracts. When a similar guarantee was not satisfied for the Osman Ghazi Bridge, the Turkish treasury had to pay the Otoyol Yaririm A.S. company more than three billion Turkish lira (USD 28 million), because the actual number of toll payments was far lower than the contract stipulated.[25] According to a report commissioned by the Ministry of Transport and Infrastructure, 54,900 ships will pass through the new Kanal in 2026, increasing to 68,000 in 2039—yet this represents a sharp reversal of current trends. Because of the increased number of oil pipelines and a decreased reliance on fossil fuels in developed countries, the number of vessels passing through the Bosphorus has fallen from 53,000 to 38,000 per year over the last decade.[26] It seems likely that the Kanal will not meet its projected number of ships.

Fluctuations in the Turkish lira, which has lost about 60% of its value against the United States dollar since 2018, reflect Türkiye's vulnerability to the logic of globalization's developmental contractions. The severe depreciation of the Turkish lira in recent years has exacerbated both contractors' and the government's liabilities while driving up service prices for consumers at facilities where such prices were denominated in dollars.[27] Turkish companies with debt in foreign currencies, particularly in the United States dollar, also saw an increase in their service costs. Foreign firms are reluctant to enter the shaky and institutionally fragile Turkish market by investing in the country's infrastructure megaprojects.

Where investment in Turkish megaprojects has occurred, it was based on the expectations that the Turkish economy would continue to grow at least 5% annually and that foreign funds would continue to pour in, keeping foreign exchange rates more or less stable. However, these expectations have not been realized. The economy began to slow in 2014 due to global economic fluctuations, an increase in the interest rate, and a decrease in foreign capital flows to Türkiye.

Currently, Turkish megaprojects structured as public-private partnerships have a contract value of USD 155 billion, which includes USD 77 billion in investments and USD 78 billion in commitments from the government as periodic payments or rents.[28] The risks involved in committing such high investments cast a shadow over grandiose construction projects such as Istanbul's new airport and the Kanal. The presence of foreign currency guarantees, often for up to 25 years, has reduced their value to Türkiye. A loss similar to that experienced by the lira in recent years "sharply raises the fiscal implications of state guarantees for these major projects."[29]

Thus, the Kanal project poses a grave threat to Türkiye's economic security. A 2019 government report estimated costs of approximately USD 13 billion, excluding likely cost overruns. Building the needed bridges and tunnels will require billions of dollars more in hard currency, which is not readily available. Turkish bankers are increasingly wary of investing in the Kanal due to "environmental concerns and investment risks."[30] Erdoğan's 2023 reelection as president strengthened his commitment to improving Türkiye's economic profile through massive investments in megaprojects. Yet Türkiye's debt is already sizable. How future governments and citizens will pay the debts incurred every day by the country's megaprojects is a difficult challenge.

Two main options are available for financing the Kanal, which may be more financially prudent than increasing the Turkish sovereign debt: securing China's participation and issuing infrastructure bonds associated with specific project components. These options would surely increase Turkish sovereign debt and will not reduce the likelihood of the nationalizations of megaprojects (already taking place) that are hemorrhaging financial losses, such as the new airport, one of the three key projects for Vision 2023.[31] China may be willing to invest in Türkiye's megaprojects to increase its clout in the region as part of its Belt and Road westward investment offensive. China has already financed the third bridge over Bosphorus and the tunnel beneath it. China may condition its funding on reduced Turkish control and greater Chinese control over the Bosphorus Strait. This situation would exacerbate existing tensions between Türkiye and its Western allies, as well as Russia, with potentially far-reaching consequences.[32]

Regarding the issuing of bonds to finance the Kanal, this mechanism is used when there is reduced global liquidity, which makes capital markets less appealing. The Turkish Capital Markets Board published a communiqué on project bonds regulating project-based securities (PBS) and project financing funds on July 17, 2021. The communiqué states that PBS will be issued to finance public sector projects such as transportation, infrastructure, energy, industry, technology, communications, or health.[33] Turkish banks and investment institutions need the approval of the Capital Markets Board to use project finance funds. A PBS may be issued through a public offering in the United States to qualified investors without a public offering or through a private placement with a specified nominal value. The PBS can also be made available in international markets.[34] Experts have noted that the success of project bond financing is far from guaranteed.

Contractors are fully aware of the financial fragility of the megaprojects. To protect themselves, contractors for seven of Türkiye's megaprojects have signed deals worth USD 17.2 billion in foreign loans. Ambitious megaprojects may not become symbols of Türkiye's grandeur but rather burdens that future generations of taxpayers and visitors to the country will find difficult to bear.[35]

Istanbul's limits on growth are being tested, with megaprojects dominating urban space and new infrastructure. There are serious concerns regarding the long-term viability and appeal of development plans based on massive infrastructure and construction investments. Dogan and Stupar argue the following:

> The anticipated expansion toward the north (Kanal Istanbul, new airport and new highway) is intended to set up new hubs of global competitiveness; however, the problems of jeopardized natural resources, property rights, and expected migration influx could violate the exiting urban fabric and deepen social friction … Therefore, the neoliberal attractiveness of new projects could easily disappear in the near future, especially when faced with the latest globally promoted imperatives of environmental awareness and social cohesion launched by many competing cities in the global hierarchy.[36]

Sydney's Barangaroo Megaproject

In the State of New South Wales (Australia), megaprojects are classified as "state-significant projects": these are typically sites that the Planning Ministry designates as having broad social, economic, or environmental significance for the community, such as universities, hospitals, employment precincts, and major residential developments. A list of state-significant sites in New South Wales (NSW) includes the Sydney Opera House the UTS Ku-ring-gai Campus site, and Central Park in Chippendale. The Central Park, famous for its cantilevered reflector and hanging gardens, is a major development across the road.

A major precedent for the Barangaroo megaproject was the 2000 Sydney Olympics: the projects were developed using similar legislative and planning frameworks. The Olympic Coordination Authority (OCA) Act of 1995, the Sydney Organizing Committee for the Olympic Games Act of 1993, the State Environmental Planning Policy No. 38–Olympic Games and Related Projects, and the Sydney Regional Environmental Plan No. 24–Homebush Bay Area are some of the many acts passed by the Australian government to consolidate state planning authority for the Olympic Games and undermine local government control of the developments.[37] As claimed by Owen, "This legislation consolidated and centralized planning powers with the State Government and with the statutory authorities that were responsible for that government, particularly the OCA." The legislative changes removed planning power from local governments, vesting consent authority for "all development for the purpose of an Olympic Games project or an OCA project' with the Minister for Urban Affairs and Planning (UA&P)."[38]

The dearth of grassroots representation in the OCA's decision-making processes stemmed from the absence of effective planning power at the municipal level. To oversee Homebush Bay's development into a top-notch Olympic Park, the OCA "excluded the Local Council [authorities] that previously had authority over the property from the regeneration process." As one Auburn Council Officer observed, the Local Council was neither involved "in the planning of... physical infrastructure at the Park or... activities/events in it, which was driven by a lack of cooperation and interest," thereby "reducing their power within local environmental decisions that may affect them."[39] This, in turn, sparked worries among residents that only those with ties to political figures would be able to obtain the money and influence needed to bring about changes in their neighborhoods. Because the state government dominated the Local Council, there was a "lack of accountability" in the neighborhood.[40] Some aspects in the planning framework have very recently been modified in New South Wales. However, the lack of significant mechanisms for effective metropolitan planning has been a key criticism of the Barangaroo megaproject since 2010.

Australian decision making around megaprojects continues to take place at the state-level; cities, metropolitan authorities, and municipalities lack any authority to decide on "state-significant projects," which renders them irrelevant in negotiations among the large variety of stakeholders that are usually involved in megaproject planning and implementation processes. Intergovernmental decentralization has not accompanied institutional changes in Australia. Nor has there been debate on the urban level regarding the merits of the global city approach.[41] In contrast, in the EU, there was a concerted effort to increase urban competitiveness via the metropolitan "rescaling" of urban regions.

Barangaroo

Barangaroo, an urban waterfront development and iconic urban megaproject (IUMP), is a major redevelopment of Sydney's foreshore city (Fig. 3.2). The Sustainable Sydney 2030 initiative includes both Central Park and Barangaroo as key sites. The Barangaroo megaproject has several components. Barangaroo Reserve, also known as Barangaroo North, is a "companion" to Bennelong where the Opera House is located. Central Barangaroo is now in an advanced phase of development and includes a new metro station. Barangaroo South is a mixed commercial and residential development managed by the Lend Lease management company.

Fig. 3.2 Aerial view of the Sydney Harbor with the Barangaroo Reserve and megaproject buildings in the foreground. Source: iStock.com

Securing the necessary approvals to develop the headland into public space required the direct involvement of key government officials, including the governor of the New South Wales (NSW) state and the prime minister at the time, Paul Keating. Their public statements emphasized the Barangaroo site's importance to local Indigenous people and were supported by the NSW Aboriginal Land Council, which also issued statements that the megaproject would benefit Indigenous culture and heritage in NSW. The Barangaroo Delivery Authority

(BDA) was created to manage city waterfront development, to deliver buildings that would be world-class benchmarks in urban design, and to ensure the project's environmental management and sustainability—all while maintaining a commitment to the cost-effective completion of the megaproject. The BDA coordinates state government's involvement in the Barangaroo Reserve, Central Barangaroo, and Barangaroo South.

According to several Australian scholars and my own observations in 2016, Barangaroo reflects trends visible in megaprojects around the world, supporting the conception of megaprojects as a global urban form and, more importantly, that their disruptive capacity is a global urban norm. Megaprojects embody the tension of competing agendas in urban development: economic growth versus environmental protection, public versus private interests, and conservation versus renewal.[42] They reflect the adoption of international development models and a move to market-led infrastructure provision, thus shifting the focus from public to private interests. More than any distinguishable local Sydney identity, Barangaroo embodies an international metropolitan brand, the privatization of public space, and privileging private over public interests.

The southern third of Barangaroo has been designated a "write-your-own-rule make-your-own-deal" development zone. Proposals that support the public interest have neither been articulated nor supported, which results in a lack of transparency and distortion of public processes, as well as the public's exclusion from the planning process. The Sydney Harbor Foreshore Authority and the BDA, the two institutions in charge of the project, have consistently disregarded the needs of the general population.[43] Yet all the levels of government in NSW have shown a lack of strategic investment in the public sphere by (1) eliminating advisers knowledgeable about the needs of local citizens, (2) effectively privatizing waterfronts, and (3) outsourcing urban projects that then do not hire local contractors and construction workers: this all leads to a failure to create public spaces for the good of the residents. Although such public spaces are essential to civil society, proposals for buildings to serve civic or communal uses are conspicuously lacking.

Critics point out that the governmental authorities utterly lack what may be described as "public imagination" and are failing to imbue projects with cultural vitality and liveliness as well as with values of equity and sustainability. Rather than conceiving a more holistic understanding of the Barangaroo project, the authorities and stakeholders have approached waterfront revitalization as distinct and separate from the city in its form, scale, and expression. Instead of connecting with Sydney's western flank to complete the longest waterfront in the city center, the project is located apart from the city. Sydney's center indeed serves as the center of many of the city's interests, including politics, culture and the arts, religion, and memorials, and Barangaroo ought to reflect this character.[44]

Iconic megaprojects, because they serve as catalysts of urban change, generate both political and economic contention and disrupt local, regional, national, and international arenas.[45] The development of Sydney's Barangaroo has been controversial from its inception. Critics in the international and local press referred to Barangaroo as a "monster development in Sydney Harbor," "contemptuous," and a "shocking breach of the public's trust." Scott Hawken describes in detail how Barangaroo's initially optimistic designs for the six- hectare headland park at North

Barangaroo were subverted, raising concerns about their authenticity—although they did retain the plantings' extensive botanical diversity as a distinguishing feature. The involvement of Crown Resorts in the Barangaroo project and its plans to build a high-roller casino there have generated extensive public criticism. The project did have a few supporters; for instance, Bruce Mackenzie describes Barangaroo as a "great contribution to a modern Sydney context through the use of design thinking, vegetation, soil science, and teamwork."[46]

Yet, the shortening of the planning process, with the tacit approval of local politicians, caused great ill will and criticism.[47] Professor Mike Harris from the University of New South Wales (UNSW) Sydney states,

> Generally, the sense here among researchers, practitioners and government (besides the few "winners") is that the process of Barangaroo was heavily flawed and created a lot of bad blood in the industry. A senior planning official described this as a slow-moving train-wreck. It has been astounding to observe how the plan changed to accommodate vested interests, with very little due process, in an advanced Western democracy. This operated far outside of any understanding of the usual development practice.[48]

Although Australian megaprojects share several characteristics with other international megaprojects, they also have several distinctive features: the dominant influence of the real estate industry, less strategic coordination with other land assets and transportation projects, and decreased public benefit outcomes, adherence to legally mandated planning frameworks, and engagement with local knowledge. The Barangaroo megaproject incorporated all these elements.[49]

Political Impacts

The development of Barangaroo is aimed at ensuring Sydney's status as a globalizing city at the center of financial services in the Asia-Pacific region. In the process it has contributed to increased complexity in local and state political processes. An "erosion of intent" regarding its initial aims and goals has generated multiple controversies and led to a highly controversial and complex political landscape.[50]

Johnston and Clegg use the concept of "circuits of power" to describe the processes of public management in the Barangaroo development; they focus on strategic agenda setting, rules-based leadership, contract relations, no-cost-to-government policies, organizing by precedent, and stakeholder management. They demonstrate that, at each stage in the process, these processes were characterized less by a rhetoric of public management and more by a distortion of this rhetoric caused by the flow of complex politics through distinct circuits of power. Their major finding is that public management in the context of a large economic infrastructural development, especially when the government is attempting to position a city globally, is far more complex and political than the prevailing rhetoric of the new public management scheme of considered rationality.[51] This confirms the "disruptive complexity" hypothesis I use in this book to explain the dynamics and impacts of iconic megaproject development in globalizing cities. For example, Stickells describes the mismanagement of the tender process for the Barangaroo redevelopment initiative, in which the international developer Lend Lease was selected to develop and create the first stage of the project, replacing a team of architects who won the project's design competition in 2006.[52]

Like most megaprojects, Barangaroo has exhibited a significant lack of transparency and accountability in governance.[53] A series of adjustments were made to increase development profitability while diminishing public benefit, beginning with replacing the plans created by a team of architects with the international developer Lend Lease, as just mentioned. This change occurred without any discernible independent reviews or public participation, though former Prime Minister Keating's investment in and influence over the project were very visible. It was seen as "perverse" that such a large enterprise should be subject to less scrutiny rather than more.[54]

The original master plan had a generic nature that could have been easily exploited to allow an "anything goes" development in the future.[55] Even one of the architects who designed it betrayed his cynicism and knowledge of how public assets are produced in NSW by referring to the design as "Sydney-proof," which meant that no matter how poorly it would be implemented, its legacy of streetscapes and modestly proportioned buildings would persist.[56] The team's participation in the project was then terminated. The government soon realized that it did not have the resources to develop the entire area and its base infrastructure of streets and parks, so instead turned to selling smaller development packages to the market.

Mismanaging public expectations, involvement, and communications can be argued to be the outcome of prioritizing the requirements of the property and leisure industries over the needs of citizens. As long as governments place more emphasis on cities as profit makers rather than the well-being of urbanites, they will continue to generate distrust of politicians and the places they force on the public (with the help of the wealthy residents).[57] As an example of the "erosion of intent" proposed by Harris, Barangaroo demonstrates how promises of "public benefit" were supplanted by the state's interests in reducing risk and boosting economic returns.

Megaprojects may thus be a tool for "financialized municipal entrepreneurialism": they certainly create "anchor capital" by remaking the built environment to the benefit of investors and developers, but they also use public land speculatively to generate substantial financial returns for state governments. As MacDonald has argued, Barangaroo development shows that urban development need not primarily seek to benefit private developers. By acting as a quasi-public firm driven by speculative ambitions, public authorities may turn the goal of "public benefits" into a strategy for increasing public revenue.[58]

Socioeconomic Impacts

Innovation is one of the main factors influencing the development of the Barangaroo. The adjective "innovative" has been used to characterize sustainability, technology, and the design performance of megaprojects. The emphasis on these marketable aspects appeals to a worldwide audience because they can be easily compared using metrics for sustainability and images that showcase improvements in technology and design. However, innovation rarely addresses the socio-economic dimensions of urban sustainability within a location. On the contrary, innovative megaprojects cause disruptive impacts associated to high-tech gentrification. One of the main goals of the plan and design of the Barangaroo is that it meet the global standards of an innovative megaproject: one that is sustainable, technologically advanced, and designed according to global standards. These marketable aspects appeal to a worldwide audience and can be easily compared to projects in other countries. However, that does not mean that they would necessarily improve the well-being of Sydney's residents. The NSW government has a track record of delivering development motivated by an external locus that aims to improve not urban living but rather Sydney's competitive position within regional and international networks. The privileging of economic motivations over social and cultural aims inevitably limits the types of institutions and spaces that are planned, resulting in a patchwork of development.[59]

Barangaroo's disruptive complexity is also reflected in the contentious nature of the negotiations around its central foreshore location in Sydney, a prime site for redevelopment. As in most other iconic megaprojects that are at the core of urban redevelopment schemes, this political contention centered on one key issue: the right of citizens to urban public space versus the private interests of developers. The prominent political ambitions of local officials and the pressures from globalization (as perceived locally) drive cities to make extraordinary efforts to align with international trends, even at the cost of jeopardizing many local citizens' standards of living and exacerbating urban inequality.

Iconic Barangaroo development has remained contentious during construction, clashing with the City of Sydney's attempts to find a solution to extremely high housing prices. In recent years, many residents have left the city in search of affordable housing. According to the most recent plans, "key worker housing"—for nurses, teachers, and first responders who are vital to the city's well-being—must only occupy 2.3% of the residential floor space in Barangaroo South, a much lower percentage compared to similar international projects. When new housing projects are built that are only financially available to a small portion of the population, social boundaries are formed: these places essentially become enclaves for a specific socioeconomic group.

Similarly, only business behemoths such as KPMG and Westpac will be able to afford the high commercial rents in Barangaroo. Consequently, retail stores, restaurants, and other entertainment-oriented services will primarily serve those who are well off. For instance, David Jones (the department store) "has been meticulously crafted to appeal to the major office community in Barangaroo." The new eateries, which are run by well-known chefs and restaurateurs, also attract customers from affluent socioeconomic backgrounds. One can mention the now apparent popular success of the area as a dining and entertainment venue and the recent completion of the public Wulugul foreshore promenade trail. All in all, it remains difficult for the less fortunate to patronize the site's residential, retail, or commercial facilities.[60]

Environmental Impacts

Private construction companies that benefited from the development of past projects, such as the Sydney Olympics, have had a significant and continuing impact on state and municipal government after the Games. As Pauline McGuirk found in 2003,

> There is a group of private sector and non-government organizations with a permanent presence in Sydney's governance landscape. In the private sector, a series of general business associations and specific business practice groups represent the major fractions of capital that dominate global Sydney: financial and business service corporations, property development and investment interests, major manufacturing corporations, and retail and tourism interests.[61]

As in many other cities, pro-growth coalitions in Sydney control the growth machine that, as stated earlier, work not at the urban level but at the state level in Australia. The Barangaroo development, the most significant globalizing initiative in Sydney thus far in the 21st century, has been shaped by the dealings of state-level politicians and Australian business leaders. The disruptive character of the Olympic mega-event is also seen in Barangaroo, although with its own distinct features.

Several organizations have voiced environmental concerns about the development of the area. Barangaroo South, like the Olympics, is built on an area where, for many years, many enterprises contaminated the soil. For example, although Millers Point Gasworks was demolished in the early 1920s, it left behind subsurface facilities, including tanks.[62] There was fear that toxins might be released into the environment in the project's development phase. The proposed five-star hotel, built on a pier projecting into the port, raised concerns that aquatic life would suffer irreparable harm.

Because of the presence of hazardous materials in some areas of Barangaroo, pollution has become a major public concern. The Barangaroo Delivery Authority (BDA) has touted the safety of other projects located on contaminated sites, including the Sydney Olympics. According to the Barangaroo Remediation Fact Sheet, Sydney Olympic Park, after extensive remediation, was made completely safe and has become a popular park. The BDA also assured the public that the cleanup process would not cost any taxpayer money and that the costs associated with these activities would be covered by the developer of Barangaroo South. Despite these reassurances, critics and many residents remain worried about the health and environmental consequences of developing contaminated land.

The Barangaroo megaproject's buildings are made of carbon-neutral materials and receive a supply of recycled water from a storm water treatment facility powered in part by rooftop solar energy. In addition to the project's financial benefits, which are expected to enhance the NSW economy by USD 2 billion annually, the environment will benefit from the project's increased green space, biodiversity, and reduced carbon emissions, power use, and water use. However, achieving such benefits will require the significant displacement of disadvantaged populations. Thus, such developments may result in "green gentrification," in which these benefits are only available to a privileged few.

Hong Kong: West Kowloon Cultural District

The West Kowloon Cultural District (WKCD) aims to turn Hong Kong into a world-class arts and cultural destination by fostering cultural and creative industries in the city and becoming a cultural gateway for the Pearl River Delta region.[63] WKCD (Fig. 3.3) houses a very large performance center; an exhibition complex; a museum (Museum+) for visual culture; and a variety of theaters: the Center for Contemporary

Fig. 3.3 Hong Kong's West Kowloon Cultural District in the foreground and panoramic view of the city. Source: iStock.com

Performance; the Thrust Theatre, primarily designed for large contemporary dance ensembles and spoken-word dramas with amplified music; the Proscenium Theatre, designed for medium-scale theatrical and dance performances with amplified music; an outdoor theater integrated within the landscape design of the Great Park; a music center comprising a concert hall and a recital hall; the Xiqu Center containing a large and a small theater, as well as a commercially operated teahouse; and an enclosed Freespace with Music Box, a cluster of two indoor venues.

Turning a city into cultural capital is a complex endeavor. The need to change one's image and attract the attention of worldwide media adds to the typical architectural, planning, and infrastructure development challenges. Debates over divergent political ideologies and the lack of agreement between professed goals and true intentions make designing and developing cultural capital difficult. Construction-related environmental harm is a significant aspect that complicates an already disruptive, disorderly, and controversial situation. Top-down urban cultural planning and attempts to create cultural capitals have become a preferred strategy for municipal and regional power brokers in the West and other developed regions of the world over the past thirty years. Iconic cultural projects such as WKCD carry intrinsic symbolic power that affects the socioeconomic and sociopolitical material realms.[64]

Impacts on the Arts Ecology

Development of the WKCD has been driven mostly by Hong Kong's economic interests that seemingly concentrate on tourism. However, as stated by Tung Chee-hwa, the first Chief Executive of the Hong Kong Special Administrative Region (HKSAR), the development of tourism was not designed to advance culture but rather to strengthen Hong Kong as a "traditional economic base." In the midst of an economic crisis in 1998, he declared that the government intended to use megaprojects to distinguish Hong Kong as "Asia's international city" and "Asia's entertainment capital." The "promised" results of megaprojects, as shown in Bilbao, Pittsburgh, Baltimore, and other cities, were exciting to Hong Kong leaders after the Asian financial crisis in 1997–1999.[65]

Prestigious museums such as the Pompidou Center in Paris and New York's Guggenheim Foundation soon expressed interest in constructing museums in Hong Kong: Thomas Krens, the former director of the Guggenheim Foundation, referred to the initiative as "the most exciting opportunity in the world because of the scale and the location," and then French President Jacques Chirac visited Hong Kong in 2004 to express the Pompidou Center's interest in the city.[66] The WKCD project hired faculty from famous universities and curators from top-tier art museums as consultants. For instance, Lars Nittve, former executive director of the Moderna Museum in Stockholm and Tate Modern in London, was recruited for a similar position at the M+ Contemporary Arts Museum in West Kowloon.[67]

Some researchers claimed that because the strategy focused on "the bundle sale of the development of creative industries," the development proposals for WKCD would not benefit local culture. Civic organizations such as Project Hong Kong and

Citizen Envisioning a Harbor feared the project would solely benefit real estate developers and urban elites, who have a clear preference for Western culture.[68] Local artists were concerned by the deep involvement of Western cultural institutions; they feared that WKCD would prioritize the works of international artists to attract new art buyers and generate income for the city at the expense of local artists and cultural organizations.

Given Hong Kong's historical status as a sort of "culture desert," local artists sought to develop a distinctive, locally based set of projects for the city's cultural sectors.[69] They were skeptical that the new billion-dollar arts centers would display significant Asian artworks; some even contended that such buildings were archaic in the 21st century.[70] There was a call to develop "software"—institutions that support creative industries—rather than the construction of expensive "hardware."[71]

Alongside the development of the WKCD, international art fairs like Art Basel have helped Hong Kong overcome its marginalization from mainland China by solidifying its place as one of Asia's major cultural capitals. These art fairs brought new galleries to the city and contributed to the internationalization of Hong Kong's art market. Hong Kong was responsible for most of China's 25% market share in world art in 2012. The city is currently the third-largest center of art auctions worldwide. Thanks to these art fairs and the city's many cultural events, Hong Kong is now perceived as a significant art hub.[72]

For example, Uli Sigg, a former Swiss diplomat and significant private collector of modern Chinese art, was hired as the executive director of M+. In 2012, he donated and was able to secure from others 1,463 artworks worth HKD 1.3 billion, or roughly USD 165 million. These artworks were donated to M+ to avoid censorship by the Chinese government; Hong Kong still tolerates politically controversial art pieces. As a result, some local artists support the WKCD initiative despite its controversial and detrimental consequences for the local art scene.[73]

Privatization of Governance

Developers and Politicians

Populist politicians in Hong Kong denounced the early versions of the WKCD plan as too generous to developers. Tung Chee-hwa, in his 1998 policy speech mentioned earlier, announced that he planned to hand over the entire project to a single large developer. The plan evoked strong opposition because it was seen as a subsidy for the city's wealthiest developer, Li Ka-shing. Many of Tung's long-time allies among business leaders did not support this plan. It was then revised in October 2005: a lead developer would build half the residential and commercial real estate in the WKCD peninsula, and other developers could bid for the rest. The three wealthy developers that had submitted proposals under the initial rules objected to the revised rules.[74]

Because of its connections to the large-scale retail industry and its commercial and residential development interests, the WKCD was also criticized as being primarily aimed at developing commercial property, rather than cultural projects. Government funding in Hong Kong primarily supports shopping, dining, and entertainment facilities, with an expectation of increased revenue, which benefits wealthy developers. Not surprisingly, property prices began to increase in residential areas in 2010 and continue rising today.[75]

Helen Siu and Agnes Ku, sociology professors at the Hong Kong University of Science and Technology, examined the WKCD project as part of Hong Kong's attempt to make itself a "global city." They claimed that the relationship of the different commercial and retail sectors involved in the project with local institutional structures and processes was defined by "the cultural policy and institutional framework in colonial times—characterized by noninterventionism and top-down drive under a fragmentary yet centralized structure."[76] Top-down decision-making structures and a lack of voice from cultural practitioners, as facilitated by globalization, have undermined cultural citizenship and civil society. For instance, the People's Panel on West Kowloon (PPWK) was established by activists in the cultural sector and civic organizations to rethink the project with the cooperation of government, property developers, and civil society; however, this panel was not effective and yielded no significant results.[77]

A Consultative Committee (CC) and a Core Arts and Cultural Facilities Committee (CACFC) were also established to oversee the provision of cultural infrastructure under the aegis of the West Kowloon Cultural District Authority (WKCDA), which oversees the WKCD project. With the creation of the WKDCA, the government attempted to distance itself from the awarding of development contracts, with the intention "to be seen as more 'democratic' and engaged with a newly 'liberated' citizenry."[78] This separation was necessary because locals have little trust in government; widespread bribery scandals between developers and politicians have generated increased criticism of the project. A corruption trial involving former Chief Secretary Rafael Hui Si-yan in 2014 worsened the public perception of public officials. Hui allegedly received tens of millions of dollars from Sun Hung Kai Properties (SHKP) in exchange for his favoring SHKP's interests in the West Kowloon project.[79]

A total of HKD 950 million (USD 120 million) has been spent on the WKCD project since 2008, and officials have announced plans to invest an additional HKD 21.6 billion (USD 2.75 billion) in the project.[80] Even the natural allies of the project—members of the real estate industry—have questioned its success and accused the government of harming their interests through excessive regulation implemented in response to public protests in 2005.[81]

Scholarly Assessments of the Project

The WKCD project is a controversial megaproject that seeks to globalize Hong Kong as an art hub city. Professor Tai-lok Lui, one of many scholars who criticize the WKCD project, describes the proposal as an example of entrepreneurialism and neoliberalism:

> The proposal was a typical package of urban entrepreneurialism, with an emphasis on chasing after megaprojects, iconic buildings, and media visibility. However, it avoided questions concerning the substance of the entire project, consensus from below, and vision of cultural development. Its failure shows that without addressing these basic questions, city competition by means of developing global architecture, mega-projects, and fabricated urban culture is inevitably futile.[82]

Moreover, some scholars point out that the WKCD has already failed to "globalize" Hong Kong by largely ignoring its history, culture, and identity. The "Brand Hong Kong" program, of which WKCD is part, has the mission "to provide a greater focus to the international promotion of Hong Kong as 'Asia's world city'." Yet, some claim that its marketing strategy has marginalized provocative artists or those out of the mainstream, according to Stephen Chu.[83]The Donald Tsang administration was keen on developing Hong Kong into a hub for Asian creative industries but did not provide an economic base for local creative industries to grow. As long as the development of creative industries is framed by the "Brand Hong Kong" concept, vernacular hybrid cultures and spaces cannot surface[84]:

> The overwhelming emphasis on branding Hong Kong has ironically led to the loss of Hong Kong's intrinsic uniqueness: the blending of the global and the local into a hybrid emerging culture which is significantly "glocal."... It attempts to assess the implications of the Brand Hong Kong program's failure in recognizing that the distinguishing characteristic of Hong Kong was its emergent community where genuine cosmopolitanisms found the space to emerge.[85]

As Jian argues, for the WKCD project to succeed, it needs to achieve balance in five areas: (1) between public participation and the on-time completion of the facilities; (2) between representation and participation in the management board and committees of the WKCD as it moves into operation/service mode; (3) between maintaining both administrative and financial sustainability, as well as resource development; (4) a balanced approach to resource development; and (5) between meeting the expectations of existing artists and supporting new ones.[86]

Environmental Impacts

The WKCD project was expected to have a significantly positive environmental impact by (1) establishing a park filled with trees and different types of landscaping and amenities, increasing West Kowloon's green space; (2) maximizing the amount of green space within the WKCD; (3) incorporating green roofs and other sustainable features within the facilities; (4) building new, aesthetically pleasing structures that would blend in with the surroundings and using landscaping components to generate visual appeal from both up close and a distance; and (5) implementing a sustainable transportation system that prioritizes walking and is supplemented by other modes such as eco-buses and outdoor escalator systems, or *travellators*.[87]

It also took these steps to minimize the number of harmful environmental effects: (1) locating the site's vehicle network at the basement level to reduce traffic noise and increase the amount of ground-level space available for planting and landscaping; (2) whenever possible, locating developments that are sensitive to air or noise (such as residential developments) away from sources of polluted air or noise emissions; (3) locating potential sources of noise, such as the outdoor theater, away from the existing and planned noise-sensitive receivers (NSRs) to avoid potential operational-phase noise impact; and (4) moving outdoor venues away from the Western Harbor Crossing to minimize its potential impact on air quality.[88]

Because sustainability is a crucial component of the project concept, the following ecologically friendly designs were part of the planning process: (1) systems that use renewable energy sources, such as wind and solar photovoltaics; (2) a district cooling system that uses significantly less energy than traditional air-cooled chillers to provide chilled water to WKCD facilities; (3) green roofs as a component of thermal regulation for WKCD structures; (4) bicycling lanes and a vast pedestrian network to complement the green transportation system; and (5)water-saving measures, including rainwater collection and reuse of the condensate from air conditioning systems when possible.[89] Despite these efforts, WKCD has caused serious environmental damage.

Degradation in Surrounding Water Quality due to Reclamation and Sewage

The WKCD was constructed on what is Victoria Harbor. It was initially a port and eventually became a tourist spot and an important transit path between the Kowloon Peninsula and Hong Kong. It is a scenic asset, supporting Hong Kong's claim to being the "Pearl of the Orient." In recent years, 55% of the tourists coming to Hong Kong have visited Victoria Harbor.[90]

Reclamation of the harbor has intensified since the 1980s. The amount of land reclaimed from 2000 to 2005 was equivalent to the amount reclaimed in the last 100 years; 250 hectares were reclaimed for container terminals in the port. The West Kowloon Project reclaimed forty hectares of land around Victoria Harbor. The environmental impacts of this reclamation were well known and even outlined in the WCKDA's 2013 Annual Report, which projected that, during the construction phase, there would be "construction site runoff, pumped groundwater discharge, drainage diversion, sewage effluent and accidental spillage that contain high levels of suspended solids and chemicals such as oils, solvents, and cement-derived materials."[91]

Dredging was required to remove underwater obstacles and replace large volumes of sand, exacerbating water pollution in Hong Kong. By removing plant life from below the surface, dredging negatively affected the marine ecosystem, which is dependent on plant materials.[92] It also caused the loss of filter-feeding organisms, including scallops, mussels, and oysters, which are essential to the vibrant seafood industry around Hong Kong. There is also evidence that dredging threatened Chinese white dolphins, which are an endangered species unique to Hong Kong: dredged materials compromised the well-being of the dolphin sanctuary, which is near the East Sha Chau contaminated mud pit where the dredged, contaminated mud was disposed of. Only 85 of the 300 Chinese dolphins survived, and those that survived the initial dredging were not expected to survive further reclamation.[93]

Poor Air Quality during Construction and Operational Phases

The WKCD is located next to the Western Tunnel Crossing, which connects West Kowloon to Hong Kong Island and is one of the busiest locations in Hong Kong. It has also has the worst air quality, with an average particulate concentration around the tunnel crossing at 63 micrograms/cubic meter (mg/m^3) which far exceeds the World Health Organization's (WHO) air quality maximum threshold of 25 mg/m^3.[94] Construction activities, such as excavating, stockpiling materials, moving vehicles, concrete batching, and loading and unloading operations, decreased the already poor air quality in the area.[95] The building of major infrastructure, such as the Western Harbor Crossing portal and additional roads and car parks within the WKCD, also reduced air quality. Because the newly reclaimed land area would attract a much larger number of people during concerts and shows, traffic congestion and density would exacerbate the existing air pollution problem.[96]

Reducing the Aesthetic Appeal of the Victoria Harbor

The project promised to build scenic landscapes in and surrounding the WKCD, adding ornamental plants, piazza, parks, and avenues through innovative design.[97] However, the construction of the metro station, the Elements Mall, international commercial centers, and large apartment buildings will dominate the area. The future construction of skyscrapers and other very large complexes in the WKCD will further reduce the beauty and access to the views of Victoria Harbor.[98]

Impacts on Public Space

Even though master plans and urban visions always emphasize the establishment of inclusive public spaces, in practice, these areas are often only considered at the very end of the project: even then, they are seen mostly as leftover spaces between buildings and development. Public buildings and spaces are frequently interpreted out of context and treated as architectural "landmarks," particularly in the case of cultural endeavors.

Instead, the void—public open space—should be seen as equally or even more significant than the solid (buildings): the two should be mutually beneficial and interactive, rather than seen as a collection of isolated items. In consultation with the developers of the WKCD, Yui in 2018 examined the problem of public spaces in the context of cultural districts and proposed how they could be effectively used to strengthen the community.[99]

Contrary to many other cultural initiatives that either have a strong architectural heritage or local memory, the WKCD was built on reclaimed land that necessary lacked an urban history. This provided both a chance to create a new artistic identity for the area but also the danger that the area would be turned into a generic location without any association to local culture. WKCD's future public life and the character of its public spaces are still being developed. The bustling street life of the core Kowloon areas, which include the harbor front and the older urban fabric, makes them some of the most popular public spaces in the city, and the WKCD may find inspiration from the harbor- front Cultural Center's well-defined public square.[100]

Yet, the WKCD is isolated from its surroundings and is inaccessible. The West Kowloon waterfront and its potential iconic architecture produce a highly visible image that can be seen across the harbor. However, when arriving by subway, the project itself is concealed between megastructures with no notable entrances and public spaces. Currently, it is very challenging for first-time visitors to locate museums and theaters, which discourages regular and frequent trips. Some theatergoers were not even aware that the park was open every day. Proximity and accessibility are important factors for encouraging daily use; they should also be incorporated into the design of public spaces to allow unscheduled daily activities.[101]

Impacts on Tourism

The Guangdong-Hong Kong-Macao Greater Bay Area (GBA), which spans nine cities and the two special administrative regions of Hong Kong and Macau, has a total area of about 56,000 square kilometers surrounding the Pearl River Delta. The GBA designation started to be widely used around 2017. The "nine + two" GBA cities are further broken down into four core cities—Hong Kong, Shenzhen, Guangzhou, and Macau—which support national economic growth and development, and seven node cities: Zhuhai, Foshan, Dongguan, Zhongshan, Jiangmen, Huizhou, and Zhaoqing.

At the end of 2019, the region's population was over 72 million, its GDP was USD 1,679.5 billion, and its per capita GDP was USD 23,371, which was higher than the country's per capita GDP of USD 10,261.68.[102] The aim of bringing these cities and together is to enhance regional collaboration and capitalize on Hong Kong's position as a hub for international transportation. The GBA is intended to serve as a national tourism demonstration zone that will capitalize on its advantageous coastal locations, natural endowments, and well-integrated transportation network using a three-pronged approach that focuses on the sea, outlying islands, and coastal cities.[103] Aligning tourist resources with urban and rural areas is the primary tactic for boosting cultural tourism with other GBA cities. The region's well-developed infrastructure enables visitors to embark on multicity and multicultural package tours.

The West Kowloon Cultural District should not be viewed in a vacuum; it will have a significant impact on its surroundings and even Hong Kong as a whole. Its success will hinge not only on how the land is reclaimed but also on how well it blends with adjacent older neighborhoods. Instead of being gentrified or relocated, existing parks, structures, heritage sites, and communities should be integrated into the new cultural area. The WKCD project offers the government a chance to create a comprehensive and proactive policy for cultural development and conservation that addresses both tangible and intangible aspects of cultural heritage and the physical environment in a manner that facilitates cross-sector and interdepartmental cooperation.

As Lam et al. have suggested, in order to optimize synergy and ensure an intermodal tourism sector, the tourist industry needs to be revitalized in the short term along with a step-by-step implementation schedule, including: (1) enhancing cooperation between Hong Kong, Macau, and Guangdong to fully utilize the combined strength of the Greater Bay Area; (2) ensuring sufficient and sustainable funding that supports district development, such as place- making, village revitalization, and thematic linkages; and (3) gaining high- level policy support to foster the momentum of tourism development after the planning phase is completed.[104]

It is not impossible to foresee some challenges, including both exogenous and endogenous difficulties, related to the WKCD and Hong Kong's tourism development. Endogenous problems include an aging population, and risks of gentrification and displacements; exogenous difficulties refer to natural disasters, climatic risks, and disease. These shocks can be managed using a networked, resilient approach with cross-departmental cooperation and improved collaboration. The public's views on tourism development should be considered in future improvements.

Conclusion

This chapter focused on megaprojects in Istanbul, Sydney, and Hong Kong, showing that they are disruptive and contentious enterprises that trigger significant economic, sociopolitical, and environmental challenges that are often difficult to resolve for the benefit of urban communities. The list of negative impacts of megaprojects is long: cost overruns, environmental risks and negative environmental impacts, gentrification risks, drawbacks of top-down cultural engineering, neglect of local cultural identities, uncertain economic success, population displacements, spoiling of cities' visual and structural uniqueness, increased congestion and overpopulation, political gridlock, and existential threats to entire ecosystems, *inter alia*. In Chapter 6, we shall see that the impacts of megaprojects are incompatible with sustainability. In fact, they work against any sustainable conception of the urban environment. This is particularly true when the combined simultaneous, interrelated, and interdependent impacts produce cumulative and cascading effects in the urban areas and populations affected.

It is questionable whether better management of megaprojects—based, for example, on limiting the influence of human cognitive bias on the decisions made by planners, developers, and managers—would greatly reduce the disruptive impacts that emerge after implementation. Like most megaprojects, the cases discussed in this chapter illustrate the privileging—through autocratic rule (Istanbul), erosion of intent (Sydney), and misrepresentation of targets (Hong Kong)—of various political and economic interests at the expense of the well-being of most urbanites. Any application of optimal rationality that disregards contextual conditions, as well as potential ecological, socioeconomic, and equity impacts, will not prevent a megaproject from negatively disrupting affected communities in major ways. This is particularly true when communities and citizens are excluded from stakeholder deliberations in the planning process and from participation in their management and governance after they are built.

The study of controversies and disputes in urban megaprojects (UMP) development contributes to understanding the main challenges for urban governance in cities seeking to gain global visibility and attract investment and visitors by creating megaprojects. It also helps identify the main participants or sets of participants driving urban policies in cities, as well as the most suitable governance arrangements for supporting and enabling sustainable urban policies.

Istanbul's megaproject investment process triggered local citizen resistance, political confrontations, and the realignment of political options along the lines of an allegedly environmentally friendly Istanbul under Vision 2023. Earlier massive investments in rebuilding Istanbul had already caused significant environmental damage. Kanal Istanbul is an urban megaproject under construction with potentially significant geopolitical consequences, great financial risk, and significant environmental harms. The Barangaroo development is intended to ensure Sydney's status as a globalizing city at the center of financial services in the Asia-Pacific region. It has led to political contention around issues of the right of citizens to urban public space versus the private interests of developers and the risk of the project fostering urban inequality in Sydney. Hong Kong's WKCD was planned and promoted as an economic stimulus, rather than as a cultural district reflective of local heritage, thereby excluding local artists. It attracted foreign artists and investors who wanted to make the city an arts epicenter by using top-down governance to privilege wealthy property developers. Dredging associated with the WKCD in Hong Kong has adversely affected the marine ecosystem.

This exploration of the impacts of UMPs on Istanbul, Sydney, and Hong Kong strongly suggests that they pose risks to the bioeconomy and biodiversity. Enhancing urban resilience and livability may require (1) mediating between private and public sector interests; (2) promoting the greater involvement of local and regional governments for projects implemented in their communities; (3) engaging citizens more proactively in consultation and direct involvement before, during, and after UMP development; (4) maintaining local heritage as a central component of any UMP agreement; and (5) preserving the environment in the air, water, and land. These are the main challenges for urban leaders and policy makers who share the goals of improving the economy, environment, and equity in urban governance.

In view of megaprojects' negative impacts, it seems reasonable to think that better development pathways exist for cities. Yet, all too often, these pathways are not prioritized given the still predominant neoliberal urban discourse and its focus on growth, international visibility, and competition in the global arena. Research-based policy needs to highlight some of the keys and preconditions for urban leaders (and urbanites themselves, as creators and recreators of the places they inhabit) to shift from neoliberal development models based on megaprojects and work to organize sustainable urban ecosystems from an ecological, environmental, sociopolitical, and economic perspective.

Complexity and Sustainability

In the current "green capitalism" regime characterized by the systematic use of environmental sustainability strategies for private profit, megaprojects are presented as "sustainable." However, it remains unclear under what conditions these allegedly sustainable megaprojects can foster equity, livability, and shared prosperity while mitigating ecological damage. Evidence instead shows that they symbolize the grandiose vision of political and economic leaders in their attempt at nation-building to effectively compete on the global scene. Will these megaprojects yield economic benefits to the cities that host them and the urbanites in those cities, and not only to the vested interests of the megaproject stakeholders?

There is growing evidence that their benefits could be very limited, as is the case with most Olympic mega-events. Growing skepticism about spectacular and large-scale urban development, as well as the multiple impacts of the Great Recession of 2008 and the ongoing coronavirus pandemic, is triggering a rethinking of urban revitalization strategies in the West, and new paradigms, such as "smart cities," have come to the fore. However, advocates of the "smart-city" idea, who rightly stress the importance of digitalization and technology for the management of urban areas, seem to overlook the fact that it is governance models and practices, not technological tools, that need further discussion and rethinking.

Jane Jacobs rightly reminded us almost more than sixty years ago that cities can be seen as "organized complexity" and cannot be addressed as a conventional problem of hierarchies and visual or mechanical orders planned exclusively by leaders and experts. Jacobs understood the city as a complex of interrelated factors in an organic whole; she recommended that urban planners and architects show respect for the intrinsic order of the city and discard the demiurgic, spectacular, and traumatic interventions that they often put into practice.[105] Cities have the capacity to promote creative and dynamic growth and simultaneously reduce the destruction of resources. They are known to be more efficient than other types of human settlement in terms of energy consumption, resource consumption, and greenhouse gas emissions.[106] This is because urban ecologies are organized through exchange networks in spatial proximity, whose synergies have positive and multiple effects. In other words, cities are complex, dynamic, and variable human clusters that can favor efficiency, competitiveness, and ecology.

We also know that networks, proximity flows, and casual encounters in defined spatial environments foster multiplier effects, which explains the impact of creativity on the economic strength of cities.[107] I am not advocating that a creative class should serve as the engine for urban prosperity, but rather we should value and promote the ecological character of the city as a place for the exchange of knowledge, information, experiences, and affections. These exchanges need public or semipublic spaces, and hence the crucial importance of preserving squares, sidewalks, parks, terraces, coffee houses, and other meeting places where economic rationality does not prevail. Online connections can supplement but not replace the primary network of face-to-face human exchanges.

Cities, just like any network, benefit geometrically from the number of existing connections. If the economic disparity between urbanites condemns certain citizens and neighborhoods to socio-spatial segregation, the prosperity of the city as a whole will be compromised. For reasons of social welfare, economic prosperity, and, increasingly, for survival of the planet and our species, it is essential to advance the planning of integrated cities on a human scale that respects the close and multiple interactions of their neighborhoods. These constant human interactions enable urbanites to shape their own identity by endowing their environment with meaning, a fundamental factor that contributes to individual and collective well-being and indirectly to prosperity.

What is truly important for urbanites is to ensure that their leaders (and the urbanites themselves, as creators and re-creators of the places they inhabit) work to organize urban ecosystems that are sustainable from ecological, environmental, sociopolitical, and economic perspectives. The good design and efficiency of a city are not sufficient to guaranteeing good urban life. Improving macroeconomic magnitude is insufficient if there is no effort to achieve higher levels of social welfare. Urban visibility (megaprojects) and urban connectivity (smart cities) strategies present as many risks as potential benefits for urban populations.

These ideas, which flesh out the urban order as an organized complexity, become the driving force behind sustainable urban development, from planning to implementation. Urban ecologies must be organized through exchange networks in spatial proximity. Planning should aim to advance integrated cities on a human scale that respects the close and multiple interactions of their neighborhoods. Evidence shows that advancing these goals would have positive effects on urban communities, which would conceivably be in a better position to build the necessary anti-megaproject coalitions and slow down urban growth machines. This community imperative would contribute to transforming megaprojects into assets that work for the welfare of urban populations. At a minimum, it would contribute to the task of transforming megaprojects from disruptive and contentious enterprises into feeble artifacts with relatively controlled negative impacts.

Chapter 4
Emblems of Capitalism

Introduction

Disruptive impacts can be observed in many types of megaprojects. This chapter presents cases of attempted urban regeneration through cultural megaprojects. It also shows the drawbacks and failures of these same megaprojects and the so-called *Bilbao Effect*—a model of urban revitalization based on iconic architecture commonplace among urban managers worldwide after the alleged success of Bilbao Guggenheim since its opening in 1997. Cultural megaprojects, it is argued, are vehicles for a property-led urban revitalization strategy. Yet, the impacts of these iconic buildings on the urban fabric are often unintended and negative. The relative success of Bilbao has been difficult to replicate elsewhere because urban leaders in most cities have overlooked the intrinsic limitations of iconic buildings that affect urban socioeconomic regeneration. Urban leaders elsewhere failed to examine and understand the specific socioeconomic and political context in Bilbao. There, the Guggenheim megaproject was just one element among many in an overall local revitalization strategy. Nevertheless, seeing the alleged success of Bilbao Guggenheim, urban leaders were eager to be a part of what came to be called the Bilbao Effect. Yet, with the failure and fading away of the Bilbao Effect, the Guggenheim Foundation's dream of museum franchises around the world—controlled by the Foundation and paid for by host cities—has not been realized. This chapter discusses the Bilbao case along with other similar efforts in urban reimaging through iconic cultural projects: Denver, Helsinki, and Abu Dhabi. Such iconic projects are called into question in three main ways: (1) iconic architecture is no longer a hegemonic visual discourse in urban revitalization; (2) the franchise model imposed by Guggenheim means that local officials have no autonomy to make major decisions on matters ranging from exhibition calendars to budgets and investments; and (3) local cultural identities are usually neglected in a foreign global art model.

G. del Cerro Santamaría, *Contesting Megaprojects*,
https://doi.org/10.1007/978-3-031-98382-5_4

Bilbao's economic performance after the opening of Guggenheim has broadly followed the ups and downs of economic cycles, a clear indication of both (1) the embeddedness of iconic megaprojects in cities at multiple scales of socioeconomic action and (2) the limited power of architectural icons to explain development, competitiveness, and urban economic change. Indeed, it is institutional contexts, specific policy instruments, and territorially grounded social dynamics that define distinct patterns of iconic megaproject development and explain the degree to which such megaprojects succeed or fail.

The fading away of spectacular architecture as the hegemonic architectural discourse in the West leads us, in the second part of the chapter (Architecture After Spectacle), to discuss which new paradigms might replace it, or at least coexist with it. Among these new paradigms, we see a renewed interest in linking aesthetics to topography through tectonics and place as source material for architectural production. We also see the intent of rebuilding resistance through the use of time and history. In sum, we observe a fuller embrace of complexity by architectural theory and practice, away from commodification, and the crossing of disciplinary boundaries, particularly when dealing with issues of urban ecology and sustainability.

The Fading Away of the Bilbao Effect

Urban Museums as Emblems

The influence of neoliberalism on urban areas, under pressure to attract investors and become visible, has led to a relatively uniform approach to urban policy across political and geographic boundaries.[1] A recipe for urban economic improvement has focused for the past decades on property-led regeneration even when such strategies have been presented as cultural regeneration.[2] Cities have been encouraging this strategy through zoning and megaproject-based investments in areas that can easily be redeveloped and are often populated by low-income and minority communities. In "entrepreneurial cities," heavy manufacturing has been gradually replaced by niche real estate, service sector employment, tourism, and culture.[3]

Urban megaprojects are intended to attract new investments, tourists, and foot traffic to underutilized areas, thus increasing property values and triggering widespread gentrification. An increasing number of cultural attractions, especially museums, are now central to urban development strategies for the inner city. These projects pursue an important economic aim centered around a flagship or iconic building.[4] Museums allow cities to promote themselves as cultural and art centers. Museums also tend to attract wealthier individuals. In this way, neoliberal public spaces are vehicles for economic development, benefitting some populations to the detriment of others.[5]

Museums, such as those discussed in this paper, have been a part of urban planning strategies for redevelopment for some time. The Museum of Modern Art in Manhattan (Edward Durrell Stone, 1939) epitomized the "white cube" model of museum architecture and the International Style. MoMA established the influence of modernism on cultural architecture. MoMA continued to be the preeminent model for art exhibition; however, the beginning of the sculptural form of museum design was Frank Lloyd Wright's Guggenheim, also in Manhattan, on Fifth Avenue. This building—both its exterior and interior curving galleries—inspired a line of sculptural museums in contrast to MoMA, which, interestingly, includes Guggenheim Bilbao designed by Frank Gehry.[6]

Following Wright's Guggenheim, the next major museum in this category is Centre Pompidou (1977), designed by Richard Rogers and Renzo Piano. Hoping to renew Paris as a leading city of culture and art, Pompidou shifted the museum's role away from a purely educational institution. It was a key strategy of the Parisian government's attempt to redevelop the historic neighborhood of Marais, and it represented the beginning of the era of iconic museums as a force of urban regeneration.[7] In 2007, when Rogers won the Pritzker Prize, the jury said the Pompidou "revolutionized museums, transforming what had once been elite monuments into popular places of social and cultural exchange, woven into the heart of the city."[8]

Thus, when the Guggenheim Foundation and Basque planners agreed to build a Guggenheim Museum in Bilbao in 1991, emblematic projects and cultural megaprojects had already played an important role in the revitalization of cities in Europe (e.g., Paris, London) and the United States (e.g., Pittsburgh),[9] and the key role of iconic architecture in changing a city's image was discussed and well understood among urban planners. Bilbao became a game changer both because of the early impact of the Gehry building in architectural circles, as well as the sociopolitical and socioeconomic contexts (political unrest, deindustrialization, and severe economic downturn) in an unknown city, adding up to a good journalistic story that swiftly traveled around the world. This is how the "Cinderella story" concept—a museum places the city on the map and rescues an urban economy in crisis—spread misleadingly (Fig. 4.1).[10]

The Bilbao Effect in the World

The Bilbao Effect can be succinctly defined as the attempts by a significant share of urban elites worldwide to build icons in their cities, largely based on a superficial and media-based understanding of the Bilbao case, which led many to firmly believe that a city in economic difficulties could be turned around just by iconic architecture. The Bilbao "Cinderella story" captivated many urban leaders. The Guggenheim Foundation received numerous offers from cities on five continents to pay the full cost of a building designed by Frank Gehry. Thus, the Bilbao Effect transcended the discursive realm into the material world, and many cities worldwide thought it possible to affect a transformation like in Bilbao by simply building a Guggenheim Museum ("build it and they will come" became a catch phrase).

For some time, the model of the global museum franchise envisioned by Guggenheim was becoming a reality. Many urban officials contacted the Guggenheim Foundation with firm plans, although only a handful went beyond the initial conversations—specifically, Rio de Janeiro, Vilnius, Salzburg, Guadalajara, and Taichung—and with the rest negotiations ended before reaching an agreement.

Fig. 4.1 Aerial images of the Abandoibarra redevelopment (including Frank Gehry's Guggenheim Museum) in downtown Bilbao, in two different years: 1996 (left) and 2017 (right). Source: Courtesy of Bilbao Ria 2000

Nevertheless, the Guggenheim Foundation was able to advertise itself openly as a global art organization with a new and successful vision for the 21st century.[11]

The Guggenheim Foundation planned for a large museum on the waterfront in Lower Manhattan and engaged Frank Gehry as an architect. His nearly complete designs for the building were showcased in 2001 at the Fifth Avenue Museum, but these plans were disrupted by both the economic downturn of the early 2000s and the September 11, 2001, attacks.

Two outposts by Guggenheim did open—Berlin (1997–2012) and Las Vegas (2001–2008)—but neither achieved much success and both have since closed. More recently, Abu Dhabi and Helsinki became successful contenders. Guggenheim Abu Dhabi—designed by Frank Gehry—has been built, is nearly complete and was orginally expected to open in 2017. Now, after several delays, it is, at the time of this writing (January 2024), expected to open in 2025. The Guggenheim Helsinki project unveiled the winning design in June 2015 following an international competition. However, in the fall of 2016, the City of Helsinki voted to cancel the project due to increasing controversy and civic opposition.

Plans seemed complete, or near completion, in 2016 for other new cultural hubs centered on museums in Saudi Arabia (Mecca), Australia (Perth), Albania (Tirana), and Brazil (Belo Horizonte). And ultimately, in what has become Eastern Europe's largest museum, Mystetskyi Arsenal, opened in Kiev, Ukraine, in 2011 with 50,000 square meters (540,000 sq. ft.) of exhibition space.[12]

The Bilbao Effect has been used to explain events whose causal relationship with the success of Gehry's building has not been easy to prove. For instance, the alleged success of the Bilbao Museum has been credited with an increased wave of museum construction, extension, and reform in the United States over the past 15 years. Wilson argued that the museum in Bilbao "was widely credited with having sparked an economic boom in northern Spain,"[13] which other cities aimed to replicate. She quoted architecture critic and Editor-in-Chief Robert A. Ivy of *Architectural Record*, who proclaimed that "Gehry's Bilbao has conflated cultural, economic, and political interests, alerting all to what a dazzling object in the cityscape can accomplish."[14]

The Guggenheim building in the Basque capital was also perceived as the beginning of a new era in which museums were thought of as monumental sculptures, potentially becoming the most important work in an institution's collection. Museum plans were being developed in the United States, including in New York, NY (a new Guggenheim by Gehry and the Whitney extension by Rem Koolhaas); Philadelphia, PA; Atlanta, GA; Austin, TX; Bellevue, WA; Boston, MA; Charlotte, NC; Cincinnati, OH; Denver, CO; Hartford, CT; Kansas City, MO; Milwaukee, WI; Saint Louis, MO; San Francisco, CA; and Savannah, GA.

According to a study from the University of Chicago, "between 1994 and 2008, 725 new art facilities were built in America at a price of more than $US 15 billion."[15] Even the 2008 recession couldn't stop the boom: according to the *Art Newspaper*, from 2007 to 2014, USD 8.9 billion was spent on museum expansions worldwide and USD 5 billion in the United States alone.[16]

> To tour the museums and art galleries of the last two decades is to take in a whimsical menagerie of iconic creations: curls of Frank Gehry–built metal rippling through Cleveland and Seattle; neo-futuristic Zaha Hadid monuments alighting in Azerbaijan and Guangzhou like so many glossy alien motherships; Daniel Libeskind shards poking out of heritage buildings from Dresden to Toronto.[17]

Cultural megaproject construction was not limited to wealthy global cities, such as Paris, New York, and London. Regional and small cities such as Biloxi, in Mississippi, or Roanoke, in Virginia, also attempted Bilbao-like urban transformations into cultural hubs via iconic museums. According to Joanna Woronkowicz, one of the authors of the University of Chicago study, "the people behind these buildings all had something in common: they had read Richard Florida and absorbed his message. Building a large, eye-popping museum was not an act of hubris; it was a civic duty."[18]

Nevertheless, the Bilbao Effect faced significant criticism and skepticism among numerous architectural and art connoisseurs. Chicago Tribune critic Blair Kamin identified the beginning of the "starchitect" trend in the 1976 Houston Pennzoil Place, dubbed by the residents "the milk cartons." He noted that the fashion spread to other cities, such as Chicago, in the 1980s, and remarked that the rise of "starchitects" poses a broad set of questions about the impact of globalization on an art that is ultimately local, and contended that architects must respect the urban context and history of the city where their design will be built:

> Should 15 or 20 starchitects be designing all the world's great buildings? What does it mean if every city has Gehry, Koolhaas, and Calatrava? Are the backers of these buildings simply seeking known commodities rather than taking genuine artistic risks? Can stars tailor their style to a vast cross-cultural array of functions and places? [...] They design eye-catching creations that would enhance a building's marketability [...] There is something [...] to be gleaned from starchitects, but only if they are willing to look deeply at [a city] and to adapt their work to the city's essence and its economics.[19]

Architectural critic Witold Rybczynski asked whether cities commissioning new museums by starchitects could become equal to Bilbao in terms of visitors. He noted that attendance at the Experience Music Project in Seattle—designed by Frank Gehry for Paul Allen in 1996—decreased by a third, 18 months after the museum opened, while the number of visitors to the local art museum increased by more than a third during the same period. Despite featuring unusual architecture consisting of colorful, rounded forms inspired by electric guitars, being a museum of rock music, and housing Jimi Hendrix memorabilia, the Experience Music Project has not proven to be a success; recently a portion of the building was converted into a science fiction museum.

Rybczynski was "skeptical that designing at the full glare of public competitions encouraged architects to produce better buildings. The charged atmosphere promotes flamboyance rather than careful thought and favors the glib and obvious over the subtle and nuanced."[20] More recently, Rybczynski has argued that "perhaps the Bilbao Effect should be called the Bilbao anomaly," since "the iconic chemistry between the design of a building, its image and the public turns out to be quite rare, and somewhat mysterious."[21]

> Herzog & de Meuron's design for Beijing's Olympic Stadium is ingenious, for example, but instead of the complex engineering, it was the widely perceived image of a "bird's nest," a nickname that did not originate with the architects, that cemented the building's international iconic status. The woven steel wrapper symbolized both China's ancient traditions and the rush to modernization. However, for every bird's nest, there are scores for building failures that are not only costly but also fail to spark the public's imagination. [...] failed icons do not disappear, which is problematic. Since the Bilbao Effect mistakenly teaches that unconventional architecture is a prerequisite for iconic status, clients have encouraged their architects to go to greater lengths to design buildings that are unusual, surprising, and even shocking. However, the shock will inevitably wear off, and 100 years from now, most aspiring iconic constructions will resemble a cross between a theme park and the Las Vegas strip.[22]

The Case of the Denver Art Museum

Both the Denver Art Museum (DAM) and its critics attribute the origins of the Hamilton Building to the opening of Bilbao Guggenheim. DAM's director Lewis Sharp attended the opening in 1997 and was captivated by what he saw; he "wanted one for Denver."[23] John Hickenlooper, the mayor of Denver, predicted in 2000 that the jumble of metal-clad and faceted geometric forms of architect Daniel Libeskind's design—which opened in 2006— would "put us on the map as a world-class destination city."[24] His prediction remains unrealized.

Snyder (2016) and Lindsay (2013)[25] have both discussed DAM's Hamilton Building (Fig. 4.2). There are abundant formal similarities between the two museums—Bilbao and Denver—including their material choices (titanium cladding), the selection of a star architect, and the construction of a daring sculptural structure. In addition to architectural similarities, the Hamilton Building was intended to play a role in developing the long under-used Golden Triangle neighborhood to its south. The use of an international architect to garner attention at the Ponti Building at DAM suggests that the influence of Guggenheim Bilbao on DAM is indisputable.[26]

The Hamilton Building project was sold to the public, not just as a building, but also as a symbol and an event. The power of architecture as an icon is dependent on the "circulation of visual images, combined with an embodied, performed a set of tourist practices on the part of architects and their professional critics and journalists."[27] That is, iconic architecture has a strong media presence and attracts tourists for visual admiration. According to architectural theorist Charles Jencks,

> an iconic building … need not be a great work of architecture, but it must be a captivating one. It has to move your viscera, whether you like it or not, and stay around as a memory image that attracts other thoughts into its orbit.[28]

As Lindsay remarked, using a building to represent a place is not new; the Eiffel Tower, Sydney Opera House, and even the Gothic Cathedrals are all icons that draw people to visit them. The Hamilton Building defies gravity with its angles and cantilevers, it stands out from nearby structures, it violates historical conventions for civic centers and art museums, and it uses a novel material for cladding.[29] It is a break that was vetted successfully in Bilbao and then copied by cities across the United States. The selection committee in Denver cited the Jewish Museum Berlin and Gehry's Bilbao as a successful strategy.[30]

Both local and global elements influenced the planning and design of the Hamilton Building. It was an urban event aiming for international recognition. It would expand the global reach of Denver. Libeskind's design was simultaneously a tool for local urban planning and a symbol of regional innovation. Fundamentally, the museum contributes to Denver's urban fabric. Specifically, the Hamilton Building links the Civic Center to the Golden Triangle. Its height descends nodding toward a residential area—the Golden Triangle—and gesturing toward the area with its prow.[31]

Gentrification is one of the disruptive impacts of such buildings. Urban leaders in Denver tried to transform the civic center of the city "from a place of homeless people into a place where the upper classes spend time." To date, the success of the

Fig. 4.2 Daniel Libeskind's Hamilton Building, Denver Art Museum in the city of Denver (Colorado, United States). Source: Courtesy of Daniel Libeskind Studio

building is only partial. Property prices in the area have not significantly changed because of building-led gentrification. A portion can be attributed to timing; despite Denver's housing market not collapsing as severely as in other cities in the United States—Las Vegas or Phoenix—it nevertheless did experience a housing slowdown during the 2008 crash.[32]

Criticism Toward the New Building

The museum received substantial criticism about Libeskind's architectural style and his work's decreasing quality, and the ability of the building to work as an art museum and its failure as a space to exhibit art. One argument made against sculptural museums, in general, is that both the architecture and internal gallery areas may prevent visitors from fully experiencing the art they house. Firstly, because visitors often come to see the building rather than art, and the architecture ends up overshadowing the art inside. Secondly, controversy and critique may discourage attendance by the public. Hamilton Building is a target of this criticism.[33]

Yet while Hamilton Building received mostly negative criticism concerning Libeskind's style and its failure at displaying art, it has received favorable comments regarding its relationship to the urban space. Critics referred to the Hamilton Building as a "surprisingly sensitive shaper of urban space," a "contemporary spin on urban context," and a "tour de force on urbanistic grounds alone. It succeeds brilliantly in weaving together disparate parts of the cityscape."[34]

Commodification of iconic buildings is a common feature of sculptural museums in urban neoliberal strategies. Just as cultural regeneration cannot be separated from property-led regeneration, the intention of urban leaders when using iconic museums in their revitalization plans is mainly to increase economic returns, growth, competitiveness, and global visibility. In this context, the role of art is at best secondary and is often neglected. Previously, museums were recognized by the collections they housed; today, this is done by architecture; in other words, the container rather than the substance is the main image.[35] Furthermore, one must not forget that when museums are intended as catalysts for urban development, there are usually many negative impacts on accessibility and quality of life in urban areas. New museums often attract visitors and energize neighborhoods, but residents— particularly low-income residents—are all too often disregarded. Victoria Newhouse writes:

> During the last two decades this role has become increasingly frequent [...]. Because major renewal projects normally spearhead gentrification and dislodge low-income populations, they cause tremendous social and economic upheavals and tend to be highly controversial. [36]

According to Snyder, the Hamilton Building was "an integral part of Denver's attempts to transform its downtown, and particularly the Golden Triangle, into habitated "urban villages" intended for upper income residents."[37] Downtown as a whole saw a dramatic increase in development, particularly in the LoDo district and neighboring northwest districts. The Golden Triangle had some development, but not as much as other downtown neighborhoods. Denver's development in the Golden Triangle and Civic Center has led to further intriguing studies on the use of arts and culture in municipal policy.[38]

Overall, the Hamilton Building in Denver did not turn the city into a world-class tourist destination. One reason for this is the intrinsic limitation of urban icons, which may trigger significant economic changes in cities albeit not the intended ones. The other is a change away from spectacular architecture in design theory and from using spectacular architecture as a tool for urban revitalization. Snyder contends that "there has been a shift in Denver's rhetoric and planning away from monumental buildings like the Hamilton Building toward the cultivation of smaller arts institutions or street culture."[39] From iconic buildings to cultural districts to sustainability, there has been a clear evolution in planning and urban strategies over the past ten years. With this development, the meaning of DAM in Denver has shifted, and its influence as an urban revitalization tool has decreased. Regarding concerns of public space, Denver's planning and construction in the Civic Center and the Golden Triangle from 2006 to the present demonstrate a change in the city's conceptualization of "arts and culture," and with it, the Denver Art Museum's declining authority as the city's cultural icon.[40]

Helsinki's Failed Guggenheim Museum

In short, the Helsinki Guggenheim Museum was a failed project; it will not be built. In June 2015, the Guggenheim Helsinki International Competition announced that the winning project would go to the Paris-based firm, Moreau Kusunoki Architectes (Fig. 4.3). However, by fall 2016, the City of Helsinki voted to cancel the project due to increasing controversy and civic opposition. This project was perceived as an onerous economic expenditure for Helsinki. The project came with a $130 million price tag and a hefty $27 million licensing fee to Guggenheim for the privileges of using the brand,[41] not to mention the cost overruns that the project would have likely generated, as most megaprojects do. Kaarin Taipale, the co-author of *In the Shadow of Guggenheim*, a book that analyzed project challenges, stated the following:

> If [Guggenheim] is a private enterprise, so it's simply not part and parcel of Finland's welfare society to support that kind of organization with public funds [...]. Everything would be financed by the city – the land, the construction, the upkeep of the building, the salaries, the license fee, everything.[42]

The economy of Finland had been in a recession since 2008, and it was difficult to justify the financing of a megaproject. The country is currently considered to be in a "lost generation" due to losses in industrial development and EU sanctions imposed on Russia, Finland's biggest trading partner.[43] The cost of the project was, indeed, the primary reason it was denounced by opponents. Paavo Arhinmäki, the Cultural Minister of Finland at the time, stated that the Ministry would not fund the project on the grounds that public funds for theaters, orchestras, and museums had already been reduced and the economic climate was extremely challenging. There was simply not enough money in the budget for culture to be devoted to the cost of building (above $70 million) and operating expenses (around $1 million annually).[44] The assumption of the Guggenheim project was that the money would be provided by external sources, such as franchising payments, which would cover the $30 million that would be paid to the Guggenheim Foundation for the rights to use

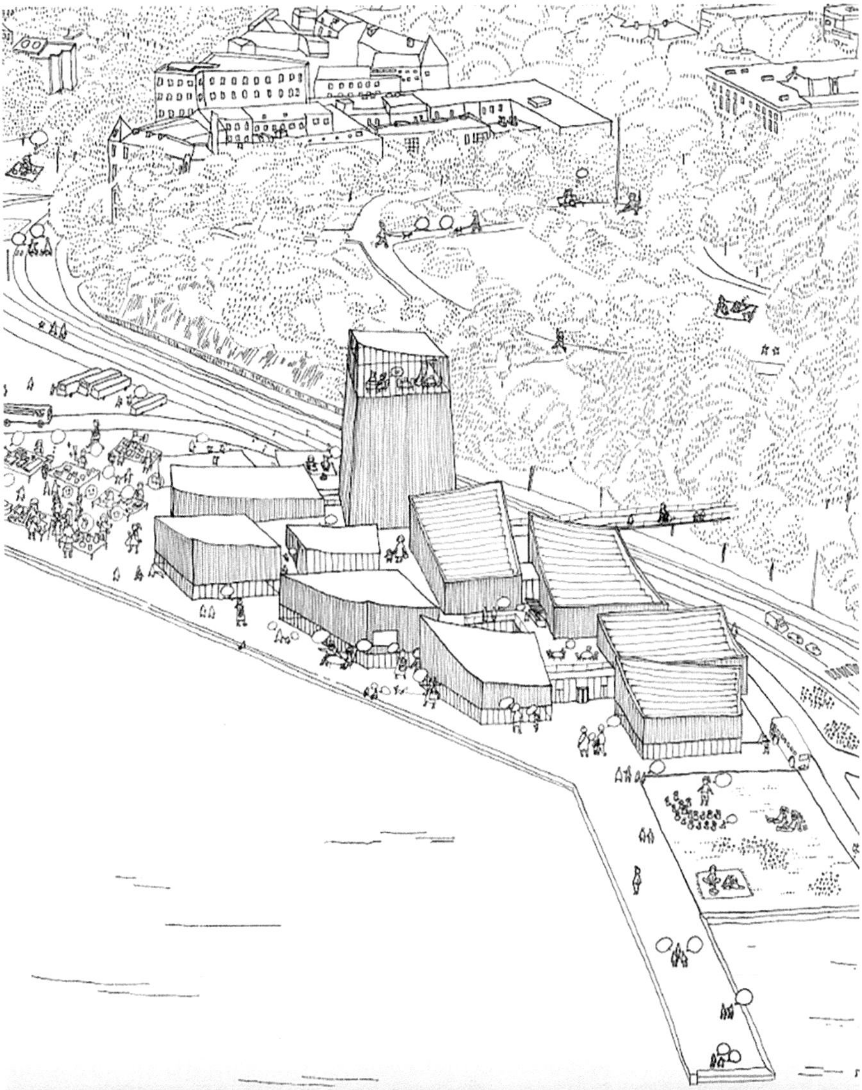

Fig. 4.3 Hand-drawing of the Guggenheim Helsinki winning design by Moreau Kusunoki Architectes. Source: Courtesy of Moreau Kusunoki Architectes

the name and brand. Arhinmäki also stated that all other payments were to be funded by taxes.

> Now, it looks like everything else is supposed to come from taxpayers. [...]. This means that when the National Board of Antiquities (which funds museums in Finland) makes cuts, museums have to scale their operations back. Therefore, this issue should be assessed honestly. This would mean that big cuts would have to be made elsewhere in the culture budget.[45]

Opposition to the Project

In the Guggenheim Helsinki competition, none of the finalists—nor any participating firms—were Finnish; true for most of the artists in the entire Guggenheim collection. Raoul Grunstein, a Finnish artist, also worried that "the meaning of our local artists would be diminished."[46] Richard Armstrong, Director of the Foundation and the Guggenheim in New York, mentioned experiencing some defensiveness and a highly developed antagonism toward the Guggenheim Foundation, a worry that he said he could sympathize with; Helsinki's unique local identity would somehow be incorporated into some sort of enormous industrial complex.[47] The impression that Finland was purchasing a famous onerous foreign brand was shared by many.[48]

The project's opponents were mainly on the political left, while the right saw it as a boost for Finland, citing the example of Bilbao Guggenheim, which, according to conventional wisdom, helped transform the Spanish city into a popular art and architectural destination. Dazzled by the promise of the "Bilbao Effect," dozens of cities courted Guggenheim every year.

However, when Helsinki city councilors were asked whether the museum would benefit Finland as a whole, none agreed strongly, while almost half disagreed.[49] The Guardian newspaper reported on the views of Mark Wigley, Dean Emeritus of the Graduate School of Architecture at Columbia University, and the chair of the judges in the international design competition for Helsinki. Wigley remarked that the "genuinely dignified" status of the public in the winning design (by Paris-based Moreau Kusunoki Architectes) was "a wake-up call to the Guggenheim and architecture in general." He added, "I am so bored with 80-year-old white men getting out of their airplanes, not knowing anything about the city but pretending to love the clients and dumping one more uninteresting museum on them."[50]

The late architect and urbanist Michael Sorkin, an opponent of Guggenheim Helsinki, criticized the six finalists of architectural competition. His initial impression was that "it's just architecture, just a collection of singular structures on a particularly alluring location. No typical exceptions were not observed. The concept that the jury is trying to protect the participants' privacy struck me as absurd; two children in my studio pretty much figured them out in a matter of seconds." [51] Sorkin was the chair of The Next Helsinki (TNH), an international competition that sought innovative ideas for the improvement of the cultural and public spaces of Helsinki.

Launched as an alternative to the controversial Guggenheim Helsinki project, The Next Helsinki (TNH) "called upon architects, urbanists, artists, and environmentalists to imagine how Helsinki and the South Harbor site allotted to the proposed museum can be transformed for the maximum benefit of the city's residents and visitors." As the official TNH website states:

> We initiated this project out of a sense of both outrage and love. Outrage at the march of the homogenizing multinational brand culture emblematized by the imperial Guggenheim franchise, Starbucks' cultural equivalent, was what launched us. The feeling of love came from our mutual affection for Helsinki in the sense that it is a singular place, unique in setting, form, and culture. Understanding the impetus to acquire a Guggenheim as a pursuit of the vaunted Bilbao Effect, the idea that some gaudy global repository would put a tired place

> *on the map,* we wondered why a city so indelibly fixed in the urban firmament, so superb, would want to surrender such a fabulous site to some starchitect supermarket.[52]

The TNH competition, pitched to "attract innovative ideas about how to more fully meet the city's cultural, spatial, and sustainability needs," branded the Guggenheim project a "misguided vanity project, and a symbol of the Finnish capital selling out to an American brand."[53] According to *The New York Times*, many city residents opposed the project, and local artists expressed concern about Guggenheim, potentially absorbing the Helsinki City Art Museum.[54] The call for submissions to the TNH opened on September 9, 2014. The deadline for submissions was March 2, 2015. Over 200 international entries were submitted to TNH competition from 37 different countries on five continents. Results of the TNH competition were announced on April 20, 2015, in Helsinki, approximately 2 months prior to the outcome of the official Guggenheim Helsinki competition. The jury was most excited about entries that suggested building on existing resources and tried to capture emergent urban trends and tendencies in the city.

By contrast, the six design finalists' of the Guggenheim Helsinki were criticized as uncreative and lacking in a sustainable bioeconomy focus—something that Finland leads in and values highly.[55] Finalists' final designs were repeatedly reminiscent of modern architecture, such as "light as a beacon, industrialized portal frames with a modern twist with piano roofs, and one entry that resembles a lotus flower or something resembling flora that had been so overdone."[56] Critics pointed out that architecture needs to be responsive and be considered as a form of "critical regionalism," which has been lacking in a significant number of projects. Even if the finalists' renderings were praised for their technical architectural competence, criticism was directed toward a perceived lack of attention to bioeconomy and sustainability issues, which was highlighted in the competition brief. Another set of critiques stressed that the designs did not incorporate any elements from the local context, culture, or heritage.[57]

Abu Dhabi and Dubaization

Most Arab world cities compete to imitate Dubai in its unprecedented effort to build the tallest, biggest, and largest ever built architectural and urban statements. This phenomenon can be best described as "Dubaization," the process of urbanizing a city with futuristic, pioneering architecture.

Dubaization, qualitatively similar to the Bilbao Effect, is part of the "icon project" trend and has spread to other cities beyond the Gulf area such as Istanbul and Vancouver. Dubaization triggers crucial questions: What are the consequences of this urbanization strategy on the future of Arab cities? What kind of social life has emerged from this development? Is this an elite-driven process of constructing, reconstructing, and deconstructing identities along with the urban landscape of Arab cities? Is there any future for sustainability in the development strategies of Arab and Middle Eastern cities?

Dubai is a model of urban development primarily based on images and icons rather than sustainable concepts and processes.[58] Major conflicts result, including failure to adopt sustainability, limited interpretation of globalization, and degradation of locality. Arab cities must consider a holistic approach to sustainable strategic development.

Architecture as a domain and creative reflection of local culture can be used as a vehicle to maintain local culture and interact with the global appetite for knowing "the other." The main condition for these architectural products to be exposed to the other is that they should come from a deep and original local vision rather than being exemplars of a globally crafted strategy. The Dubaization of Abu Dhabi includes a new Guggenheim Museum designed by Frank Gehry—originally set to open in 2012; then in 2017; and as of January 2024, projected to open in 2025.

The Guggenheim will be one key element in the new Saadiyat Cultural District (Fig. 4.4), together with the Zayed National Museum (also to be completed in 2025). The new Cultural District will also include the Natural History Museum Abu Dhabi, featuring a research and teaching center, and the teamLab Phenomena Abu Dhabi, a cutting-edge art installation collaboration with Tokyo art collective teamLab. These will join the already-completed Louvre Abu Dhabi, the event space Manarat Al Saadiyat, and the music conservatory Berklee Abu Dhabi.

The Guggenheim in Abu Dhabi is twice the size of the museum in Bilbao and 12 times the size of the Frank Lloyd Wright Guggenheim in New York. Carol Vogel in *The New York Times* refers to this Gehry design as "a graceful tumble of giant plaster building blocks and translucent blue cones."[59] The multiple controversies and disruptions associated with the Guggenheim Abu Dhabi indicate errors and failures in planning, policy, and implementation.

The original cost of Guggenheim was approximately $800 million out of the $27 billion budget for the cultural district.[60] Ramin Salsali, honored as a patron of the arts by the UAE vice president Sheik Mohammed bin Rashid Al Maktoum, said that the project would not stimulate the local economy nor contribute to cultural development because Abu Dhabi does not need the museum to enhance its reputation. Salsali expressed skepticism about the need for the Bilbao Effect in Abu Dhabi. It is a city already with a global reputation. Therefore, since the Guggenheim project would clearly be of most benefit to the Guggenheim Foundation, it made sense for the foundation to pay the city if the project should go ahead.[61]

Abuses on Foreign Workers

Foreign workers in Abu Dhabi comprise 85% of the workforce, mostly originating from countries such as Bangladesh, Pakistan, Nepal, India, and Sri Lanka. According to a PriceWaterhouseCoopers compliance report,[62] 86% of the workers paid illegal recruitment fees to agents, while 92% paid their own relocation fees, including visa, accommodation, and travel expenses. Many of these workers were young men who were led to believe that they would work high-paid construction jobs. While others,

Fig. 4.4 Model of Master Plan for Saadiyat Cultural District in Abu Dhabi, with the Guggenheim Museum in the foreground. Source: Urbanmyth/Alamy Stock Photo

according to Human Rights Watch, were under the impression that they would be working as hotel employees or waiters only to find out upon arrival that they would be working far less lucrative construction jobs.[63] Many of these foreign workers live and work in dire conditions. Therefore, in conjunction with NYU Abu Dhabi and the Louvre, Guggenheim insisted on guaranteeing workers' rights to prevent discord from board members and donors. Despite this assurance, *The Guardian* released a 3-month investigative report on work and living conditions and found the following:

- Companies withheld the passports of migrant workers, trapping them in the UAE;
- Thousands of workers lived in substandard conditions elsewhere in the UAE, in an apparent breach of the TDIC's—a developer of major tourism destinations in Abu Dhabi—pledge to house them all in its model Saadiyat accommodation village;
- Dozens of workers had been deported in 2014 for striking over pay and conditions;
- Mobile phone video footage of a riot showed dozens of men roaming the camp armed with metal spears and planks spiked with nails, with men jumping out of the windows; and
- One worker, who claimed that he had lost his legs while building luxury villas, was forced to live on the top floor of a migrant camp for a year. He only received a prosthetic leg months after the injury and was reliant on the Red Crescent for medical support. His claims for compensation and requests for ground-floor accommodations have been rejected. [64]

Religious and Political Censorship

In Abu Dhabi, and specifically in Saadiyat Island—the location of the new Guggenheim—purely ornamental art is welcome[65] while modern art to be exhibited in the Guggenheim Museum is generally regarded as "largely liberal, tolerant, and exuberant."[66] This contrast between local culture and modern art creates a societal clash where "cultural biases threaten to divide the project philosophically, while practical differences suggest that the users and providers also disagree about the eminence of economic or aesthetic interests."[67] Verena Formanek, Senior Project Manager of the Guggenheim Abu Dhabi, admitted that the process of compiling contemporary art was difficult because it was managed by government officials with government money; they had to proceed with caution to avoid shocking people and triggering a backlash against the new museum.[68] However, she also noted that the delay in the construction of the museum gave the government time to educate the public and prepare them for art that would be exhibited in the museum. Rita Aoun Abdo, Saadiyat's Cultural Director, noted that museums need to develop organically within the local culture; otherwise, their survival is compromised.[69]

Apart from religious censorship, there were also concerns about political censorship in Guggenheim following the 2011 Arab Spring Movement. Although the political movement largely skipped the UAE,[70] the country did introduce Internet restrictions in 2012, specifically on the use of social media to organize protests after a large group of Islamists were imprisoned and charged with plotting a coup in 2013. One example of political censorship is the sacking of the Jack Persekian as the Director of the Sharjah Biennial in 2011. He planned to display a picture considered offensive to the ruling Sheikh in Sharjah.[71] Consequently, Arab Spring was not predicted to take root in the UAE due the government's strong societal control.[72]

Environmental Impacts

To accommodate the Guggenheim Abu Dhabi area, which includes a satellite of the Louvre and an accompanying performing art center, all built on Saadiyat Island, new lands and breakwaters have had to be created.[73] Breakwaters are known to stagnate the flow of ocean waters and can lead to environmental problems. For instance, where flow is reduced, waste cannot flow out and nutrients cannot flow in, deteriorating the shore environment. Also, turbidity and sedimentation levels along the shores of Saadiyat Island could increase, which would further harm aquatic life due to insufficient oxygen dissolved in the water. Multiple flooding events have also occurred during these reclamation projects. Due to variations in reclamation activities and locations, the flow over the pond is continually changing. Therefore, careful planning is required to best manage the weir boxes and maintain the channels and scum booms. However, as the project has progressed, material is accumulating in the settling pond. Due to the continuous flow of water over the pond, excess material cannot be removed, and settling capacity is reduced. The settlement pond, which was incorporated into the island's architecture, needed reclamation. As the reclamation process continued, however, less space was left for settling, and the height of the weir boxes could no longer support a sufficient capacity to guarantee the tailwater quality. A closer examination revealed that the trench depth was still outside the range required by the screen.[74]

Sensitive marine communities, including mangroves, sea grasses, and coral communities, are adversely affected by dredging. Moreover, sedimentation rates greater than 15 mm per year smother aerial roots where gas exchange occurs. Despite the models of sediment accumulation have indicated little risk of mangrove root smothering, mangroves located inside the boundaries of reclamation areas are likely to suffer from both reclamation and dredging. Coral communities are also being harmed, threatening the area's biodiversity.[75] While mangroves and sea grass are very sensitive to the salinity and acidity of the surrounding water, coral communities are likewise very sensitive to the sediment load carried by surface water. Impervious surfaces, such as roadways, increase runoff rates and carry water mixtures containing pollutants; these waters are eventually discharged into coastal waters, negatively impacting the coral.

The Fading Away of the Bilbao Effect

As mentioned earlier, capturing a share of the world's mobile wealth is fundamental to justifying megaprojects. From an urban-spatial and design perspective, this assumes that cities need to become "visible" and attractive to international capital. The "icon project," that is, the widespread construction of architectural icons in globalizing cities around the world, accomplishes these goals.[76] The construction of iconic urban megaprojects (IUMPs) has grown into a standard policy choice for urban and regional elites in globalizing cities. Politicians, business leaders, and others in local and regional growth machines fulfill their personal and professional ambitions by investing in and promoting IUMPs, aspiring to reach global status and positive economic change for their cities.

Many urban elites worldwide have been greatly influenced by the so-called "Bilbao Effect," and the perception that the Guggenheim Museum Bilbao resulted in a "Cinderella transformation" of the northern Spanish city and economic capital of Basque Country. In the dominant discourse concerning architecture's surrender to capitalism's commercial goals, Guggenheim Bilbao is mistakenly and repeatedly portrayed as the model "catalyst" for a city's radically successful transformation from industrial powerhouse to regional service center. The author hypothesized around 2007 that iconic architecture could be replaced as the hegemonic discourse of urban revitalization, and with this development, the Bilbao Effect would fade away.[77]Along these lines, architectural critics, such as Michael Kimmelman, have come to understand that the Bilbao Effect has always sat on shaky foundations:

> The truth is, the Bilbao Effect is largely a myth. Frank Gehry's museum alone did not turn around the city. It has capped decades of civic renewals. Flashy, even brilliant buildings, rarely rejuvenate neighborhoods or guarantee crowds and cash simply by virtue of their design [...] Sadly, museums, like cities, have squandered fortunes praying to this false idol. They still do.[78]

The example of the new Ordos Art Museum in Inner Mongolia—beautifully designed by MAD, a prestigious firm of Beijing architects—suggests (not too surprisingly) that building a terrific museum is not sufficient to ensure success. The city of Ordos has sprung up quickly and is relatively wealthy, thanks to discoveries of oil and gas, but the museum has no collections and few plans for exhibitions. No wonder it is devoid of visitors.

Even before the actual construction, the Guggenheims in Helsinki and Abu Dhabi attracted significant criticism, as shown in this chapter. These projects can be questioned in three main ways: (1) iconic architecture is no longer a hegemonic visual discourse in urban revitalization; (2) the franchise model imposed by Guggenheim means that local officials have no autonomy to make major decisions on matters from exhibition calendars to budgets and investments; (3) local cultural identities are usually neglected in a foreign global art model. Moreover, the environmental impact of such megaprojects is not negligible. The Abu Dhabi project has also been controversial in terms of workers' rights and labor conditions. IUMPs have triggered many controversies, including cost overruns, negative environmental impacts, gentrification risks, the drawbacks of top-down cultural engineering, neglect of local cultural identities, and uncertain economic success. None of these externalities bode well for cities that count on instant icons to salvage them during times of economic malaise. Despite mounting criticism, if the new Guggenheim Museum in the United Arab Emirates results in even half the impact of that of Bilbao's, the term Bilbao Effect will continue to carry weight on both sides of the debate.

Despite the media success of Bilbao Guggenheim, the Bilbao Effect has proven difficult to replicate in most places, even in the case of other designs by Frank Gehry. On the other hand, some architectural icons, such as Gehry's Stata Center at MIT, work well with no Bilbao Effect—most MIT scientists working in the building praise its playful and inventive feelings, as the author has personally witnessed.[79] Cooper Union alum Daniel Libeskind's jagged edges, sharp angles, and complex geometries—as seen in DAM's Hamilton Building, the Royal Ontario Museum in Toronto, and the Danish Jewish Museum in Copenhagen—do not have nearly the universal acclaim of his Jewish Museum Berlin, an illustration that success, impact, and visitor attraction are not necessarily functions of a building's spectacular design. Many works by Shigeru Ban and Tadao Ando are excellent examples of highly admired and successful architecture in stark contrast to iconic buildings designed to stun.

The jury is still out today regarding not only Gehry's highly anticipated Guggenheim Abu Dhabi but also the massive West Kowloon Cultural District (WKCD) in Hong Kong. Both have been among the most prominent cultural megaprojects in recent years. As we saw in Chapter 3, the WKCD is a project of such scale and ambition that it could "define the nature of the public realm in the 21st century," according to Rem Koolhaas' hyperbolic statement.[80] WKCD met with significant criticism from the planning and construction phases. Although Guggenheim is not part of the project, it replicates all the expected controversies associated with IUMPs, including cost overruns, negative environmental impacts, gentrification risks, drawbacks of top-down cultural engineering, neglect of local cultural identities, and uncertain economic success.

Woronkowicz and others (2015) analyzed new cultural facilities built in the United States between 1994 and 2008, a building boom for museums and other arts institutions.[81] According to their analysis, the Bilbao Effect did not materialize in most cases. There has been a decrease in poverty rates and an increase in property in communities where new cultural centers or expansions were built, but poorer residents have also suffered displacement in those areas. In addition to the usual gentrification effect caused by cultural megaprojects, researchers show that supply may have outstripped demand, and cities have been left with the responsibility to maintain or even pay for cultural centers that they do not entirely need.

Furthermore, the analysis shows that expansion and new museum projects do not have spillover effects one way or the other (positive or negative) for nearby cultural institutions. In addition, the authors explain that leaders involved in building cultural megaprojects depend on "inside knowledge" obtained from their own experiences, and those of their collaborators' experiences. "What tends to be absent in their thinking, however [...] is 'outside knowledge' regarding the distribution of projects that did *not* go as planned," the book continues.[82]

Cost overruns and project delays are representative of cultural megaprojects. In the cases analyzed in Woronkowicz and others (2015), 91% of performing arts centers built during the study period went over budget. Of the projects studied, 54% featured lower revenues than projected, while 59% featured higher expenses. What the research shows is a gap in cultural facilities management lacking level-headed analysis that leads to misplaced expectations and mistakes, resulting in budget shortfalls and cost overruns.[83]

In the summer of 2017, Centro Botín was opened in Santander, Spain. Located just an hour from Bilbao, the contemporary art center was designed by Renzo Piano, a "starchitect" who is no stranger to flashy, monumental design. As Hune-Brown explains:

> The museum is an elegant, striking structure, but its creators have been eager to tamp down any Bilbao-related rhetoric on civic transformation. According to the Botín foundation's president, the museum was built for the people of the city, not to "create an icon." It is nearly invisible from within the city itself—a "self-effacing" building according to one architecture critic. The building feels like a public affirmation that the heady days in which we discussed architecture as a savior of cities are over. When Piano was asked about his design approach, he did not mince the words. "I suppose our strategy was the opposite of the Guggenheim," he said. "How many Bilbao Effects can you have after all?"[84]

Learning from Bilbao

In summary, UMPs often rely on iconic architecture for success. Iconic UMPs (IUMPs) have come to play a central role in standard urban policies designed to gain global visibility, attract visitors, and invest in cities. Bilbao is widely considered a successful case of image reconstruction via iconic architecture; however, Guggenheim Bilbao, like most other urban icons, presents lights and shadows when the focus of the analysis is on economic and cultural impacts. Guggenheim Bilbao triggered immediate and lasting worldwide interest among tourists, artistic circles, the architectural profession, journalists, and the educated public based on the iconic architectural style of the building as well as local, contextual, economic, and political conditions. Far from being the trigger for and prime mover of revitalization, the museum postdated it. To date, there has been no substantial foreign investment in the Basque city or any significant positive outcomes in the job market due to the museum.

Denver, Helsinki, and Abu Dhabi, the cases examined in this chapter, add to the list of cities around the world that have attempted urban transformations like Bilbao. The Helsinki project has been canceled and will not be built. With the other two, there continue to be numerous controversies and civic opposition triggered during the planning and design phases. In Denver, urban planning developments in the years since the opening of the Libeskind-designed Hamilton Building in 2006, put the Denver Art Museum on the sideline of local improvements efforts. In Abu Dhabi, the new Guggenheim will play in a complex environment within the context of an increasing diversification of the local economy, and if its fate is like Masdar City's eco-experiment on the outskirts of Abu Dhabi (discussed in Chapter 5), the success of the new museum is far from guaranteed.

Even if Bilbao itself is considered a successful case of image reconstruction via iconic architecture—something that by itself did not solve many of the structural and socioeconomic problems of the city—the Bilbao Effect was largely a failure and has faded away. The fading away of the Bilbao Effect is due to the limitations of existing political rationality and decision-making processes at times when globalization puts pressure on urban leaders to redevelop and become globally visible.

There is also a poor understanding by outsiders of the true reasons behind Bilbao's urban revitalization success. Indeed, such success has little to do with iconic architecture. Instead, it can be attributed to a sound economic policy by the wealthy and financially autonomous Basque government and a well-crafted and comprehensive revitalization plan in which the Guggenheim was just a very small and ad-hoc component. Furthermore, while Bilbao has been more successful than other cities in terms of branding and image change, the realities of urban life in Bilbao remain subject to the complexities, contradictions, and shortcomings of the neoliberal model of urban development, including precarization of the labor market and severe socio-economic inequality. Iconic architecture has brought tourists to Bilbao, but it has not contributed to solving the structural problems of Bilbao's urban economy.

Bilbao's economic performance after the opening of Guggenheim has broadly followed the ups and downs of economic cycles—a clear indication of both the embeddedness of IUMPs in cities and the very limited power of architectural icons to explain development, competitiveness, and urban economic changes. Institutional contexts, specific policy instruments, and contextualized social dynamics inform distinct patterns of IUMP development and contribute to explaining the degree to which such megaprojects succeed or fail.

Likewise, should the star of Bilbao Guggenheim begin to dim and visitor attendance begin to dip, the consequences for this Basque city would not amount to a significant economic decline. The museum represents just 2.2% of the Bilbao economy in terms of GDP—per the author's calculations. Cities are complex. A spectacular building alone is not usually capable of shifting fortunes in fundamental ways.

Urban context matters in megaproject planning and implementation. In addition, not every city is well positioned to be "put on the map," especially second- or third-tier cities—comparable to Bilbao in terms of size—but located away from main thoroughfares. Bilbao is in a country, Spain, that remains one of the top three tourist destinations in the world, a key factor in the museum's spectacular ability to attract visitors. Spain receives approximately 85 million visitors annually, of which approximately 4.5 million tour Basque Country. Approximately one million of those visit the Guggenheim Museum in Bilbao.

Bilbao's place on the world stage has irreversibly changed after Guggenheim. Unequivocally there is potential for of IUMPs to transform a city's image in times of globalization. The global media impact of Guggenheim Bilbao alone represented a turning point that significantly enhanced debate about the commodification, commercialization, and replicability of megaprojects in large-scale redevelopment schemes. Megaprojects are infrastructure units that obey specific planning, design, and management priorities; and they are expressions of larger socioeconomic forces. The dramatic increase in the interest of urban elites to surrender to the promise and discontent of UMPs has become one of the keys to explaining prominent aspects in the deployment of contemporary globalized urbanization.

Cities should not expect to replicate Bilbao's success simply by implementing a fashionable urban policy marketized via appropriate global media discourses. Each city has a unique local history, a specific geography and political makeup that influences the local decision-making processes. Cities and regions adhere to their own logic of development, and each may require localized strategies for redevelopment. Applying standard elements in the revitalization mix, including IUMPs, may be unavoidable because of the rapid and acritical adoption of policies from the center to the periphery. However, expecting to replicate another city's success by blindly adopting such a strategy is often a recipe for disappointment.

Twenty-six years after the opening in 1997 of Guggenheim in Bilbao, the Bilbao Effect has faded away. Inflated claims of this age of urban icons no longer hold. When every city appears to boast a spectacular museum, the strategy of building cultural megaprojects no longer yields positive outcomes. Instead, we should work to favor other approaches such as traditionally overlooked synergies between research-based evidence, management, and urban governance for holistic sustainability, and make those a priority.

Afterword: Architecture After Spectacle

Iconic architecture is not fading away altogether, but rather has expanded to areas where it previously did not exist. To be sure, this globalization of emblems—the icon project[85]—might appear to contradict the argument about the fading away of the Bilbao Effect developed earlier in this chapter.

However, recall that the fading away argument centers on how the diminishing returns of iconic architecture directly correlate with the expansion of this mode of urban development. The Bilbao Effect is tied to context and uniqueness, runs against serialization, and increases scope and reach. When every city seems to boast of an architectural icon that defines its identity, the strategy of using megaprojects to attract investment and tourists has diminishing returns. Catchy looks and curvaceous shapes—the formal architectural requisites of most emblems—continue to be popular in the many parts of the world where municipalities and corporations desire to promote their brand globally. These iconic buildings are intended to become drivers of economic renewal and agents of economic development that would appeal to a universal audience, create a positive image, and reinforce self-esteem. The availability of capital (e.g., in the Persian Gulf), the massive scale of construction and development (e.g., in China), the wide availability of high technology (software, visualization techniques), and the Internet's crowded global marketplace of imagery ("digital monuments," as Broot puts it)[86] contribute to the resilience of an architectural paradigm whose expansive thrust paradoxically erodes its effectiveness over time.

Architecture is a political act that works by mapping space and place. As discourse, practice, and form, architecture operates at the intersection of power, production, culture, and representation. The politics of emblematic and spectacular architecture are intimately tied to digital globalization. In fact, it is legitimate to question whether we could even have an icon project without a digital economy that enables, on a global scale, the visible image of cities and icons (through branding and reimaging) traveling by means of urban policy mobilities (e.g., the Bilbao Effect, Dubaization, Manhattanization) in a society of spectacle and entertainment. The consequences of global visibility can be said to be urban image change and the shaping of an urban global identity, as well as overexposure, vulnerability to overseas contractions, and their concomitant risks, such as tourism gentrification.[87] Furthermore, this top-down strategy is the hidden side of achieving and maintaining political hegemony. At the time of writing, globalizing strategies such as digitalization may have passed their golden years in a new context of technological competitiveness driven by geopolitics. While cities are still building icons to attract tourists and investments, an increasing preference for proximity in supply networks and localized advanced manufacturing production, together with a new focus on urban technology (see Chapter 5) and urban sustainability (see Chapter 6), are beginning to restructure urban economies.

During this neoliberal period, there has been a general trend of questioning many tenets and practices of architecture. This is not surprising given that both architecture and urbanism have been exhausted by the effects of neoliberal globalization.[88] Urban emblematic projects have been at the service of global capitalism along with the known effects of cost overruns (coupled with strategic misrepresentation), negative environmental impacts, gentrification risks, drawbacks of top-down management decision-making and cultural engineering, neglect of local cultural identities, uncertain economic success (compounded with optimism bias), population displacement, spoiling of cities' visual and structural uniqueness, rise of congestion and overpopulation, political gridlock, threats to ecosystems, and other impacts.[89]

As non-Western regions of the world catch up to the game of emblems, the fading away of spectacular architecture as the hegemonic architectural discourse in the West leads us to discuss potential new or coexisting paradigms. Among these new paradigms, we see a renewed interest in linking aesthetics to topography through tectonics and place, as source material for architectural production. We also see the intent of rebuilding resistance through the use of time and history. In sum, we observe a fuller embrace of complexity by architectural theory and practice, away from commodification, and also the crossing of disciplinary boundaries, particularly when dealing with issues of urban ecology and sustainability. The remainder of this chapter explores these new paradigms and ideas and contrasts them with the ideas behind iconic megaprojects. While this section is markedly different from the kind of analysis developed in the previous pages of this chapter, our discussion about the emerging ideas characterizing *architecture after spectacle* is a necessary afterword or epilogue to our discussion of iconic megaprojects.

Tectonics and Materiality

Tectonics—the craft, science, and art of construction—is concerned with the nature of the creation and modeling of materials and how this focus forms a creative force in construction, structural features, and architectural design. It is closely related to the work of structural engineers and the natural context of buildings, their topography, and climatological conditions. It rejects the focus on visual effect and the effacement of history, which are two features of iconic architecture.[88] In *Studies in Tectonic Culture*, Kenneth Frampton provided a framework to tie architectural theory to the essence of construction and experience. [90]

Frampton believed that architecture is about placemaking and the passage of time as much as it is about space and form. He was explicit that space is architectural when the tectonic process is front and center of architectural projects. The focus on the local context allows the tactile, tectonic, and hybrid to create an "architecture of resistance at the interstices of freedom" in direct opposition to the commodification of architecture into a global consumer product. Projects are not only designed to reimagine cities, but they are also engineered within and for the urban built environment, with lasting and sometimes irreversible consequences. As a subject that encompasses all aspects of design, structural engineering encourages reflection on scale, constructability, aesthetics, and the technical, social, and environmental contexts.

The focus of tectonics is akin to the notion of *materiality*. The built environment is an inescapable material reality to be grasped from the outside, through "the observation of concrete materials, not the workings of the mind in isolation," according to Richard Sennett.[91] Jane Jacobs observed that buildings, streets, and neighborhoods work as dynamic organisms, changing in response to how people interact with them.[92] Materiality aims to reveal a structure's influences, results, and consequences, not by defining the objects, but instead by becoming sensitive to matter itself. In this vein, French sinologist François Jullien stated that "a wise person does not have ideas independent of matter."[93] Materiality can be seen as a theoretical extension of the notion of *affordances*.[94] Affordances, as applied to both things and places, are properties that allow a person to do some things and not others but are not fully constraining. The idea of affordances thus refers to the properties of matter, these properties being what influences how humans interact with things.

Time and Memory

A city is a reflection of the complex relationship between time, space, and memory, strongly influencing human society. All locations contain layers of history: some visible, some hidden, and some partially erased. It is something that must be discovered in each city and that partly explains in part the enormous fascination that human beings feel about cities.

Urban architecture has become a language that deliberately erases, preserves, rewrites, communicates, transmits, hides, and restores memory. For example, Berlin's architectural quest for authenticity, an elusive use of media simulation and historical citations, implicitly suggests the positive use of iconic architecture in the service of time and memory. These uses were collected both in official institutional buildings (the Reichstag, the destruction of the Palast der Republik, and the reconstruction of the Schloß facade) and commercial experiments (Potsdamer Platz, Hotel Adlon, and Friedrichstadt-Passagen). Projects that glorify the contemporary globalized economy, the restoration of forgotten but highly symbolic buildings, and the destruction or surgical removal of socialist icons have created an indecipherable palimpsest, composed of a multitude of layers that point to the complex past, present, and future of the city.

Peter Eisenman designed the Holocaust-Mahnmal (Holocaust Memorial) to preserve critical memory in Berlin, a reflective and contemplative memory that evokes the structural social conditions of a certain urban and European past. The Mahnmal is built in the heart of the city next to the Brandenburg Gate, the Reichtag, and the Tiergarten, and its 2,711 blocks of different heights evoke memory in a subtle and abstract way through various architectural procedures: repetition, temporary experience, ground ondulations and inclines, and the grid that the blocks form. Repetition promotes abstraction but also preserves memory. In this framework of preserved memory and recovered and monumentalized historical identity, a trip to Berlin is increasingly connected to the experience of a vibrant and authentic place where layers of history can and must be sought, discovered, remembered, or recognized. Part of the architectural renewal that the city underwent was related to tourism and the opportunities offered by this type of urban consumption. A greater focus on authenticity and sustainability drives the increasing diversification of visitor experiences.

Space and Place

Tectonics, time, and memory are resistance strategies. So is the practice of place. Walkers in the city continually invent spaces in the built urban environment. According to De Certeau (1984), walking defines *spaces of enunciation*. Practices of resistance become, then, "spatial stories," and through them there occurs "a constant transformation of places into spaces and of spaces into places." Thus, spaces are open to human creativity and action. De Certeau believes that spaces can be more easily liberated than Foucault imagines, because individual practices "spatialize" rather than localize in repressive grids of social control.

Therefore, space is not simply a metaphor for a site or container of power.[95] Resisting means that marking boundaries for the symbolic creation and recreation of spaces is an act of partitioning and differentiating. In this sense, it also constitutes an act of foundation, "of creation of a field that authorizes dangerous and contingent social action" in a "fragmented," "miniaturized," and "polyvalent" form. Resisting (spatial stories, practices, and operations) also involves transcending the frontiers and crossing bridges. By privileging a "logic of ambiguity," the spatial stories of resistance represent "a departure, an attack on a state, the ambition of a conquering power, or the flight of an exile; in any case, the 'betrayal' of an order," the "tour over the state," and narrativity "in its most delinquent form."[96]

De Certeau conceives narrative in a quite broad way as creations of spaces (as opposed to established places), as description (as opposed to theorization), as an art (as opposed to discourse), and as a private knowledge that remains "on the margins...of scientific or cultural orthopraxis." It is "the status of a know-how without discourse." Not only all manifestations of popular culture (ordinary language, tales, games, legends), but also any kind of "spatial practice" (walking, incarceration, railway navigation), and also reading and believing constitute objects for narrativity. The basis of spaces and the avoidance of totalization are represented through narration (which, as we have seen, challenges the unifying thrust of places). In fact, without narration, a group or individual regresses into the unsettling, fatalistic experience of nighttime, hazy totality.[97]

This "poetics of space" (to use Bachelard's term) is enacted in the practice of place as resistance within and amongst the interstices and multiplicities of capitalism, leading to the well-known Foucauldian conception of space as "heterotopy" (1967), where the concern with intimate spaces is replaced with a focus on spaces of crisis, deviation, exclusion, and illusion.[98] Here, power and history establish a link between the individual and the collective, indicating that resistance from within architecture also needs to use socioeconomic categories. Sorkin puts it as follows: we need "the political demand for redistribution, a Ghandian refusal to play ball, a celebration of the informal, the demand for a space of occupation, new tactics of the *flâneur*, a celebration of the possibilities of the visionary, a return to socialism, a demand for a larger state."[99]

Complexity Beyond Spectacle

Tectonics, time and memory, and the practice of place require alternatives to icons and images as representational tools: the paradigm of complexity is becoming an alternative. In some respects, complexity is a component of iconic architecture, allowing the design and construction of eye-popping structures. The current and future potential of complexity lies in the notion of nondeterministic diversity and fractal self-organization (as the basis for the analysis of risks, resilience, and urban sustainability). This is akin to the intrinsic, pluralistic order of cities (disorderly and chaotic), as Jacobs taught us,[100] and is also similar to the idea of a "jumping universe," according to Jencks.[101] Jencks argued for an architecture based on deep nature, one that keeps changing in slight variation (it is self-similar rather than repetitive), and one that follows fractal patterns found in natural, nonhuman life. In this sense, the complexity of architecture is essentially urban, as it shows a concern with information and entropy, such as those generated in urban environments, as well as a focus on interaction, feedback, collaboration, and integration. Thus, complexity is wicked and messy. It must face the limitations of human cognition as well as grapple the difficult whole and embody "the difficult unity of inclusion rather than the easy unity of exclusion."[102] These ideas help us classify cities and megaprojects into two categories of complexity. Both are complex adaptive systems, yet the natural order of cities self-emerges, to a certain degree, from spontaneous responses to prototypical and unique situations, while megaprojects are artificially contrived and invented. In megaprojects, we can observe the intent of an iconography of order exerted by power, which is ineffective because of the stochastic environment defined by uncertainty and risk. The consequence is the production of disruptive complexity in the form of impacts (materialized risks or the consequences of risks) in the urban arena.

The paradigm of complexity needs to deal with creativity and the fundamentals of architectural imagination for it to become human and incorporate both perception and cognition as attributes of the flow of consciousness. Re-enacting time and memory and the practice of places are part of an ecology of creativity. Citizens are co-creators of places and buildings, after all, and they should be invited to collaborate in the quest to re-center the ideas of public value and common good. As Michael

Hays has argued, architectural imagination lies between sense data and understanding, and schematizes with the use of diagrammatic procedures, thus producing complex knowledge.[103] The schematic imagination becomes diagrammatic when it accommodates newly conceived issues of multiplicity, potentiality, virtuality, and becoming, as well as various materialist tendencies. This is consistent with the thrust we observe in architecture after neoliberalism, where architecture is responsible for emphasizing its material embeddedness in heterogeneous networks of other forms, forces, and interconnected constellations.

Architecture is the making and remaking of the world: territorialization and re-territorialization of the concrete assemblage through a particular diagram. The diagrammatic imagination accounts for the entanglement of architecture within a complex of social, technological, and historical forces that are deep-seated, shifting, and contradictory. It is these forces that detailed formal readings of architectural projects seek to reveal.[104]

Transdisciplinary Urbanism

A complex, nonreductionist framework necessitates boundary breaching across research disciplines. Architecture is prone to such crossings. Architecture is the art form most dependent on and intertwined with outside forces, including physics, capital markets, client whims, and the passage of time.

Reality is socially constructed, as shown in Berger and Luckmann's model.[105] The ontology of reality, especially urban reality, cannot be completely understood by the social sciences alone. Leading urban researchers such as Manuel Castells, Janet Abu-Lughod, and Saskia Sassen have recognized that the reality of the city cannot be understood from a single disciplinary perspective. This point was brought home during two joint conferences of the British and American Sociological Associations held in 2001.[106] It follows that hybrid modes of inquiry should be a part of the knowledge landscape as urbanism engages, both as a discipline and as a profession, with broader societal concerns (e.g., situated knowledge, participatory design, and everyday activities).[107] Transdisciplinary knowledge production requires merging academic and nonacademic knowledge, theory and practice, discipline, and profession. It fulfills an important role by raising awareness of the need to co-create knowledge in the interstices between disciplines.

In the context of disciplinary boundary crossing, the contributions of architecture to urbanism and urban studies are well known. Furthermore, social scientists have benefitted from the understanding of space in the urban realm by architects, and architects have long tried to understand the architectural project within capitalism.[108] Architect Michael Sorkin was a major force in blending architectural and social-scientific concerns.[109] From a more substantive perspective, the sociology of architecture (both macro and micro) has been developed with some success.[110] Approaches such as science and technology studies (STS), and actor-network theory (ANT)[111] have in common their suggestion to tackle urban issues not according to predefined ideologies or (critical) theories, but to study them as a problem of the outside—as situated, complex gatherings of all sorts of agencies (human, nonhuman, nature, technology, engineering, the social, facts, values), where the notion of transdisciplinarity can be applied meaningfully.

There is a growing interest in explicitly approaching urbanism as a meeting point where architecture is integrating other disciplines—including the social and environmental sciences—into design thinking. In *Ecological Urbanism*, Mostafavi and colleagues demonstrate a variety of disciplinary interests, from humanities, literature, and cultural theory to engineering, biology, and computation, with a focus on natural complexity.[112] Architecture's encounter with natural complexity, which goes back to Patrick Geddes and his protégé Lewis Mumford, interprets the environment as context, climate as a resource, and forms as emergent rather than produced or invented.[113] Unlike other studies attempting to infuse design with complexity science in a systematic way, *Ecological Urbanism* focuses on conceptual ties and transactions across disciplines, and much less so on specific strategies that would link science, design, humanities, and other relevant disciplines.[114] By focusing on concepts, practices, cases, and technologies (with environmental planning, landscape ecology, and biodiversity at the core of the proposal), this project shows that urban environments can promote ecology and sustainability more effectively than nonurban environments, leading to a more detailed understanding of the ties between complexity, disruption, sustainability, and transdisciplinarity.

Conclusion

This chapter showed that megaprojects function as emblems for global capitalism and that their disruptive impacts are significant. By analyzing four cases of urban reimaging via iconic cultural megaprojects, this chapter unveiled the controversies and disputes surrounding contested iconicity in Bilbao, Denver, Helsinki, and Abu Dhabi. We saw that the success of Bilbao Guggenheim is difficult to replicate and that the Bilbao Effect is fading away due to the diminishing returns of serialized megaprojects and urban icons worldwide. The chapter also discussed the main disruptive impacts of emblems—related to the need for visibility, the focus on spectacle and entertainment, and the traveling of icons via urban policy mobilities. We then outlined a path for architecture post neoliberalism. We suggest a focus on tectonics, matter, memory, and the practice of places as well as a shift toward a fuller embrace of complexity and transdisciplinarity by researchers and practitioners. By embracing these tenets and their thrust of resistance, we believe that architecture and urbanism would be better prepared to fulfill their promise of improving human settlements. To be sure, the nature and shaping of projects is due to the developmental, institutional, and geographic contexts in which they emerge. However, as we saw in Chapter 3, urban megaprojects, regardless of their geographical location and socioeconomic context, constitute contentious enterprises that produce disruptive complexity, controversy, and criticism, which is clearly observable if we focus on impacts rather than risks. This is also the case in cases of *smart destructive creation*, which we examine in the following chapter, High-Tech Megaprojects.

Chapter 5
High-Tech Megaprojects

Introduction

This chapter analyzes high tech megaprojects, that is, megaprojects characterized by processes of high-tech innovation in urban settings. We focus on four such megaprojects: (1) urban technopoles, (2) manufacturing megaprojects (factories of the future), (3) innovation districts (innodistricts), and (4) smart-city megaprojects such as data centers. We discuss our findings from empirical evidence gathered about projects in Barcelona, Boston, London, Masdar City, and Toronto.

The chapter also discusses the pervasive expansion of the smart-city paradigm, and that paradigm's gradual evolution into a nascent artificial intelligence (AI) urban regime. An AI regime is characteristically disruptive, and it includes distinct features that impose valuation and devaluation strategies over entire neighborhoods and urban areas. These strategies reinforce marked inequalities and a dual pattern of benefits and high risks in already socioeconomically divided cities.[1] By focusing on high-tech development and use, particularly new kinds of automation capabilities and AI, this chapter explores and critically analyzes the smart-city paradigm along with the promises and discontents of urban sustainable infrastructure based on these new technologies.

The paradox of AI urban environments lies in the promise of attaining a certain type of sustainability at the risk of further jeopardizing urbanity, equality, and liveability. Clearly AI contributes to the improvement of megaproject management. However, if a relational concept of megaprojects highlights *embeddedness* within the wider economy and society,[2] it is less than certain that better megaproject management by itself can contribute to fostering urban sustainability. In other words, AI alone does not necessarily foster urban sustainability even if specific AI innovations can. Furthermore, in the case of innovations that foster sustainability, both the benefits and risks of the innovation strategies must be factored in. The key question is, therefore, how to craft and build a framework for sustainability that can

G. del Cerro Santamaría, *Contesting Megaprojects*,
https://doi.org/10.1007/978-3-031-98382-5_5

discriminate and identify which innovations contribute to the desired sustainability outcomes, an idea related to recent arguments and warnings about AI's existential risks.[3]

Smart Destructive Creation: An Era of Disruptive Innovation

We are witnessing the movement of advanced manufacturing and high-tech business activities to urban regions years after the creation of producer service complexes in cities.[4] In their worldwide study of high-tech development, Castells and Hall (1994) emphasized the significance of technopoles for the growth of urban and regional areas. Contemporary equivalents of technopoles, such as factories of the future and innovation districts, are new megaprojects that profoundly alter the urban fabric and socioeconomic structure of entire neighborhoods in distinct, specific ways.[6] Simultaneous and significant reorganization of production is occurring in many cities around the world. Using new technologies—particularly *smart* innovations and AI—in old technopoles, new factories of the future, and innovation districts, is a common strategy. Many other countries have resourced infrastructure development via megaprojects in the form of industrial corridors (e.g., India) and reterritorialization via megaregions (e.g., China).[6] As Zukin (2020b) put it:

> Since the 2008 economic crisis, city governments have aggressively pursued economic growth by nurturing these ecosystems. Elected officials create public-private-nonprofit partnerships to build an "innovation complex" of discursive, organizational, and geographical spaces; they aim not only to jump-start economic growth but also to remake the city for a new modernity.[7]

It should be noted that high-tech urban investments of the kind Zukin writes about originate in an umbrella process of urban financialization that encompasses different kinds of massive infrastructure investments in large projects—with a focus on culture, iconic architecture, green, and high-tech. In any case, corporate policies—especially those relating to the use of high technology, such as AI—have a direct impact on urban sustainable development insofar as these new centers of corporate activity are located in cities, megacities, megaregions, and urbanized areas. In this context, the new spatial reorganization of production is fundamentally urban and presents formidable challenges for sustainability; it enables it and constrains it to an unprecedented scale.[8]

Despite the diffuse, diverse, and controversial meaning of the smart-city concept, its proponents assume that a smart city is capable of better addressing the challenges of urban resilience and sustainability. Smart cities are presumably more liveable, sustainable, and resilient. They are presented as being better prepared to respond to urban challenges than the more traditional and transactional forms of urban governance. Therefore, it is vital to evaluate the advantages and the risks of smart cities as their development along with the widespread application of AI techniques, processes, and devices suggests that a new AI urban regime will be expanding in the coming years.

Although incremental, the disruptiveness of smart-city implementation is undeniable. This disruptive AI urban regime—described in this chapter as *smart destructive creation*—generates a critical mass of negative externalities in the development process of smart cities, stemming initially from the contemporary hegemonic nature of technological innovation in urban socioeconomic processes. Such a disruptive thrust represents a risk for the centrality of public spaces and civic friction between humans that should occur, according to Jacobs, in the core of the urban as self-organized complexity.[9] Technological gentrification (interrelated with green gentrification) is an emerging feature in spaces adjacent to innodistricts and other tech neighborhoods. There are other risks and ethical dilemmas brought about by the smart-city strategy that have to do with control, privacy, and security. For these reasons, *smart* megaprojects are indeed landscapes of disruption, like those examined in Chapter 3, albeit exhibiting their own genesis, development, and impacts. The intelligent disruption—destructive creation—produced by artificial intelligence systems is no longer avoidable, although we can debate what kind of sustainability we want for ourselves and future generations. Existing and forthcoming analyses and discussions would gain by focusing on the notion of "smart and just sustainability" as a normative horizon setting the framework to decide what technological innovations and what AI applications we need and want to see implemented. We discuss these matters in Chapter 8.

Starting in the 1980s, a new round of global capital accumulation began to focus on the extraction of value-add from innovations in science and technology (and digital information technologies) that took place in the United States, around Silicon Valley and Boston Route 128.[10] In the following decades, the information revolution expanded to other areas within the United States and around the world, particularly to the Global North (North America, Europe, Asia-Pacific), but also in some areas in the Global South.[11] While emerging market economies continue to increase investment in R+D and innovation, advanced economies have experienced a slowdown since 2000 relative to both prior periods and to other less advanced economies. In contrast to newer revolutionary technologies such as AI, machine learning, and automation, experts generally agree that the impact of information and communication technologies (ICTs) is waning. Significant technological advancements may be difficult to maintain over time, but persistent disparities allow emerging countries to catch up through adoption as well as their own adaptation and invention.[12]

Globalization and innovation have widespread effects on productivity.[13] Through cross-pollination, the worldwide dissemination of information and technology produces advantageous network effects because it enables technology-receiving countries to advance their research and development (R+D). Such dynamics have played a significant role in China's explosive growth in R+D spending and in the accumulation of international patents in South Korea. These nations have joined traditional leaders in industries such as electrical and optical equipment and—in the case of South Korea—machinery.[14]

The diffusion of knowledge and technology promoted by the functioning of a global economy operating in real time on a planetary scale accelerates and strengthens the incentives to adopt new technologies and to innovate.[15] With globalization, new ideas and technologies can spread quickly across the world. Increased exposure to a highly diverse cultural world results in the blending and hybridization of established ideas.

This drives process and product innovation.[16] One of the benefits of globalization has been its promotion of productivity-enhancing technology transfer and innovation worldwide. The April 2018 World Economic Outlook by the International Monetary Fund highlights freer trade, increased foreign direct investment, and the international use of patents and copyrights as some of the pillars of the increased transnational exchange in trade, goods, services, capital, and information characterizing globalization.[17] Certainly, the hyper-globalization characterizing the neoliberal period from roughly 1990 to 2020 is currently severely undermined, even if the pace of technological innovation shows no signs of slowing down, and a perennial need for more inclusive growth still characterizes global capitalism.[18] Living in the era of the so-called knowledge-based economy also has major implications for the ways in which states, cities, and even supranational political units, are spatially planned, governed, and developed—suggesting that knowledge-based *economization* can be understood as a geopolitical process that produces territories of wealth, security, power, and belonging.[19]

What is known today as the *Fourth Industrial Revolution* represents the evolution, transformation, and maturation of the information technology revolution around digitalization, biotechnologies, materials and nanotechnologies, and environmental and energy technologies. Schwab proposed the following classification of new processes and technologies in the Fourth Industrial Revolution:[20]

- Extending digital technologies, such as ubiquitous computing, mesh computing (distributed computing across many devices on a network), the Internet of Things, and technologies that increase the possibility of a two-way flow of information between humans and the environment.
- Reforming the physical world: AI and robotics, advanced materials, nanotechnology, additive manufacturing, and multidimensional printing.
- Altering the human being–biotechnologies, neurotechnologies, and virtual and augmented realities.
- Integrating the environment–energy capture, storage and transmission, geoengineering, and space technology.

Urban Technopoles

The coming of an *industrial renaissance* in the West can become a major development in the global political economy because of the end of hyper-globalization and the return of geo-politics to the world stage. Coupled with this, risk materializations in global supply chains under pandemic conditions have shifted foreign, commercial, industrial, and economic priorities among Western countries toward a realignment of trade blocks based on political affinities. Semiconductor manufacturing is an example of this trend, with plants being built in Western countries to reduce dependence on Taiwan, the world's top manufacturing country.[21] In this context, two major smart megaproject types seem to emerge that must be discussed: urban technopoles and high-tech manufacturing plants—or *factories of the future*. Urban technopoles can constitute central knowledge districts (CKDs) in cities, whereas factories in the future can be part of those districts or be in industrial districts in the surrounding metropolitan areas.

The primary drivers of the urbanization of technopoles in CKDs are new reindustrialization plans, support for urban development, and the establishment of synergies between actors and stakeholders. Metropolitan regions are considered the best locations for the growth of new industries, and technopoles serve as new Marshallian industrial districts that specialize heavily in a few key industrial sectors.[22] Building an urban fabric and luring a knowledge community have taken place simultaneously, and there has been much conflict among the interests of investors, scientists, and entrepreneurs. The nature and perception of proximity have changed, and various technopoles place differing emphases on various forms of territoriality and proximity—organizational, cognitive, geographical, cultural, and social.[23]

Additionally, we have observed the rise of so-called *open innovation* and the absorption of ideas in campus architecture, synergy management, connections between production and R+D, and the development of dynamic company ecologies. From inflexible structures to dynamic and unstable ecosystems, recent urban technopoles have arisen in the context of extremely dynamic and loosely connected ecosystems. Furthermore, synergy management requires a deeper comprehension of how collective leadership processes work in relation to the concepts of innovation ecosystems and districts.[24] Because of the necessity for coordination between levels of government to suit the needs of the industry, governance has played a more significant role in determining the emergence of technopoles in the past 30 years.

STEM subjects increasingly serve as anchors at universities for technological activity. They also continuously nourish innovation ecosystems. As a result, corporate innovation has evolved from merely technology development to proactively building and fomenting start-ups. According to the characteristics of interactivity, interdependence, hybridization, and reciprocity, entrepreneurial universities are specifically designed to develop complex network innovation ecosystems. As emerging platform technologies arise, technopoles have great potential to act as change agents. Consequently, technopoles are responding to the great challenges of the 21st century—resource scarcity, global warming, urban inclusion/sustainability, access to energy, and civic security—as they become more integrated into national technical innovation networks. As the idea of *living labs* has been extended to the level of urban fabric, they are also becoming more integrated inside existing cities.[25]

Urban technopoles and science cities are closely related to urban development, and, in some instances, the two labels designate the same phenomenon that locates science-based economic development within established metropolitan areas. In the

2000s, the U.K.'s national government declared six English cities to be *science cities*. The Northern Way, a strategy to revitalize the north of England, was tied to the 2004 announcement of first three—Manchester, Newcastle, and York. Then in 2005, Birmingham, Nottingham, and Bristol were added representing the Midlands and southwest regions. London and the southeast, including Oxford and Cambridge, were intentionally left out because the policy was primarily intended to spur growth in underdeveloped regions. Even though the cities had to have a solid scientific foundation, these English science cities were more about rebalancing the nation's economy than they were about recognizing current scientific capabilities. In France, several science city-style proposals are currently being pursued and are centered on new and relocated university campuses. This is especially true for Plateau de Saclay in southern Paris, which is becoming a significant science hub with a concentration of universities, public research facilities, and business R+D centers. Grenoble, which has several well-known public labs and significant corporate R+D centers, is a recognized center for science-based development in another area of France.

Brisbane, Australia, benefited from a state-led strategy in the 2000s that had all the characteristics of a science city strategy, although it was not known by that term. With Australia's major ICT and biotech hubs, as well as a state-funded synchrotron, Melbourne has also recently developed an urban science strategy with the help of the state government.[26]

High Tech Manufacturing Megaprojects: *Factories of the Future*

Megaproject Structure

Manufacturing plants are megaprojects devoted to the production of goods. The factory of the future is a vision for how manufacturers should enhance production by making improvements in three dimensions: plant structure, plant digitization, and plant processes. This is a worldwide trend driven by competitiveness, and it is primarily observed in plants located in the metropolitan areas of the top ten manufacturing countries: China, United States, Japan, Germany, India, South Korea, Italy, France, U.K., and Indonesia.[27]

New manufacturing facilities are more adaptable, multidirectional, modular, and environmentally friendly. Products are placed on autonomous delivery systems and individually steered through production by communicating with production machines in multidirectional factories of the future. For example, there is no fixed conveyor in the factory near Heilbronn, Germany, where the Audi R8 is made. A laser scanner and radio frequency identification technology embedded in the floor are used to guide driverless transport systems, as they carry vehicle bodies through the assembly process. Thus, it is feasible to rapidly change the assembly layout using these systems.[28] Indeed, future manufacturing facilities should include reconfigurable production equipment and interchangeable line modules. Toyota has installed "thin and simple" production lines in its plants in China and Mexico. The automaker uses a modular conveyor constructed on the factory floor rather than in a pit, providing workers with more freedom to modify the length of the line and move the equipment located on the line side.

In general, new manufacturing plants are built to produce goods in an environmentally friendly manner, prioritizing the effective use of resources, such as energy and materials. For example, to dramatically lower energy use, Webasto has fitted its plant in Arad, Romania, with LED lights and an active overnight cooling system.[29] Most factories of the future will implement some or all of the following plant processes:[30]

- Intelligent robots installed to gather data from each item of the generated work and automatically adapt their behavior to its features.
- Use of collaborative robots without barriers of protection.
- Additive manufacturing, such as the use of 3D printing, to produce tools and parts.
- Use of augmented reality allows employees to view information as an overlay on their visual field using augmented reality—such as smart glasses.
- Use of simulations for production to streamline workflows and material flows.
- Immersive training sessions to teach employees in real-world setting.
- Decentralized production steering implementation using cutting-edge technology to make it possible for people, machines, and work items to communicate with one another; thereby, establishing autonomous manufacturing processes.
- Use of analytics and big data applications to automatically evaluate massive volumes of data.

Lean management and the improvement of plant operations will still be significant by 2030. Customer centricity and continual improvement are two crucial aspects of lean management that further enhance digital technology. The value chain, which consists of suppliers, component manufacturing, press, body and paint shops, final assembly, and the client, will be completely integrated into the factory in the future, erasing conventional distinctions. The thorough integration of IT systems and accessibility of all necessary production data will enable manufacturing along the entire value chain. This integration improves ties between R+D, production, sales, and other departments within an organization.[31]

Manufacturers can boost *pull* in production, thereby lowering waiting times and the amount of work in progress, by using production simulators. By showing operational instructions, augmented reality—such as smart glasses—can assist operators in carrying out assembly and maintenance tasks. Manufacturers will have a much better understanding of each stage of production after evaluating production data using sophisticated big data algorithms, enabling them to continuously enhance production procedures.[32]

Producers would need to solve issues relating to three areas—strategy and leadership, personnel skills, and IT infrastructure—to actualize the vision of the factory in the future. Businesses integrate a factory-of-the-future strategy into their business strategies and modify their leadership philosophies to accommodate new working practices. The development of a workforce with the new set of skills needed to carry out industrial operations that are heavily reliant on technology is another area of concern. Finally, businesses must establish an IT infrastructure that facilitates communication along the entire value chain while maintaining data security. Industrial companies increasingly need to combine the creation and manufacturing of physical objects with the use of algorithms and AI to parse expansive, interconnected datasets and thus be able to make strategic connections that would otherwise be hardly possible. Competitive advantage seems to be increasingly rewarding those with the most robust, data-driven insights rather than the most valuable assets.[33]

Lights-Out Manufacturing

The first *dark factories*, or lights-out factory, began to arise in Japan in the 1980s, when businesses began to utilize advancements in robotics and automation technologies to avoid the high expenses of human labor. At the time, business analysts predicted that dark factories would grow more commonly as technology advanced, and skilled people grew increasingly difficult to find and expensive to support. However, in the interim, the expansion of manufacturing to underdeveloped countries with cheap labor momentarily halted the emergence of dark factories. Additionally, the 1980s saw a failed attempt at automated manufacturing at General Motors, which resulted in a quality reduction and a corresponding sales decline.[34]

However, there have been numerous successful examples of dark factories worldwide. In the United States, many machine shops operate either fully, or partially, unattended. Despite GM's early failures, robots are now a standard in auto business. Amazon.com heavily uses robotic systems in its distribution centers. In 2022, it paid $775 million in cash to buy Kiva Systems, the technology manufacturer. FANUC Robotics runs a lights-out factory in Japan where robots are used to create other robots.

A specialized research business named FANUC CNC was founded in Japan in 1956. It is the largest professional manufacturer of Computer Numerical Control (CNC) systems worldwide, with a 70% market share. The first electrohydraulic motor was introduced by FANUC in 1959; subsequently, a hardware-based open-loop numerical control system was created and refined. FANUC successfully created FANUC CNC system 5 in 1976. This was followed by the SIEMENS CNC system, which was created at an advanced level of 7. Since then, FANUC has grown to become the leading professional maker of CNC equipment worldwide.[35] Based on its compact size, low cost, and ease of composition, mechatronic systems for tiny, affordable machines, system 9 is a condensed version of system 6. System 9 is intended to strengthen the foundation of System 6. These systems are particularly well suited for processing complicated and expensive aviation parts, which require highly dependable, powerful, and multi-axis CNC machine tools. By altering the software, systems can be customized for a wide range of applications.

Canon, a major electronics manufacturer in Japan, recently disclosed plans to phase out human labor at multiple factories. Philips also makes electric razors in the

Netherlands in a plant with 128 robots and nine human, quality-control workers.[36] Lights-out manufacturing results in significant energy savings and, with widespread adoption, would be environmentally advantageous. Nevertheless, analysts question whether displacing human laborers with computers, machines, and robots is ultimately beneficial in terms of social impacts. For many, the growth of automation will be more likely to create significant economic division, further dualization, and widespread conflict.

Benefits

The key advantages of light manufacturing include improved productivity and efficiency, lower costs, superior talent use, and energy efficiency. However, in some situations, human replacement is impossible. Manufacturers strive to identify automated procedures. Consequently, light-out applications are typically found only in parts of manufacturing plants.

However, experts have noted that completely automated manufacturing is more feasible than ever.[37] Unsurprisingly, there is some controversy surrounding the concept of light-out manufacturing. On the one hand, it boosts productivity while requiring less investment in labor, thus increasing cost-effectiveness. However, it eliminates jobs because factories often require only a small portion of the staff needed when running traditionally.[38] Recent lean manufacturing initiatives have reemphasized the significance of human factors after decades of machine advancement, although many manufacturers use a variety of robots, conveyors, and other forms of automation. This is especially true for the production of large, sophisticated goods such as trucks, tractors, airplanes, vehicles, boats, locomotives, printing presses, and refrigerators. Ironically, some in the corporate world still believe that people are "the most programmable, efficient machinery for most assembly activities."[39]

Semiconductor Manufacturing

Making chips is a high-stakes, high-cost game. Building a cutting-edge "fab"(a factory for semiconductor manufacturing) can top $30 billion, while designing a single chip costs hundreds of millions. In the United States, Intel received $8.5 billion direct public investment for its projects through the CHIPS and Science Act, and TSMC received $6.5 billion from the United States government for their $65 billion investment in manufacturing facilities in Arizona, followed more recently (March of 2025) by a $100 billion investment aimed at building *fabs* to manufacture the most advanced 2- and 4-nm (nanometer) chips (which China is not yet capable of manufacturing), semiconductor "packaging" plants (using metal, glass, ceramics or plastic materials) and an R&D center.[40] In addition to public subsidies, TSMC has chosen Arizona for several reasons: available and cheap land, available advanced infrastructure, abundant talent, absence of natural disasters, and tax incentives. It is evidently a strategic decision that seeks to protect production from the Chinese threat, although there is no evidence at the moment that TSMC seeks to move all production and talent to the United States.[41]

Economic statecraft and strategic public support for the industry is a key component of the innovation ecosystem to ensure global leadership. Chip-manufacturing is a key geo-strategic industry on which other advanced manufacturing industries depend. To thrive in this fiercely competitive industry, companies must scale up to fund these massive investments and access global markets to recover costs and fuel future innovation.[42] Nvidia's Eric Breckenfeld suggests more public investment on innovation, collaboration with geopolitical partners to capture key markets and expand innovation ecosystems, a focus on understanding and improving the transition from basic science into technology production, and a systemic approach where no technology innovation field is neglected given that any breakthrough in any field can have synergistic and cascading effects in the broader ecosystem.[43]

As part of the CHIPS and Science Act, the Biden-Harris administration, through the United States Department of Commerce's Economic Development Administration (EDA), announced in October of 2023 the designation of 31 Tech Hubs in regions across the United States. This was an economic development initiative designed to drive regional innovation and job creation by strengthening a region's capacity to manufacture, commercialize, and deploy technology with the aim of advancing American competitiveness. The program invests directly in burgeoning, high-potential United States regions and aims to transform them into globally competitive innovation centers in approximately 10 years, "while catalyzing the creation of good jobs for American workers at all skill levels, equitably and inclusively." The active Tech Hubs focus on semiconductors, autonomous systems, quantum technologies, biotechnology, energy transitions, critical minerals and materials manufacturing.[44]

The Coming of Innovation Districts

Due to the deployment of advanced technologies that are accompanying the Fourth Industrial Revolution, cities and regions have been witnessing the relocation of high-tech corporate activity to urban areas.[45] This activity tends to be located in so-called *innovation districts*—a highly visible process in the Global North and one that is spreading around the world in a "planetary Silicon Valley" process.[46] Many urban megaprojects (UMPs) concentrate on innovative corporate functions and activities such as global management and control functions. Innovation districts concentrate on mature start-up ecosystems and high-tech manufacturing activities and constitute urban megaprojects in terms of their significance and impact. These newly developed urban areas significantly alter the urban fabric and socioeconomic structure of entire neighborhoods.[47] Innovation districts, like most megaprojects, are "privileged particles of the development process." They are planned to be *trait-making*, that is, to significantly modify the structure of society and cities: they are strategic urban projects.[48] R+D and expert knowledge exchanges are salient features of *innodistricts* as are other forms of innovation in inherently multicultural settings, such as cities. They foster social learning, the formation of collective intelligence, and the shaping of new epistemic cultures.[49]

Ongoing socio-spatial transformations in the global economy represent a formidable challenge for businesses and society in the 21st century, particularly in developing countries. The rise and continued expansion of megaprojects worldwide is one of the most significant changes affecting the global economy. Megaprojects and innovation districts have multiplied worldwide as an urban response to pressure for development, competitiveness, and innovation in the context of globalization and neoliberalism. The narrative of international competitiveness and the rhetoric of economic survival embrace most megaprojects. In this context, innovation districts work by fostering urban transformations that would enhance the city's position in the global economy and within a fluid sociopolitical division of labor. The agenda is to align urban initiatives with the real or perceived requirements of global production, consumption, and deregulated international neoliberal economic systems.

Over the past 50 years, the innovation landscape has been dominated by regions such as the Silicon Valley—spatially-isolated, corporate, suburban corridors accessible only by car, with little emphasis on quality of life or the integration of work, housing, and leisure. However, in recent years, a new complementary urban model has emerged—the so-called *innovation districts*. According to the Brookings Institution, these districts are "geographic areas where state-of-the-art institutions and companies are grouped and connected with new businesses, business incubators, and accelerators."[50] Compact innovation districts that are accessible to traffic, have high-tech infrastructure, encourage open collaboration, promote talent pooling, and offer attractive places to live. With increasing frequency, start-ups, incubators, and technology accelerators worldwide have been grouped around these innovation districts. By creating shared value and promoting *placemaking*, this emerging geography of innovation in cities has been attracting the attention of scholars and experts.[51] Innovation districts can play an important role in an integrated strategy designed to attract, retain, and cultivate talent, improve networks and communication flows among innovators, and make the district an attractive destination. Innovation districts have begun to occupy a preeminent place in which culture and tourism have occupied urban revitalization strategies for three decades.

Therefore, the geography of innovation has changed. Google, for example, has taken Silicon Valley's core R+D and innovation activities in several cities and urban cores over the past 10 years. The company's presence in the Tech City of London, the Chelsea District of New York City, and Bakery Square of Pittsburgh shows Google's calculation that being in the cities increases the company's admission to the growing ecosystems aimed at technology, superior lookup institutions, giant talent companies, and several regional economic specializations.[52] Barcelona, Berlin, Copenhagen, London, Medellin, Montreal, Seoul, Stockholm, and Toronto include emerging or established innovation districts. In the United States, the most iconic districts can be found in the city facilities of Boston, Atlanta, Cambridge, Detroit, Philadelphia, Pittsburgh, and St. Louis.

In each of these innovation districts, there is a combination of high-quality research universities, scientific and technological agencies, firms, and complexes that trigger business expansion as well as commercial and residential growth. Movers to innodistricts include high-value and research-oriented sectors, such as life sciences, and creative fields, such as architecture, design, theater production, advertising and marketing, and advertising and marketing. We have even seen a return to small-scale and personalized manufacturing activities made viable by 3D printing and robotics. Much of this activity displays an indispensable rethinking of corporate activities around the concepts of innovation, open innovation, entrepreneurship, sustainability, and requirements for increased competitiveness.[53]

Barcelona, Boston, London, Masdar City

Barcelona

Barcelona is credited with creating in 2000 the first innovation district, 22@ Barcelona, Districte de la innovació (Innovation District), or just 22@ (Fig. 5.1). The 22@ project is perceived as a success and has become the pioneering model for other innovation districts, including Boston's. In the area of El Poblenou, where the innodistrict is located, approximately 70% of the industrial land has been redeveloped and remodeled, including 141 individual plans for upgrading real estate and facilities. According to Barcelona City Hall, approximately 72% of all employees in 22@ have a college education. Since 2000, approximately 4,500 companies have opened in the district, employing approximately 56,000 workers.[54] Since 2000, the population of this area has increased by 130,000. Landmark companies in 22@ include the MediaTic Barcelona Growth Center, an innovation center built through a public-private partnership. As expected in an innodistrict, several incubators and accelerators have been developed in 22@, such as the Biomedical Park, the MediaTic building, and Barcelona Activa.[55]

The 22@ project also requires continuing education of the community in information-oriented activities such as coding, product design, and training in IT services. Many universities have also established their presence at 22@, such as Pompeu Fabra University, the University of Barcelona, Polytechnic University of Catalonia, and Open University of Catalonia. The most important takeaway in 22@ Barcelona has been the validation of innovation districts as a sustainable urban economic regeneration tool. Through strategic programming and a robust planning model that emphasizes the physical, social, and economic spheres, Barcelona has created one of the most successful innovation districts in the world.[56]

Boston

The South Boston Waterfront Innovation District (SBWID) is the first officially labeled innovation district created in the United States (Fig. 5.2). In May 2010, former Boston Mayor Thomas Menino announced plans for the city to develop 400 hectares of land in the South Boston Waterfront. The SBWID is the fastest growing

Fig. 5.1 An east view of the 22@ innovation district in Barcelona, with the Glòries Tower (formerly the Agbar Tower) in the foreground and the Mediterranean Sea in the background. Source: iStock.com

area in Boston today and has stimulated significant economic development in the city. Since the district's origin, 5,000 new jobs have been created and more than 200 new businesses have been formed. According to the Boston Mayor's Office, 40% of companies located in the SBWID share joint workspaces.[57] More than 1,100 housing units have been built, including 300 innovative micro-units. The increase in rental prices in SBWID has raised concerns that rapid real estate development in the area discourages entrepreneurs and emerging businesses, organizations, and people

Fig. 5.2 Aerial view of the South Boston Waterfront Innovation District, showing the historic Fort Point Channel neighborhood in the foreground. Source: Courtesy of Aaron Fedor

that the district's design intends to attract. In just a few years, rents increased by 43% at the seaport, and the trend continues today."[58] In the SBWID case, the revitalization effort was an attempt to aggregate international and local firms in a more concentrated geographic environment with the hope of fostering increased collaboration. Collaboration among the city government, education institutions, and private sector has contributed to the success of the district.[59]

Many other cities in the United States are also pursuing the development of innovative districts—New York exhibits, among others, the new Cornell Tech campus on Roosevelt Island and the IBM Watson Center, next to The Cooper Union in downtown Manhattan.

London

The Kings Cross Development, which is transforming an old railway station in downtown London into a science and business hub—or a *knowledge quarter*, is an innovation district, a center for London innovation in the digital economy. Central Saint Martins of the University of Arts, London, became the first occupant.

Kings Cross adopted a redevelopment format that spatially combined knowledge creation and the usability of ideas in an entrepreneurial economy. For example, The Francis Crick Institute, hosting approximately 1,250 biomedical scientists, opened the site in 2015. The newly built European headquarters of Havas (the global communications group) opened at King's Cross in February 2017 uniting its 24 U.K. agencies. And now Google has decided to move its U.K. headquarters to King Cross Central. An iconic building in the area, The Landscraper, will be the U.K. headquarters for Google, and as of April 2025, it is near completion (Fig. 5.3).

Indeed, the London innovation ecosystem stands out for its exceptional access to affluent consumers and powerful corporations, a solid funding landscape and ambitious government initiatives. The Kings Cross is in a knowledge-intensive, mile-wide area that comprises Bloomsbury and Euston. The area includes over 500 research centers, 3,000 scientists, 12,000 academics, 50,000 staff members, and 77,000 students. Within a mile radius, there are 21 museums and galleries, and 27 libraries and archives. It represents one of the greatest knowledge concentrations in the world.

Kings Cross includes an *Impact Hub*—part innovation lab, part business incubator, and part community center. Impact Hubs are spaces with co-working areas in various locations designed to promote communication, networking, and exchange of ideas. They are an invaluable resource to catalyze impact in a wide range of fields.[60]

King's Cross Central (KXC) is currently the largest city center redevelopment area in Europe. It was a brownfield site—that is, a former industrial use site—with excellent public transportation links. Buildings under construction at KXC range from one to nineteen stories. St. Paul's protected views from Parliament Hill and Kenwood House will not be affected. At least one-third of the site (25 acres/10 hectares) is dedicated to new public routes and open spaces. The developer proposed the creation of 20 new major routes and 10 new privately-owned public spaces. Five of these are major new squares—Granary Square, Station Square, Pancras Square, Cubitt Square, and North Square—which together total eight acres (3.2 hectares). In addition, the proposals include 6.5 acres (2.6 hectares) of new public space along the Regent's Canal (the Gas Holders Zone and Coal Drop Yard) and within a new area, called Cubitt Park. Argent LLP, the property developer, invested in

Fig. 5.3 Aerial view of King's Cross rail hub and The Landscraper in London's Camden and Islington districts. Source: Shutterstock.com

improvements to the canal corridor, notably lighting and access from Granary Square. The mile along the towpath to the west is the Camden Market, and beyond that, Regent's Park and London Zoo; to the east is the Islington Tunnel and then Upper Street, a busy retail and entertainment area. Camley Street Natural Park, Old St. Pancras Church, and Somers Town are now connected to KXC by a footbridge, one of three new crossings over the Regent's Canal.[61]

Fig. 5.4 Visualization of Masdar City, located next to Abu Dhabi International Airport. Source: Courtesy of Foster + Partners

Masdar City

Masdar City is an example of a highly questionable urban innovation project. A planned ecological city on the outskirts of Abu Dhabi, Masdar City, was designed for clean technologies and renewable energy (Fig. 5.4). The construction of Masdar City began in 2006. It was planned and designed as one of the most prominent attempts by the government of the United Arab Emirates "to diversify its economy in anticipation of the future depletion of oil reserves. Mubadala, a government company in Abu Dhabi dedicated to investing in major strategic projects, financed the Masdar City experiment at $22 billion."[62] Günel (2019) studied the origins and realization of Masdar, the environmental management policies in the new city, and the perceptions of the highly qualified professionals who designed it.[63] On the positive side, Masdar's leaders developed many innovative projects. They even planned a new *energy currency*, called Ergos.

> Numerous paradoxes emerged during the development of the energy currency project. The currency would work by issuing 'a balance of energy credits as a means to define and regulate an assigned energy budget,' which sounds dangerously similar to the 'social credit' system implemented in China. In fact, although Ergos's purpose was to reduce energy consumption in Masdar, some project designers recognized that it could become a tool for a 'technocratic dictatorship'.[64]

Another notable project within Masdar is the driverless personal rapid transit system (PRTS). It was intended to cover the entire city, but soon these plans were abandoned. PRTS began to take people to parking lots. Many wondered why they did not travel such a short distance by bicycle or foot. Günel (2019) shows some of the failures of techno-futurism, where expectations and realizations are often clearly disconnected.[65] Masdar's leaders devoted much effort to formulating policies on a technology called *carbon capture storage* (CCS), which works by injecting carbon dioxide waste into the soil—a controversial climate change mitigation technology. Masdar leaders even made a proposal for the United Nations—presented in the context of the Framework Convention on Climate Change (Ferguson, 2018). As argued elsewhere, Masdar, the entire city, functions as an innovation district.

> [B]ecause the innovation component is at the center of its conception, both in design (planned and executed by Foster and Partners) and in content, Masdar can be seen with a multidimensional lens, as a "new city" and as a gated community. It resembles a technopolis similar to Tsukuba Science City in Japan or King Abdullah University of Science and Technology (KAUST) in Saudi Arabia. It can also be seen as a special economic zone (SEZ) for the development of renewable energy and clean technologies.[66]

Like the most recent ecological city project in Dubai, The Sustainable City, Masdar can be seen as a prestigious megaproject aimed at promoting a brand, a paradigmatic case that applies the familiar principles of neoliberal urbanism and shows all the shortcomings of such an approach to urban development. Masdar City also represents neoliberalism's high-tech approach to sustainability, a controversial endeavor since its construction requires massive amounts of energy, land space, and water resources. Its location in the desert, with temperatures reaching 47 °C in the summer, results in very limited water resources, especially for drinking. Therefore, Masdar is highly questionable as a paradigm in innovation districts and urban development.

Innovation District Controversies

Requisites and Constrains

Innodistricts represent the urbanization of the knowledge economy around links between science, technology, innovation, and sustainability with the promise that these ideas and practices will be held and honored. They are becoming widespread in Western cities and are usually embedded in regional and national innovation systems. In some cases, such as 22@ Barcelona, these districts are based on government-led planning and investment efforts, built on the premise that innovation districts can become both effective urban regeneration and economic development tools. Thus, the creation of an innovation district is usually "an attempt for the city to leverage its strengths and resources to emerge as a hub of innovation in the knowledge economy."[67]

Innovation districts show a wide typology depending on leadership, cluster type, and firm support programs. The Brookings Institution's three district models include the following:[68]

- Anchor Plus–an innovation district where large-scale mixed-use development is centered around major "anchor institutions" and a rich base of related firms, entrepreneurs, and spin-off companies involved in the commercialization of innovation.
- Reimagined urban areas–an innovation district rooted in proper transit access, historic building stock, proximity to downtown areas, and a rooted anchor company.
- Urbanized science parks–an innovation district intentionally developed from suburban office parks grounded in the work done there.

There are several requirements and constraints surrounding the development of innovation districts. Perhaps, one of the main barriers to innovation district development is the current structure of incentives for investment-privileged finance and property-led urban regeneration. A gradual shift of focus is required from built environment investment to socioeconomic output, and from financial, real estate, and property investments to innovation.[69] Marginson (2016) argues if the rewards for investment in financial assets are higher than the rewards for investment in knowledge—intensive industry innovation—the latter will be neglected. Marginson (2016) believes that this is a serious problem in the U.K. economy, where finance generation often seems to be the main game.[70] However, as Zukin (2020a, b) has documented in New York, the trade-offs between finance and innovation are at the core of the rapid development of the Big Apple as a high-tech hub.[71]

Close collaboration between stakeholders and investors seems to be a requisite for successful innodistricts. Porter's well-known model uses four categories of private sector drivers. These include:

- Human and natural resources;
- Related and supporting industries reflect the role of supplier industries in traditional clusters such as the steel industry
- Demand conditions consider the size and sophistication of home markets and
- Strategy, structure, and rivalry examine how companies are managed, how they compete, and the business culture

In addition, Porter highlighted the role of government as an external facilitator of cluster performance, noting that chance factors are important. Chance factors may include having the right skilled people available when a new technology takes off. However, pure chance should not be overemphasized.[72]

Today's most powerful innovation originates from collaboration, sharing of ideas, and mashing of radically different disciplines and technologies to create new solutions to new problems or upgrade traditional industries. Effective multidisciplinary, open collaboration requires intellectual density (concentration of skilled actors), diversity, tight proximity, strong networks, and partnerships among citizens, businesses, laboratories, academic institutions, and investors.[73] All innovation districts include physical, economic, and networked assets. An innovation ecosystem results when these three assets combine in a context and culture that are prone to risk-taking. Such an ecosystem often follows a triple-helix model, whereby entrepreneurs connect with universities and research centers to promote innovative thinking and practices supported by government funding. The role of universities is important because they have the potential to increase revenues by fostering opportunities for R+D.[74] Since innodistricts draw a range of activity communities to an innovation area to work, learn, live, and play, mixed-use design techniques are often helpful for activity-based integration. Additionally, recent research shows that smaller development plots, dispersed spatial configurations, and a variety of land uses lead to improved socioeconomic integration because they increase the chances of interaction between various income groups.[75]

Capital, technology, and built environment constitute tangible assets in the development of innovation districts. Intellectual density, impact innovation, and social and economic networks are intangible assets (IAs). The physical proximity and density of key actors can be created in an urban environment to foster collaboration. However, creating an entrepreneurial spirit also demands social, cultural, and behavioral aspects of crucial importance. These have to do with intangible, long-term, societal processes, such as quality of education, leadership formation, and business ethics.[76] The move toward these innovation hubs reflects the growing importance of the geography of innovation in urban areas. It also shows how the development of industrial clusters can deliver economic growth, employment, and community regeneration. Brookings has called for local decision-makers, global companies and financial institutions, and government "to 'unleash,' 'embrace, ' 'support and accelerate' innovation districts. The result: a step toward building a stronger, more sustainable, and more inclusive economy."[77]

Drawbacks

For detractors, innodistricts are useless ecologically, uncertain economically, and counterproductive in terms of real estate values; simply, they harbor false promises in an already uncertain and precarious era. For example, the location and residential appeal of new neighborhoods can create new housing pressures that exclude disadvantaged populations. Social mixing is not guaranteed by residential diversity, and classical sociological studies show that neither social homogeneity nor heterogeneity guarantees that individuals will socialize.[78] Broadly speaking, it is difficult to add high-tech manufacturing spaces to the urban fabric. As Zukin (2020b) argues:

> Embedded industries and social communities want protection from expanding tech companies and the real estate developers who build for them. City council members, state legislators, and community organizations oppose the city government's attempts to satisfy big-tech companies. While a city's density magnifies conflicts of interest over land use and labor issues, the covid-19 pandemic raises serious questions about the city's ability to both oppose Big Tech and continue creating tech jobs.[79]

Specifically, if the inno/ecodistrict allows exclusively self-referential synergies, its urban development and sustainability objectives will be in question, as in Masdar City.[80] High-tech, eco-friendly, and master-planned utopias (like Songdo, South Korea) tend to fall short of expectations or become somewhat disfunctional.[81] Neom (Saudi Arabia), the world's first mega-smart-city, will exhibit large-scale AI systems, robotics, and automation applications; however, it already shows some of the same problems that led to the failures of Masdar and Songdo.[82] Innovation strategies in urban areas often develop alongside the pursuit of unrealistic growth and sustainability targets through major capital-intensive projects (e.g., megaprojects). Instead, cities in developing regions should benefit from recognizing their local context, history, and culture; they should concentrate on their strengths to address priorities such as affordable housing, accessibility to public services, and education. Thus, they can create resilient, long-term positive and exponential impacts.[83]

Competitiveness vs. Sustainability

Interestingly, promoters of innodistricts, along with some commentators, present them as major alternatives to conventional megaprojects, property-led urban regeneration, and neoliberal urban strategies. This alternative, represented by innodistricts, purportedly enhances urban quality of life and sustainability while preserving economic development, growth, and prosperity.[84] If this is the case, then unlike other urban megaprojects, innodistricts represent two distinct narratives that develop simultaneously and are potentially opposed—a) that of innovation-based global competitiveness and b) that of sustainability and quality of life.

Regarding global competitiveness, innodistricts are multiplying around the world as an urban response to pressures for development, competitiveness, and accelerating innovation, and are increasingly linked to strategic spatial planning for urban innovation.[85] The narrative of international competitiveness and rhetoric of economic survival pervade the rationale for megaproject development and implementation.[86] In this context, innovation districts are intended to work by fostering urban transformations that would enhance the city's position in the global economy and within a fluid sociopolitical division of labor. The agenda is to use the transformative potential and strategic advantage of urban technological innovation in the real or perceived requirements of global production, consumption, and competitiveness in a highly globalized transnational economic system.[87]

Regarding sustainable development, innovation districts present formidable challenges, like other megaprojects. Rather than simply fostering sustainability (as is commonly argued), they should enable it in urban areas to an unprecedented scale. The paradox of AI urban environments lies in the promise of attaining a certain type of sustainability at the risk of further jeopardizing urbanity, equality, and liveability, as discussed earlier. Many innovation districts underscore their eco-orientation and design for environmental sustainability. However, the approach to ecological relationships within a district often reflects a ghetto structure that operates in isolation from the external environment. Synergies must be projected in a reticular manner according to the external conditions of the urban environment.[88]

Disruptive Innovation Environments

Smart Cities and Intelligent Environments

Unlike innodistricts, a smart-city is not territorially bounded. The concept (and practice) of smart cities is expansive, encompassing, and relentless. It is not static; there is no absolute definition or endpoint. Rather, it is a process or a series of steps. Smart-city applications have been developed to manage urban information flows and allow real-time responses. Despite the fuzzy, diverse, and contested meaning of the concept of *smart-city*, its proponents believe that a smart-city is capable of better addressing the challenges of urban resilience and sustainability. Smart cities are said by some to be more liveable, sustainable, and resilient.[89] They may be better prepared to respond to challenges than traditional and transactional forms of urban governance.[90]

Production, adoption, and governance are three salient aspects of smart-city developments:

- On the production side, expertise in the development of smart technologies, programming, and coding tends to concentrate in specific urban areas (in innodistricts termed TAMI (technology, advertising, media, and information), such as the Flatiron District, DUMBO, Brooklyn Tech Triangle, and Lower Manhattan/Wall St in New York City; and King Cross, Tech City, and Canary Wharf in London.

- Regarding implementation and adoption, some urban areas are more technologically advanced and use different electronic methods, voice activation devices and methods, and sensors to collect specific data. Similarly, it is difficult for the smart-city to reach all urban neighborhoods. A smart-city divide separates early adoption areas from the rest.
- From the perspective of urban governance, smart-city technology gives city officials direct access to both community and city infrastructure, allowing them to keep track of what is happening in the city and how it is changing. ICT and Internet of Things (IoT) are typically utilized to improve the quality, effectiveness, and interactivity of urban services, cut costs and use of resources, and foster greater citizen-government interactions.[91]

The ideas and practices of the IoT have formed the foundation for the development of smart-city initiatives. The proliferation of IoT methods, processes, and devices is made possible by the decline in the cost of manufacturing environmental sensors and network devices as well as by the growth in the use of dependable mobile telephony and cloud computing. [92] The IoT is essentially an interconnecting and optimizing data approach used across multiple built environment systems and equipment.

This increases both the volume and velocity of data movement and creates new opportunities to interconnect the physical operations. Thus, the use of networked sensors, data management, and analytical platforms to improve the quality and operation of built environment systems is a frequent feature of smart-city planning.[93] As the Fourth Industrial Revolution develops, expands, and becomes a key component of the socioeconomic productive fabric, cities are increasingly characterized as *complex adaptive innovation ecosystems.*[94] In such an ecosystem, *intelligent environments* function via computers, sensors, and machine learning algorithms that are being built in roads, bridges, buildings, homes, transport systems, and other urban amenities and facilities.[95] Smart-city practices have evolved toward a more comprehensive and integrative approach, potentially enhancing the impact that high tech can have on urban environments.[96] Processes have gone from connecting various threads of public infrastructure and knitting the data collected through them to a more integrated network aimed at bringing together government, universities, industry, and citizens under a *quadruple helix* model.[97]

Using this quadruple helix model, Schiavone and others (2020) identified evolutionary smart-city processes that promote an integrative approach on three fronts: strategy, entrepreneurship, and organization. Concerning strategy, integrated smart-city processes aim to connect the physical spaces of cities with the economic and social spheres. In doing so, they foster increased coordination of entrepreneurial actors involved in further efficiency. All these benefits form an improved understanding of the organizational factors that apply to complex high-tech urban transformations.[98] With the quadruple helix model in mind, smart-city processes advance the goals of data gathering, data processing, and communication to stakeholders. The objective is to promote circular economies (e.g., London's Globechain or Envirocon), test new urban ideas (Melbourne's CityLab), and develop many other projects whether aligned, or not, with the preservation of the common good and well-being of urbanites.[99] Two features of smart-city processes are a lack of normative focus and controversies surrounding data ownership. It remains unclear how recent advances toward incorporating AI technologies into the smart-city mix can contribute to mitigating shortcomings and controversies.

Data Centers

Data centers are highly specialized structures that hold computer servers and network equipment. They are furnished with power and cooling infrastructure. Data centers are the physical representation of the cloud, taking the shape of coils of coaxial cables, fiber optic tubes, cellular towers, air conditioners, power distribution units, transformers, water pipes, and computer servers. The digital superhighway is a real, tangible place with enormous energy requirements. Building a data center, especially one of the hyperscale varieties, requires a vast amount of labor due to the utilization of economies of scale, intricate construction methods, and bespoke engineering on multiple levels. These massive projects not only involve designing and constructing the building itself but also providing the electricity needed to run massive amounts of IT equipment and state-of-the-art cooling systems to keep servers at the right temperature. Mega data center projects involve multiple component teams in addition to the teams handling the extensive mechanical, electrical, and plumbing (MEP) scopes of work. On these massive data center projects, MEP systems account for about 60% of the work. Civil, structural, and architectural teams comprise several engineers and project managers.[100]

Because data centers require a lot of power, they are situated in places with plenty of electricity, lots of water for cooling, fiber connectivity, reasonably priced real estate, tax breaks, and distant from locations that are prone to natural disasters. Companies constructing data centers can find inexpensive labor, inexpensive water and power, and loose or nonexistent environmental restrictions in the Global South. The world's data center industry is expanding quickly due to the surge in demand for data storage and consumption brought on by edge computing (servers located near end users), processing requirements for augmented and virtual reality, and the development of artificial intelligence. The United States, with about 2,500 data centers in 2024, accounts for about 40% of the global data center market. The market is forecasted to grow by some 10% per year through 2030—with hyperscale providers projected to experience some of the most significant growth over that timeframe.[101]

The four main categories of data centers are onsite, colocation, hyperscale, and edge, all with different purposes. Situated on a company's corporate campus or headquarters, an *onsite* data center is rather easy to access and manage. They can easily be scaled up or down as needed, and their close proximity to business operations—many of which are located in metropolitan areas—alleviates network troubleshooting. A data center that provides space to house servers and computer gear for businesses is known as a *colocation* facility. *Hyperscale* data centers, which can be considered megaprojects of digital capitalism, are enormous, vital commercial spaces built to effectively serve scalable applications. They are frequently connected to huge data producers like Google, Amazon, Facebook, IBM, and Microsoft. Citing energy cost savings, numerous businesses, academic institutions, and other entities running their own small-scale data centers have started moving their data to hyperscalers or cloud colocation facilities. An *edge* data center is situated between linked IoT devices and a centralized data center or the public cloud. Time-sensitive data may be processed at the place of origin by an intermediary server in an edge computing architecture that is physically close to the client. These facilities optimize bandwidth and lower latency, opening up new applications. Measured in milliseconds, latency is the interval of time between a user's action on a network or web application and when it reaches its destination.[102]

One major factor driving data center owners to locate their facilities in cities more often is the reduction of latency. Businesses are quickly moving a large portion of their workload to the public cloud, but they are also becoming more aware of the benefits of different data centers. Applications that need real-time insights with very low latencies, like autonomous driving, might work better near the data source. Edge computing may also benefit from reducing the expense of moving massive amounts of data to and from the public cloud. Regulations pertaining to data privacy and residency that mandate that specific categories of data be kept close to their original source also impact on business choices. There are other applications where a delay in the milliseconds can prove critical: online sports betting apps, trading stocks, or multiplayer games like Fortnite. In New York, 60 Hudson St. (the old Western Union building) is home to vast halls of computer servers. The One Wilshire building in Los Angeles, formerly home to a network of law offices, now oversees one-third of all Internet traffic between the United States and Asia. Amazon Web Services has been pushing mini data centers, which it calls Local Zones, in 32 cities across the United States Walmart, the retail giant, is renting out sections of its huge stores to host data centers for third-party companies.[103] Data centers in cities are stimulating the real estate market. They are also providing opportunities to reuse the heat they generate and the energy they consume in the urban ecosystems where they are located. However, data centers are problematic because of their socioenvironmental and other impacts. We address this matter in Chapter 7.

China's High-Tech Capitalist Communism

While China is engaged in a global race for technological supremacy, it keeps using high-tech innovations to ensure internal control of the population by the Chinese Communist Party (CCP). Data-driven benchmarking procedures in Chinese smart urbanism are standardized by the state, not by Internet giants. The government has invested a significant amount in big data solutions over the past years, using them to manage urban and social issues and to support government efforts in social regulation and coordination. A well-known example is the concept of *social credit*. The implementation of the social credit system demonstrates the administration's dedication to big data, its collection and its interpretation. On the other hand, the degree of citizen involvement in smart urbanism designs and solutions raises issues related to cities' underlying proclivity to technologically solutionist approaches, exacerbated expectations around platform urbanism, and seeing technology as a panacea to urban issues.[104]

The application of a high-tech neoliberal paradigm in China differs from that of other Asian nations. The dynamics of power transfer from the state to non-state sectors, from the central government to local agents, and from organizations to individuals is a salient feature of power structures and mechanisms in China. Subnational regimes are best able to impose neoliberal experiments and manage their territories.[105] This supports the decentralization of state resources and recalibrated municipal roles for local and regional innovation and economic competitiveness based on a *politics of scale*, in contrast to Jessop's concept that neoliberalism results in the hollowing out of the state.[106]

Even though they are market- or citizen-oriented, China's neoliberal subnational government entities do not conflict with top-level national designs, strategic planning, or laws. Despite being labeled with such terms as *authoritarian capitalism*, *state neoliberalism*, or *market socialism*, the political system in China is complicated and multiform, making it difficult to identify specific power relations between entities. However, the market and individual freedoms that take place in this complex scenario are watched and restricted by the Chinese Communist Party (CCP)'s formidable display of surveillance technologies everywhere.

High-tech neoliberalism in China exposes the top-down power structure of the state and its supporting institutions, as seen in the limited degree of administrative autonomy enjoyed by municipal governments. Studies reveal that such autonomy falls short of the goal of giving municipalities and the private sector independent decision-making power from the federal government.[107] For example, while municipal governments hold comprehensive transportation data sources (e.g., road networks and infrastructure data), they require robust and citizen-oriented comprehensive data from private enterprises to inform smart-transportation programs. These data from private companies are seen as comprehensive and up-to-date and are thought to contain the value that the government can use for efficient urban control and governance. Moreover, technical solutionism appears to be restrained by distinctive Chinese political and legal systems, thus confirming He and Wu's (2009) views on the neoliberal characteristics of modern Chinese urbanism. In other words, despite imitation and adaptation of technical solutions from other nations, what drives the Chinese technologically-solutionist, smart neoliberalism is not the private sector but rather the national government's systematic designs for social control.[108]

The CCP directs the governance of smart urbanization, whether at the national or municipal level. Indeed, the dominant position of local governments, as compared to private businesses, privileges the state for the purpose of effective urban governance through technology. The state controls how algorithms and urban informatics work to promote inclusion or exclusion, resulting in technocracy and tokenistic democracy. This is true even though there is some degree of distributed power and citizen participation in the design of platform urbanism.

Zhang and others (2022) contend that in urban China, all citizens are in a smart-city when they are under the watchful eye of surveillance technologies. At this point, China's capitalist communism has established political control to create a rigid social order and to quell social unrest. Even in the planning and decision-making stages, where they may play a minimal role, citizens cannot be involved outside the scope of government surveillance.[109]

Toronto: How Smart Cities Fail

What occurs, how do existing institutions react, and how are models of urban governance impacted when one of the big digital businesses joins the real estate, land use development, and urban planning fields? Carr and Hesse's (2020) investigation of Alphabet Inc.'s intentions for Toronto's waterfront (Quayside) sheds light on these issues (Fig. 5.5).[110] In-depth placemaking activities are revealed through their investigations, which span all governmental levels, the public, and networks of corporate actors. A world leader in technological innovation never before attempted to enter the field of urban planning as an urban developer.

Waterfront development in Toronto represents a significant advancement in digital cities because it relies on a single potent digital enterprise to fuel urban expansion, while enabling local public policy to run unchecked. This case reveals the tension between "smart" and "sustainable" characterizations of the urban.[111] Strategic behavior and poor communication on the part of the authorities in charge of land use and urban planning are examples of this, which is characterized by a cleft between politics and policy. Urban planners and development professionals were concerned that the limitations of smart-city models should consider Carr and Hesse's explanation of Quayside as a cautionary tale. Additionally, they explain Quayside as another example of post-political urban growth, adding to the body of knowledge in the field that conceptualizes contemporary urban studies.

Carr and Hesse argue that big tech companies, in general, and discourses on smart cities in particular, are prone to pursuing a post-political, impartial, expertise-led, technological agenda. According to post-political theorists, formal state-instituted processes and venues for participation do not increasingly provide a forum for meaningful public discourse or appropriate spaces for contest. Accordingly, Toronto citizens created new opportunities to participate in politics as Alphabet Inc. became a major player in the city's urban planning and Canadian politics.

Quayside in Toronto is an exceptional example of a large-scale, digital firm entering the field of urban planning and attempting to control growth over a specific area to further its own commercial objectives. While the urban-regional imprint created by Amazon.com—because of its parallel worlds of digital, logistical, and locational activities—and the long-term effects that large tech companies may have on

Fig. 5.5 Location of Quayside development (in purple) next to the Portlands section of Toronto's Waterfront. Source: Courtesy of Scribblemaps.com

urban-regional life can be significant, the Quayside case offers crucial insights into how public institutions react to newly established, significant corporate participants in the market, and how modalities of urban governance and planning are impacted.

In Toronto's urban planning discourse, Sidewalk Labs—Alphabet Inc.'s urban innovation leg—persistently sought hegemony by controlling political discourse. A chorus of independent political civic action organizations did, however, emerge to raise awareness of the potential drawbacks of Sidewalk Labs and Quayside. This activity was conducted in areas that were not subjected to official planning.

People did not object to Sidewalk's plan for Toronto due to concerns of architectural preservation, building heights, densities, or construction methods—often grounds for protests. It was the project's tech-first approach that irritated many people. The main reason it failed was undoubtedly its apparent disregard for the Torontonians' privacy concerns. In Canada, there is little tolerance for firms that manage public transportation and streets nor ones that collect data on people's daily lives.

Quayside might have acted as a proof-of-concept, providing a novel method of urban design for all cities had it been effective. The adoption of a sensor-rich, smart-city paradigm has made it popular in China. The Persian Gulf may have demonstrated that societies more democratic than China can adopt a smart-city model. Instead, despite spending 2.5 years trying to build a neighborhood "from the Internet up," Sidewalk Labs was unable to persuade anyone to move there.

According to the company, the Sidewalk project was abandoned by May 2020 because of "the tremendous economic uncertainties brought on by the covid-19 pandemic." However, financial ambiguity appeared at the conclusion of years of heated debate in the media about its $900 million plan for a data-rich city within the city. The new strategy from the city of Toronto, which focuses more on wind, rain, birds, and bees than statistics, appears to be a practical answer to this situation.[112] Smart cities appear to have been built to eliminate the very thing that makes them amazing because of their emphasis on the optimization of everything. New York, Rome, Cairo, and Toronto are not great cities because of efficiencies; rather, people are drawn to chaos, captivating interactions, and fortuitous events among a vastly different mix of residents.

Instead, proponents of smart cities have embraced the notion that a city is something that can be measured and managed. Jacobs adds:

> By making building technology more efficient and offering less polluting transportation alternatives to cars, smart-city technology should be able to reduce carbon emissions, accelerate the construction of affordable housing, improve the effectiveness of public transportation, and shorten commuting times. However, frequently, rather than focusing on what it ought to accomplish, supporters emphasize on what it can do. The failure of Sidewalk's Quayside indicates that these technologies must be more responsive to user needs.[113]

Toward a New AI Urban Regime

In the recent past, the biggest challenge for megaprojects has shifted from the ability to overcome engineering hurdles and workloads to administrative, management, and planning processes. In other words, it is seldom the case that builders and engineers cannot build. It is rather bureaucratic complexities and uncoordinated design efforts that prevent megaprojects from being carried out in the first place or from performing according to plans when they are implemented. Failed megaprojects are common.[114] At some point in the near future, it is expected that megaprojects will be completely designed with Building Information Modeling (BIM) systems, manufactured by means of CNC systems or 3D printers, and installed by robots. Owing to cost reduction reasons, the world will be built by machines; for critics and the public, this probably constitutes an Orwellian scenario.

The emergence of the AI regime is a sign of the development of the information economy, where maximizing data and information is a major source of economic value. Specifically, the AI regime is a by-product of the so-called Fourth Industrial Revolution—where autonomous decision-making by machines promises to enhance efficiency and effectiveness in all kinds of productive processes. The emergence and growth of a new AI urban regime in the coming years is indicated by the expansion of smart cities toward widespread use of AI techniques, processes, and devices. An AI urban regime can be seen as a step in the evolution of urban "growth machines," [115] characterized by a collection of formal and unofficial agreements between public agencies and private parties that (1) prioritize the development of smart cities, particularly algorithm-based technologies, and (2) propose the institutional frameworks required for the realization of the goals and objectives of the AI industry, which are said to trickle down for the benefit of all urban population segments.

AI is a set of computer algorithms designed to resemble human decision-making. It describes a range of computer programs and systems that can perceive their surroundings and think, learn, and act in accordance with their preprogrammed goals.[116] AI is worthless without a set of intentional goals to complement it. As a result, the first task for those who design, construct, and oversee physical systems is to choose the types of goals for which they want machine learning algorithms.[117] Datasets, communication platforms, prompt decision making, and efficient actions are generally regarded as the four essential elements of smart-city processes.

AI is also a specific technology that adds some essential features to smart-city processes: machine learning capabilities, automation processes, autonomous decision-making, and, with so-called deep learning, forecasting capabilities that can be combined with digital twinning and simulation techniques.[118] Researchers agree that AI is a key component in the creation of smart cities. AI is highly suited for providing an analytical underpinning for smart-city programs, especially when combined with machine learning. Although AI is predicted to play a vital role in expanding the number of smart-city applications, research in this area—albeit in its infancy—shows the massively disruptive potential of this technology.[119]

Governments have a direct geopolitical interest in overseeing and regulating (and even in strategically designing) AI developments, particularly in relation with defense and national security. In the United States, strategic initiatives by the federal government regarding AI technologies and applications support private stakeholders, within a well-established entrepreneurship culture. As an example, the recently announced Stargate Project (a $500 billion AI infrastructure joint venture between OpenAI, Oracle, SoftBank y MGX) receives strategic support, fiscal and regulatory benefits from the federal government but no direct funding.[120] More clearly defined developmentalist approaches in some other countries may shape up as "mission-oriented" innovation policies to begin rethinking the role of the state in the economy by identifying societal challenges that require system-wide transformation before defining specific policy pathways related to high-tech and AI.[121]

AI Urban Spaces

Artificial intelligence (AI) is already present in modern high-tech clusters and locations—mainly in districts and factories of the future, but increasingly in science cities, smart seaports and city ports, networked producer service areas, eco-industrial parks, etc. These areas have similar concentrations in advanced, cutting-edge manufacturing, and information-based sectors as the old technopoles. Some innodistricts mix AI and robotic platforms with existing high-tech enterprises and mature start-up ecosystems that have relocated to city centers in the last 10–15 years.[122] As discussed earlier in this chapter, advanced manufacturing research and innovation taking place in factories of the future—located in urban areas and metropolitan regions—are making extensive use of autonomous decision-making processes and devices to optimize manufacturing production. An example is lights-out manufacturing where production sites are fully automated and CNC machining requires no human presence on site.[123] All urban areas (for example, in London and New York) are becoming AI production spaces where programming and algorithm summits and hackathons are organized to produce code. Some innodistricts host these gatherings, which are set as *design thinking* events. The goal of a hackathon is to create functioning software or hardware at the end of an event. According to critics, sponsors take advantage of free labor to create "fictional expectations of innovation."[124]

AI in Megaconstruction

The implementation of AI technologies in several industry sectors is increasing and is now attracting the attention of the megaconstruction sector, which is plagued by problems and shortcomings including massive cost overruns, a high incidence of failure, and a myriad of disruptive impacts.[125] Sarkheyli and Sarkheyli (2022), for example, discussed the utilization of Artificial Neural Networks and Adaptive Neural Fuzzy Interface Systems for risk control and better management of mega-projects.[126] AI is mainly used as an analytical tool to quickly discover trends, patterns, and meanings in data. Planners of large-scale projects also use AI to boost their productivity. To enable project teams to model modularity and construction sequencing, Bechtel, for instance, established the Big Data & Analytics Center of Excellence (BDAC) in collaboration with data science company Miner and Kasch.[127] AI is also capable of managing scheduling, reminders, follow-ups, evaluation, risk identification, and risk reduction.

An AI method called *reinforced learning*, which relies on trial and error, can evaluate options and combinations using analyses of related projects and other project planning optimization activities. Additionally, robots can be built to learn from simulations and assist with upcoming typical activities such as on-site prefabrication.[128] Utilizing AI advancements from other industries could be advantageous for the megaconstruction industry. Commentators advise megaproject planners to get more involved and integrate into the industrial ecosystem to gain from synergies and cross-fertilization.[129] Decision-makers also need to enhance investment in R+D to foster digitalization, given the current scenario of very limited high-tech use in the megaconstruction business.[130] It remains unclear, however, how the use of the most advanced technologies in planning, construction and management can prevent the complex web of interrelated and interdependent impacts that emerge after megaproject construction and affect the urban living conditions of large numbers of citizens, particularly those who are more exposed and vulnerable.

AI Mobility Megaprojects

eHighways

Electric highway (eHighway) systems are among the most important intelligent mobility solutions because they lower greenhouse gas emissions considerably. An overhead line infrastructure (OL infrastructure) and an overhead line hybrid truck were combined in the eHighway system (OH truck). An onboard battery and overhead line are connected by a pantograph on the driver's cabin—how the OH truck receives power. After leaving the OL infrastructure, the vehicle uses battery energy, and if the battery runs out, a regular combustion engine. The goal is to create a global eHighway network that can reduce or eliminate the need for fossil fuels with the aid of AI solutions. The Federal Ministry for the Environment, Nature Conservation, and Nuclear Safety supported the German eHighway pilot project ELISA (ELektrifizierter, Innovativer Schwerverkehr auf Autobahnen [electrified, innovative road freight transport on motorways]). The ability of ELISA to generate readily available and trustworthy data required for a long-lasting artificial intelligence (AI) megaproject is essential for the project's success. This research suggests that ELISA has the potential to be developed into a long-lasting, robust, and AI megaproject if data availability and dependability can be maintained and people retain control over decision-making when significant modifications to the technology are made.

In addition, as technology advances, new mobility concepts have emerged. These concepts must be considered in data-driven analyses to enable the development of future AI solutions. One of the primary issues with data-driven solutions is that, even with a strong data-driven approach, it is possible that trustworthy, long-lasting conclusions will not come from the data themselves.[131] AI-powered data-driven solutions require a thorough understanding of human decision-making. A task becomes a specific action if it is programmed to occur in every conceivable circumstance. However, when human decision-making methods are used, decisions for ambiguous instances are unstable. "Uncertainty of the future" and "granularity of the decision" are obstacles for AI-based decisions. Uncertainty, complexity, and equivocality pose three obstacles to decision-making. For unforeseen circumstances or scenarios that are not anticipated by the designer, outcomes based on AI-based solutions are not guaranteed. In addition, because the decision-maker may be prejudiced while constructing a data-driven solution, ethics in data-driven solutions are a problem for objective decision-making.[132]

Hyperloops

A hypothetical high-speed transportation system, called a *hyperloop*, moves passengers or cargo through a vacuum tube. Although the concept of a hyperloop dates back hundreds of years, a working prototype has not been created until recently. Large-scale infrastructure projects such as the hyperloop sometimes require expenditure of significant amounts of public funds, which makes it imperative to confirm that public works projects are genuinely possible before significant investments are made. Notably, there are currently very few public investments in hyperloop technologies. The business sector and student contests are the main sources of hyperloop R+D.[133]

The present hyperloop debate began in 2013 with the publication of a white paper by a prominent billionaire that suggested building a hyperloop between Los Angeles and San Francisco. Although work on such a project has never started, the paper rekindled interest in the concept, which eventually led to the establishment of private companies intending to develop a hyperloop, academic programs devoted to researching them, and perhaps even the emergence of a new industry. Interest in hyperloops is understandable given that proponents of the technology claim that among its benefits are the ability to transport people and goods more quickly and efficiently, as well as a high carrying capacity, low operating costs, and relatively little environmental impact. However, there are also significant difficulties related to the aerodynamics, curvature, buffeting, and gravitational forces. As with autonomous vehicles and other cutting-edge transportation technologies, tremendous care must be taken to ensure egalitarian, sustainable, and practical implementation of hyperloops in today's smart-transportation systems, as well as in the cities of the future.

Currently, it seems that the vast bulk of writing on hyperloops concentrates on the technology's design, and the technology's possible social consequences are rarely discussed. The largest hyperloop prototype can only travel a few hundred meters and is not yet able to reach supersonic speed. A shortage of track has been a significant obstacle to the creation of a functional hyperloop. Furthermore, the actual cost of building a hyperloop remains uncertain. While at least one expert contended that hyperloops could be constructed and maintained at costs comparable to those of other transportation methods, other analyses contend that the costs of building a hyperloop would be enormous, possibly exceeding those of either high-speed rail or air travel and that they would also face significant political opposition. Therefore, it is important to consider the socioeconomic effects of hyperloops in urban areas. One might readily envision a time in the future where only the wealthy could afford to use the hyperloop. Additionally, more research is required to fully comprehend the security dangers associated with human-hyperloop travel, as even a slight mishap in a hyperloop might have catastrophic consequences for everyone involved. Additionally, hyperloop passengers experience strong gravitational forces, and their effects on health require further study. Finally, safety issues might be one of the biggest obstacles that a commercial hyperloop company must overcome to succeed.

Conclusion

The rise of AI has promoted the so-called sovereign AI (where countries are developing their own foundational models) and a new age of "techno nationalism," a policy approach that positions technology as a critical tool in global power dynamics and is vital to national security, economic competitiveness, and geopolitical influence. As a result, there has been a louder call to accelerate AI development. We witness an increased advocacy for an AI version of the Manhattan Project—a public-private megaproject, fusing the federal government's money and technology of Silicon Valley.

At the urban level, some New York hackathons had already taken place before the pandemic. These smart-city processes, as well as digitalization and virtualization, have intensified since the advent of the coronavirus to form a perverse high-tech regime. Some refer to this as the rise of *disaster capitalism*, a *screen New Deal*, or *tech shock doctrine*,[134] which is at times justified as a resilient solution against humans seen as biological hazards.[135] In the meantime, ever bigger megaprojects (giga- and teraprojects), promoted as sustainable by megaconstruction companies, continue to be built around the world, in spite of delays triggered by current global supply chain clogs and blockages.[136] In a global economy propelled by land values, natural resources, and intellectual property, tech companies consolidate and expand their social, cultural, and political influence, whereas high-tech innovation and AI continue to be presented as transformational tools for greater efficiency in megaproject management, economic growth, and urban sustainability.[137]

In addition to these concerns, researchers could benefit from developing analytical blueprints to account for megaprojects as socioeconomic particles that produce negative externalities and hugely negative societal impacts. Ultimately, human development and sustainability considerations should determine the technological innovations required. In other words, technological innovation and AI alone do not necessarily foster urban sustainability, even if specific innovations (technological or otherwise) do. Furthermore, the idea of a sustainable AI megaproject may be interpreted by some as a dystopian nightmare.

It is possible that some AI will contribute to improving megaproject management,[138] but if we use a relational concept of megaprojects that privileges the *embeddedness* of megaprojects in the wider economy and society,[139] it is questionable whether better megaproject management via high-tech enhancements can by itself contribute to assuaging disruptive impacts and to a situation of greater urban sustainability. Urban sustainability *for all* would need to take precedence over high-tech innovation and AI schemes, as the former ultimately aims to preserve life on Earth. Thus, rather than staying at the edges—either sleepwalking through the Anthropocene,[140] or accelerating technological innovation without checks—we could gain from working backward, from the kind of sustainability we want to the types of innovations that can make it happen.

Intelligent disruption (destructive creation) produced by artificial intelligence (AI) systems is no longer avoidable, although we can channel it with relevant debates about what kind of sustainability we want for ourselves and future generations. We need to see how the recent calls for AI regulation that started in the spring of 2023 (an initiative by experts and scientists, and by industry) will play out in the context of intense geo-strategic competition between the United States and China. Discussions around Suleyman's proposed "containment" path would need to explicitly integrate the notion of smart and just sustainability, a normative horizon from which we can decide which technological innovations and AI applications we need.

As we know from Earth system science, we live on a planet that works as a complex, self-organizational, and self-regulating system.[141] Sustainable life and development on Earth thus requires global coordination for target setting, as has been attempted with the UN SDGs. The SDGs can be seen as a rudimentary antecedent of sustainability requirements under planetary futures such as Lovelock's Novacene. In Novacene, AI minds, super-intelligent beings, and electronic life (the apex of technological innovation) meet Earth's sustainability needs and cooperate with humans to achieve them. Thus, in Panglossian fashion, one would have nothing to fear at the prospect of megaprojects manufactured and built entirely by AI systems and humans "all watched over by machines of loving grace."[142]

Chapter 6
Greening Megaprojects

Introduction

In the previous chapters, we analyzed the environmental impacts of megaprojects in several cities (including Istanbul, Sydney, Hong Kong, and Abu Dhabi). Chapter 4 documents a transition toward architectural and design paradigms and away from iconicity. Chapter 5 focuses on the rise of science, high technology, and innovation as strategies for building projects and ensuring urban competitiveness. In both chapters, we highlight these strategies' drawbacks and negative effects. In this chapter, we focus exclusively on the impacts that some megaprojects have had on the surrounding environments (in Tokyo, London, Paris, and New York).[1] The chapter also analyzes some strategies used (in Shanghai and Riyadh) to develop allegedly "sustainable" megaprojects to counter environmental damage. In our discussion of the empirical evidence, we have focused on showing trajectories of disruptive impacts rather than providing the latest developments in each megaproject, unless these developments add to or qualify our main argument.

The "greening" of megaprojects has become a widespread strategy for urban development worldwide. However, we contend that green megaprojects claimed as "sustainable" represent the logic of strategically disinvesting in issues that appear secondary to the competitiveness of cities, while investing in issues crucial for forming urban regions as strategic scales for capital accumulation. Sustainability has evolved into a vague catchphrase whose significance has changed over time depending on the context, players, and interests. Since the 1990s, the slogan has served as a potent discursive framework for public policies that prescribe particular modes of action. It has also served to legitimize—or delegitimize—particular types of actors, issues, strategies, policy tools, and, more broadly, certain ways of looking at the environment. Why are urban environmental policies frequently motivated by the goal of improving cities' competitiveness or limiting elements (pollution, traffic, urban sprawl, etc.) likely to impair competitiveness? These policies are

G. del Cerro Santamaría, *Contesting Megaprojects*,
https://doi.org/10.1007/978-3-031-98382-5_6

characterized by their thematic, social, and spatial selectivity by opting to pursue win-win goals for the environment and economy and to meet the aspirations of the middle and upper classes.[2] We contend that the approach to sustainability of "green capitalism," used in "sustainable" megaprojects, is a strategy that does not adequately address the main challenges posed by megaproject impacts. We propose some essential elements for understanding megaproject sustainability as a complex and multidimensional concept.

One of the most permanent human influences on the global ecosystem is converting Earth's land to urban usage. Urbanization impacts energy consumption, climate, and hydrologic and biogeochemical cycles. Additionally, it triggers habitat fragmentation and biodiversity loss while hastening the decline of highly productive agriculture.[3]

The fastest rates of the proposed land conversion are projected to occur in biodiversity hotspots that were comparatively unaffected by urban expansion just two decades ago, representing a direct threat to high-value ecosystems.[4] Megaprojects significantly contribute to urbanization processes and, therefore, to the environmental impacts described above. Megaprojects also generate impacts on their own because of their size, scale, and interaction with surrounding buildings. Research has found that less than 1% of building projects worldwide currently measure and report lifecycle emissions.

As of 2021, enterprises operating in the built environment sector are urged by the World Business Council for Sustainable Development to adopt a lifecycle approach to monitor greenhouse gas (GHG) emissions. Six materials—including steel and concrete—are responsible for 70% of the embodied carbon in buildings.[5]

Environmental Impacts in Global Cities

Tokyo Bay's Megaprojects

In Japan, as around the world, the urban policy environment drastically changed in the early 1980s. Inspired by Thatcher and Reagan's neoliberal principles, Prime Minister Nakasone pushed for deregulation, privatization, and budgetary restraint. Urban planning deregulation initiatives have been increasingly framed in terms of the economic significance of Tokyo being a world city. Tokyo offers a prime example of the impact of globalization on urban transformation. The central government has consistently sponsored and encouraged this unremitting invasion of life space by economic space to create a successful global city over the past two decades. Many aspects of Japan's relationship with the state, society, and market are unique because of the country's history, and it is possible to argue that global economic integration and international competition have increased the pressure to convert traditional residential areas into high-rise office spaces. The loss of planning safeguards, which inner-city communities acquired through citizen mobilization in the 1970s, is a key result.[6]

Globalization does not appear to have affected urban change in Japan through foreign direct investment in urban (re)development, but rather by making a strong case for the weakened enforcement of local planning regulations and citizen participation procedures. The literature shows that communities in inner-city areas have, on the whole, been fighting a losing battle against a resurgent property development industry that seeks to transform inner-city neighborhoods into high-rise towers ever since the peak of citizen mobilization over environmental issues and the progressive control of local governments in the mid-1970s. Communities in inner-city Tokyo continue to experience pressure, and life spaces are still being repurposed as commercial spaces. Minimal attention is focused on the quality-of-life and urban livability. This background explains both the development of Tokyo Bay and the subsequent environmental impacts, which are similar to challenges elsewhere in the developed world.[7]

The megaproject developments around Tokyo Bay drew inspiration from the megastructure movement in the 1960s. Urban theorist Zhong-Jie Lin examines the history of Tange's Tokyo Bay project, one of the first megastructure proposals for the area, and its impact on later developments.[8] Lin describes how Kento Tange, one of the major architects and influences in the Tokyo Bay project and the writer of the 1960 Tokyo Bay Plan, was influenced by Metabolists. Metabolists saw the transformation of a city as an organic process, where cities can expand in futuristic ways—toward the sea or into the sky. Tange's plan involved conceptualizing a city "as a gigantic structure with a hierarchical arrangement of circulations and programs" and creating several artificial islands "to house a highway that could facilitate communication in the area." Tange suggested an incredibly large, completely integrated network of super roads, artificial islands, and bridges. In Tange's opinion, concentric development in Tokyo up to this point was inferior to a linear model of expansion. The solution to urban development's ills—noise, traffic, and housing scarcity—was to expand the boundaries linearly. The whole Tokyo megaregion's development axis would have been completely reorganized as a result. This initiative was never implemented, but subsequent developments were profoundly impacted by its unique techno-optimism and ambition to expand into the bay.

In response to Tange's plan and the opportunities the bay offered, Yokohama built Minato Mirai 21 (Fig. 6.1), a business waterfront development next to Kannai, the downtown area, and Motomachi, a historic shopping street. Despite being first suggested in the 1960s, years passed before it was even close to being completed. Though they do not initially appear to have many parallels, there are some similarities between the two developments. Both have a desire for bay-focused development and an understanding of the need for system development. The Minato Mirai 21 development strongly emphasized establishing the transportation and rail infrastructure required for private development to complete the remaining tasks. Also, though on a far smaller scale than Tange's concept, it likewise placed a great deal of emphasis on axial growth.

The Tokyo Metropolitan Government is in charge of the development around Tokyo Bay. Over the years, megastructures and skyscrapers have been constructed

Fig. 6.1 Aerial view over the Minato Mirai, Kannai, Motomachi and the bayside of Yokohama. Source: iStock.com

in the area without clear planning frameworks. The construction boom housed many costly ornaments, a convention center or two here, a posh hotel there, and older, more familiar factories, smokestacks, and oil storage tanks coexist.[9] According to conventional Japanese wisdom, the development around Tokyo Bay benefited Japan following World War II's devastation. It has worked as a strong magnet, attracting millions of people from all over the nation and giving them access to new occupations and lifestyles.[10] Tokyo Bay is a gigantic conglomerate of office towers, and there are plans to create more megastructures, including the Shimizu Mega-City Pyramid and a floating airport. Rather than focusing on just one megaproject, we briefly discuss two main disruptive impacts that are common to all of them: the urban heat island effect and the liquefaction of reclaimed land.

Urban Heat Island Effect

Owing to the megastructures in Tokyo Bay, residents of Tokyo now suffer severe heat, especially during the summer. In 2008, the Tokyo Metropolitan Government showed that the average temperature rise in Tokyo during the 20th century was 3 °C.[11] The Japan Meteorological Agency also revealed that between 1898 and 2010, the yearly average temperature increased countrywide at a pace of 1.15 °C per century, significantly higher than the 0.74 °C increase in average global temperature.[12]

This temperature increase has been attributed to the Urban Heat Island effect (UHI). Because concrete and asphalt are used to construct buildings, the heat energy from the sun and other structures is retained, turning the city into a gigantic absorber and then an emitter of convected heat.[13] This then creates "guerrilla storms and renders the nights (always cooler than the day due to lack of sunlight) barely better than stifling days."[14] According to the Inter Press Service, the media is now citing statistics demonstrating how Tokyoites are no longer cooled during summer nights due to the massive buildings surrounding Tokyo Bay. The Shinagawa sea-front's much-admired neon glitter prevents cool air from blowing into the city after dark.[15]

As an expert in the UHI effect, Komichi Ikeda, deputy director of an environmental scientists' organization, the Environmental Research Institute Inc., argues that government policy putting the economy above environmental protection has turned Tokyo into a nightmare. He adds that building is being done "at the expense of the environment and the health of inhabitants by landowners and developers."[16] In response to the UHI effect, there are plans to build twin buildings 246 m apart, joined by a pedestrian deck. Upon completion, an ancient 12-story structure blocking the wind from Tokyo Bay would be destroyed, which is anticipated to create a cooling wind flow.[17]

Liquefaction of Reclaimed Land

The glistening lights of Tokyo Bay attracted residents, although not many people could afford the skyrocketing housing prices and had to settle in cheaper areas. However, liquefaction of land can occur, especially in an earthquake-prone country such as Japan, located at the meeting of four different tectonic plates.[18] As recorded over the years, even if the earthquake magnitude is small, substantial damage can still occur.[19]

Most of Tokyo Bay has been developed on reclaimed land. When an earthquake occurs, the reclaimed soil acts as a liquid: the sand and water from underground rise and structures on the surface sink.[20] Despite the danger of liquefaction, construction and development did not stop in Tokyo Bay; in fact, numerous new reclamation projects are in the pipeline.[21] During the 2011 Tohoku Earthquake and Tsunami event, the largest recorded Earth tremor in Japan, although no megastructures fell, soil liquefaction damaged large areas.[22] During liquefaction due to the earthquake, many courtyards were coated with wet mud. The ground level decreased at numerous locations. Manhole covers were forced out of the ground on the roadway, water lines ruptured, and utility poles began to sag.[23] Despite reclaimed land being highly vulnerable to liquefaction, many people continue to live in these areas because they are affordable.

London's 2012 Olympic Park

In London, as in many other urban areas, megaprojects have defined the city's physical form and greatly influenced its socioeconomic dynamics since the Romans built the London Wall in 60 CE. Megaprojects are privileged elements of British capital development. The London Bridge, Tower Bridge, sewerage system, underground transport system, and Thames Barrier have historically provided infrastructure allowing London to gradually grow and expand to reach its current structure and dimensions.

More recently, the Docklands redevelopment, Thames Gateway, Olympics, Kings Cross Development, East–West Crossrail, DP World London Gateway Port, Wembley Park, Royal Wharf, and London City provide examples of megaprojects that re-engineer, reshape, and modify London's morphology and provide the foundations for socioeconomic innovation and, sometimes, radical urban transformations and disruptions.[24]

The Thames Gateway is a development eastwards from London and is Europe's largest regeneration project. Starting at Canary Wharf, the project spans 40 miles (64 km) into the counties of Essex and Kent. The central government set objectives for the creation of 180,000 new jobs, 53,000 hectares (131,000 acres) of protected and improved green spaces, 160,000 housing units (35% of which were designated as affordable housing units), improvements in residents' skill levels, and access to high-quality healthcare. The plan called for both growth on new greenfield sites and renovation of inhabited or abandoned regions (brownfields in many cases). Work progressed quickly in the western region, where the Olympic Park and related venues for the 2012 Games were constructed. Other areas of the project are still under regeneration.[25]

Other megaprojects transforming the urban fabric of London include:

- The renovation of *Battersea Power Station* in London's Nine Elms neighborhood is the core of a larger redevelopment in London. The estimated USD 17.6 billion, eight million ft^2 project includes apartments, hotels, and offices. Construction has been proceeding in seven phases from 2014 to 2025. Apple opened offices there in 2023.

- *Kings Cross Development* (discussed in Chapter 5) is becoming London's main innovation district via the transformation of the area around the railway station in downtown London into a science and business hub. Central Saint Martins of the University of Arts, London, became the first occupant of the new premises. In January 2013, Google acquired a million square feet within King's Cross Central (KXC). In September 2014, Havas, the Global Communications Group, consolidated its existing twenty-four London offices into one new building at Kings Cross Central. The Francis Crick Institute, hosting approximately 1250 biomedical scientists, opened the site in 2015.
- *East–West Crossrail*, an innovative megaproject and the largest infrastructure project in Europe, is a 73 mile (118 km) railway line under development in London, running through parts of the city and the counties of Berkshire, Buckinghamshire, and Essex. The main central portion of the line between Paddington in central London and Abbey Wood in the southeast opened in 2021.
- *DP World London Gateway* is a fully integrated logistics facility comprising a semi-automated deep-sea container terminal on the same site as the U.K.'s largest land bank for the development of warehousing, distribution facilities, and ancillary logistics services. The facility is located on the north bank of the River Thames in Thurrock, Essex, 30 miles (48 km) east of Central London. The deep-water port opened in November 2013 and can handle the largest container ships in the world. The port is now linked on a weekly basis to 51 countries and more than 90 ports worldwide. "The new facility significantly increases the capabilities and efficiencies of the Port of London in handling container shipping to help meet the growing demand for container handling at Britain's ports."[26]
- *Wembley Park*, *Royal Wharf*, and *London City* are three prominent, ongoing, large-scale housing developments in various parts of the city, combining residential and commercial spaces as well as public areas.

London's projected economic growth and increasing concentration of highly skilled workers and firms within the city "pose many practical challenges for policymakers. First, there is the issue of environmental sustainability. Judicious sustainability policy initiatives can bring ample economic payoffs from attracting and retaining business and creative people in the short and long term."[27] London is planning and implementing environmental policies and strategies to accomplish its sustainability goals. In alignment with the global shift toward sustainability, the city contributes to environmental sustainability by maintaining the viability of its ecological system and seeking the best possible environmental practices. However, housing to accommodate the expanding metropolis remains a key challenge in London. This challenge goes hand-in-hand with the need to enhance an already capacity- constrained transportation infrastructure. Effectively addressing these issues requires thoughtful consideration of the changing role of London in the U.K. and global economies, especially as the effects of Brexit begin to alter the urban prospects of the city.

Despite London's privileged economic relations with North America and Asia, the decision by the U.K. to exit the European Union, as determined in the referendum on June 23, 2016, may have already damaged the image of London as a center for international commerce, particularly with London's standing as a leading European financial center and its importance in financial markets such as investment banking, asset management, and Eurobond trading. The EU crucially links London and the U.K.'s financial markets with Europe.

Additionally, the EU is an important market for U.K. goods and services across industries, accounting for 45 percent of U.K. export demand in 2014.[28] A major drawback of Brexit is that it severely limits the opportunities for London to attract international talent, which is currently one of the key economic strengths of the city. The uncertainty surrounding the form and nature of the relationship with European and international markets reflects both the risk and opportunity for the prospects and fortunes of British capital in its role as a quintessential global city.

Olympic Disruptions

London Olympic Park, also known as Queen Elizabeth II Olympic Park, was constructed during the 2012 Olympics. It is situated east of Stratford, adjacent to Stratford City Development (Fig. 6.2). The park occupies five boroughs: Newham, Tower Hamlet, Hackney, Waltham Forest, and Greenwich.[29] The park houses the Olympic Village, Olympic Stadium, London Aquatics Center, Arcelor Mittal Orbit, an observation tower, and a piece of public art. The motto of the London Olympics was "Toward One Planet Olympics," a vision for sustainable and low-carbon Olympic Games.[30] Because of implementing the vision, the London Olympics were assessed as the "greenest Olympics ever" by the Independent Commission for Sustainable London in 2012. The venues were made of at least 25% recycled materials, and the transport arrangements were successful.[31]

Cooperation with the Rio Tinto Mine

The medals of the London Olympics were made almost entirely from materials extracted and smelted at the Kennecott Copper Mine, owned by the Australian mining multinational Rio Tinto Kennecott. The mine was involved in a federal lawsuit in Utah, where the owner had been accused of violating the United States Clean Air Act for several years.[32] The Kennecott Copper Mine produced a quarter of all United States copper, as well as vast amounts of gold, silver, and molybdenum. This mine is also said to be the largest man-made mine in history. However, unregulated extraction has resulted in the mining of the largest industrial pollutant in Salt Lake. Kennecott was accused of advocating health groups endangering the health of Salt Lake City citizens by disregarding the emission standards

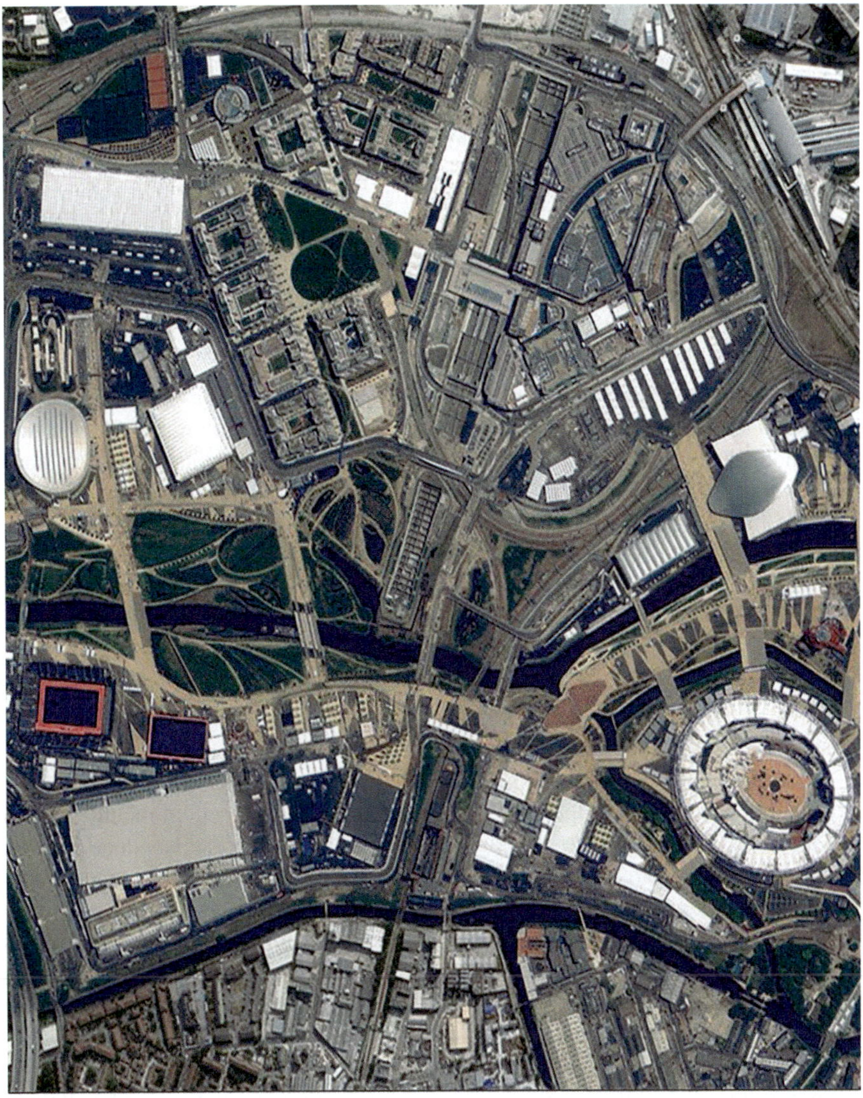

Fig. 6.2 Aerial view of Queen Elizabeth Olympic Park, the main site of the London 2012 Summer Olympic Games. Source: Google Earth

set by the United States Environmental Protection Agency, making Salt Lake City one of the worst cities for air pollution in the United States. The allegations by doctors, lawyers, and advocates stated that the Rio Tinto mines were responsible for 30% of air pollution in Salt Lake City, inducing long-term respiratory problems in its citizens, including children.[33]

Loss of Biodiversity

The construction of the Olympic Park severely affected the immense biodiversity of the area. The five boroughs produced a Biodiversity Action Plan a few years before the London Olympics, which detailed local wildlife and biodiversity and focused on key habitats and species. The report identified more than one hundred sites of importance for nature conservation. There was a vast amount of woodland, standing water, reed beds, and grassland in Hackney, one of the Olympic Boroughs. However, the London Thames Gateway Development Corporation drew a picture of "fragmented ownership" and "declining industry," which depicts North East London as "an area littered with relics of industry and rusting steel planes."[34] Later, the corporation bought land at a convenient price.

The lower Lea Valley, home of the river Lea and referred to as London's "green lung," has also been impacted. Many species live there, including kingfishers, and for hundreds of years, birds, including ducks, geese, warbles, and thrushes, have used it as a preferred migration path through London.[35] The lower Lea Valley is also an oasis for kestrels and herons. The reserve was removed to build a velodrome arena for Olympic track-cycling events. An official statement declared that significant labor was required to clean up the land and remove invasive species. The parkland redevelopment included a new wet forest habitat that promotes biodiversity and increases access to recreational opportunities.[36]

Mismanagement of Nuclear Waste

Documents obtained from the British Government via the Freedom of Information Act revealed that radioactive nuclear waste was buried under the London Olympic Park, especially under the Velodrome and Basketball arenas.[37] Even though the Olympic Development Authority (ODA) insisted that there was "no exposure to increased levels of radiation either through direct contact or dust," it was discovered that over 7000 tons of waste were buried in a radioactive storage bunker beneath Stratford International Station, transport hub for Olympic attendees. The British Environment Agency forbids the mixing of higher and lower radioactive materials, which is how radioactive waste, comprising radium, polonium, thorium, and uranium, is diluted. The ODA denied a request for radiation monitoring, and no yellow radiation signs were used to indicate the presence of radioactive waste.[38]

Significant quantities of radioactive contamination were unknowingly excavated at the main stadium and spread across other areas. Public Relations from the ODA claimed that during the earthwork program, highly sensitive monitoring was performed on the site, and as a result, a small number of low-level radioactive findings in the nearby soil were discovered. This soil, which has some minor radioactive qualities owing to historical industrial activities over the past hundred years or more, is being controlled to ensure that neither the general public nor workers are at risk.[39]

However, excavation was completed before radioactive monitoring could be conducted. According to public reports, "When radiation was detected in excavations in

November 2007, no action was taken and it was not made public. Only in February 2008 were the relevant authorities alerted and remedial and protective measures started."[40] Paul Charman, an analyst on the Citizens Intelligence Network, expressed his concern about management by the ODA:

> It has become clear that the real priority is the avoidance of negative PR and of any delays to the delivery schedule—and the less that was known the better. The ostrich-like approach led to real injustice and left the workforce and the local population vulnerable.[41]

John Large, an independent nuclear analyst, said at the time:

> The Olympic site's hurried and unplanned development may have resulted in a great deal of public harm to the local communities remaining around the site. Overall, there is some doubt about the applicability and validity of radiological risk analysis undertaken for future legacy use.[42]

Loss of Allotment and Greenspace

Public spaces were lost during the Olympics. For instance, the Manor Gardening Society was right in the middle of Olympic Park and felt the sharp edge of the compulsory purchase order being pushed into their 100-year-old grounds as a result. According to the Manor Garden Society, the displacement was considered a betrayal of thousands of people on allotment waiting lists in the local area. The society hoped for a genuine example of sustainability and response to the local needs in the Olympic legacy, and not the replacement of the old site with something far inferior in the future Olympic Park.[43] Other complaints regarding the loss of open space include the disruption of Greenwich Park and Leabank Square. In March 2012, the Save Leyton Marsh Group protested Porters Field to oppose the construction of Olympics practice basketball courts.[44]

Olympic Legacy

The Olympic Park was created as an urban destination that reimagines a historic community as a sustainable place to live, work, study, and visit. It was intended to address the effects of climate change, encourage responsible consumption and production, enhance health and well-being, encourage community involvement, and promote biodiversity and sustainable land use. The project as a whole was planned for the effects of climate change, and included sustainable parklands, neighborhoods with walkable areas, and streams. Megaprojects typically put economic growth ahead of sustainability, which has a negative social and environmental impact. Furthermore, political unrest and policy changes can impede sustainability efforts. This is seen in the Queen Elizabeth Olympic Park, where policy changes resulted in a reordering of sustainability priorities. Such sustainability priorities in "green" megaprojects come together with a need for effective governance and stakeholder engagement, the need for innovative financing mechanisms, and the importance of performance monitoring and evaluation. Nevertheless, more than these optimizing practices in design, planning, and implementation are needed to prevent the disruptive impacts megaprojects cause in the neighborhoods and cities where they are built. Specifically, the Olympic Park is yet another example that greening megaprojects for sustainability leads to gentrification, displacements and exclusion, a controversy also surrounding the recent 2024 Paris Olympics.[45]

According to *The Guardian*, 11 years after the opening ceremony of the London Olympics, only 13,000 homes were constructed on and around the Olympic site. Only 11% of these were affordable to people with average wages. At the same time, there are approximately 75,000 households on the waiting list for council accommodation in the four boroughs: Newham, Tower Hamlets, Hackney, and Waltham Forest, many of which are in abject poverty.

Additionally, thousands of former residents relocated outside the region since the Olympics. The empty storefronts below their concrete facades promote the promise of "luxury living in East Village," the former athletes' hamlet that has been rebuilt with Manhattan-scale aspirations. The private rooftop garden is safely elevated from the street and framed by the concrete facades of the buildings. Another pair of towers nearby welcomes locals to its protected "village green," which is free of dogs, ballgames, and unattended children. These projects, created by the development arm of the Qatari royal family, contained no affordable dwelling units. Many in the area describe the situation as "a massive betrayal." Although things may appear to have improved, most long-term inhabitants do not agree.

After 11 years, the legacy yielded a pleasant park filled with outstanding sports facilities and opulent mansions, with some cultural attractions. The most vulnerable and impoverished people, who still live in the poorest neighborhoods of London, have suffered. According to one researcher, it has been a complete success in terms of social engineering: "a lovely place for those who can afford it, with people living in desperate poverty right outside it."[46]

La Défense Megaproject in Paris

In 1958, the French government initiated a major urban planning initiative for the second half of the 20th century. The Public Establishment for Installation of La Défense (EPAD), a business- and industry-focused government organization, was responsible for town planning, establishing infrastructure, and selling available land space. La Défense is a major business district of the Paris metropolitan area, located in the commune of Courbevoie and parts of Puteaux and Nanterre, west of the city of Paris (Fig. 6.3). It spans through a perimeter of 760 hectares and was developed at a time when the area "had seen very little development and was filled with rundown houses, was full of small factories and shanty towns."[47] The government decided to build a modern office district bordered by the historic axis, which started from the Louvre in Central Paris, continued along the Champs-Élysées and the Arc de Triomphe, and ended at La Défense. EPAD brought into play the principles of

Fig. 6.3 View of Paris *La Défense* from L'Arc de Triomphe. Source: Shutterstock.com

"modern movement" laid down in the Athens Charter of 1933, which advocated the lack of roadways, the development of towering structures that maximize natural light, the use of paved pedestrian spaces to separate vehicular and pedestrian traffic, and the use of underpasses to connect high-speed roads that surround the area.[48]

By the 1990s, La Défense covered an area of 160 hectares, including commercial and residential neighborhoods, and served the requirements of 40,000 households and 140,000 workers in 1600 businesses (including 14 of France's top 20 firms and

15 of the top 50). With 2600 new hotel rooms, a significant transportation hub, and 26,000 subterranean parking spaces to support almost one million people annually, the concept was created with potential for domestic and international tourism. Parks, museums, and a plethora of consumption options were also promised (including the largest shopping mall in Europe at the time) but have not been particularly well achieved thus far. While La Défense consists of hundreds of different structures (a type of miniature Manhattan), it is unmistakably an UMP in the truest sense. The Grande Arche building, which serves as the center of La Défense, was designed with an iconic status in mind from the beginning.[49]

La Défense was reinvented to improve its business and consumer- friendliness after the first two disastrous decades, in which not a single right to develop was sold to the private sector. As a result, a new generation of high-tech structures was built, such as the enormous Quatre Temps shopping center (1981) and the iconic Grande Arche. The inauguration of the Grande Arche took place with much public fanfare on July 14, 1989, the 200th anniversary of the French Revolution, during the G7 Heads of State Summit. An even larger spectacle, directed by Jean-Michel Jarre, was performed in July 1990 in front of an estimated two million spectators along the major axis connecting the Arc de Triomphe and Grande Arche. Many state-owned businesses in France were privatized around this time, and major transnational corporations with French and foreign headquarters started to invest in La Défense. This highlighted the shared objectives of those who owned and controlled the largest corporations, globalizing politicians and bureaucrats, globalizing professionals, and consumerist elites, many of whom were French and typically attended the same universities.[50]

Disturbance of Skyline

The skyscrapers in La Défense have been criticized for being disruptive to the view of certain monuments, including the Arc de Triomphe. In 1931, the government developed plans to renovate the area, but with a limit on the height of buildings along the historical Triumphal Way. Towers were only allowed at the end of the historical axis, as they would affect the view of the Arc de Triomphe if allowed in the middle of the Way. La Défense's first master plan in 1964 required all office towers to follow the same rules. In this plan, all towers would be identical in size: "a base of 42 meters by 24 meters, a maximum height of 100 meters, and a floor area of 30,000 square meters."[51] The plan was to build two rows of skyscrapers of 25 stories each, but this was never implemented, as most companies started putting pressure on EPAD for taller office buildings.

Consequently, La Défense ended up with towers of various heights. The tallest of them, the CB21 Tower (formerly GAN Tower), is 589 feet tall. The public criticized the "forest of towers," in which buildings came to disturb the view of the Arc de Triomphe, as seen from La Défense. The skyscrapers were also criticized as generic buildings—towers that are considered as a postmodernist style that "could be built anywhere."[52] To counter the disruption of the Arc de Triomphe view, the government decided to build the Grande Arche, which provides a clear line of sight along the Triumphal Way. However, the architectural style and impact of the Grande Arche remain controversial.

The Grande Arche

The Grande Arche is one of several aspirational and real iconic structures referred to as *grands projets*, linked to Mitterrand and subsequent presidents.

These include the Musée d'Orsay, a former train station that was transformed into one of the most well-known and stunning museums in Paris in the 1980s by Italian architect Gae Aulenti, and the new pyramids at the Louvre designed by Chinese-American architect I.M. Pei. They also include the technologically advanced Institut du Monde Arabe (built by architect Jean Nouvel), the brand-new Bibliothèque de France (Dominique Perrault, 1998), and a sizable recreation area known as Parc de la Villette (which began in the 1980s with buildings designed by Swiss-American Bernard Tschumi and others). They are world-renowned tourist destinations that share a certain level of iconicity. Although it was built before Mitterrand took office, another well-known tourist destination is the Pompidou Center, which was designed by Richard Rogers and Renzo Piano.[53] The Grande Arche in La Défense is a marble and glass cube spanning 110 m in height by 110 m in width. In total, 300,000 tons of glass and Carrara marble from Italy cover the structure. Because of the porosity of the stone, which made it susceptible to weather and pollution, one in six marble tiles on the structure fell apart between 2004 and 2010. Fears of marble shards falling below led to the sealing off of the area on the northern side of the arch.[54]

Unequal Distribution of Natural Light

Giant skyscrapers often cause an unequal distribution of natural light. Following the first plan for La Défense, the number of towers increased. Some of these include the Areva Tower (formerly FIAT Tower), which is 184 m high and has 44 floors. Employees criticized such towers for permanent indirect lighting.[55] In 2008, the government proposed building a pyramid-shaped structure designed by the architectural firm Herzog and de Meuron. The building was named The Tour Triangle.

Architects claimed that the shape would not block sunlight and would also allow for "optimum solar and wind power generation." Green activists demanded "shadow studies," where "new designs are computer tested in a virtual landscape to see where shadows will fall at different times of the day and year." Despite efforts by Paris City Hall to limit the heights of buildings, building shapes also affect the environment through overshadowing. Critics pointed out that Paris did not limit the construction of buildings the way New York did by regulating and limiting the shape of the building.[56]

Yves Contassot, a Paris environmental activist, indicated that "the project would cast a shadow over locals and its foundations would disturb a water table below."[57] The Tour Triangle has also been criticized for disrupting the city center's skyline. Olivier de Rohan Chabot, from the Safeguard of the French Art Group, said:

> I was not against towers in principle, but within central Paris they ruined the skyline's general harmony. The Montparnasse Tower crushed the Hotel des Invalides (housing Napoleon's tomb). The monument was built as a grandiose. But what has it become? A dwarf. The tower ridicules it. In this sense, it is a veritable attack on the beauty of the capital.[58]

Discussions surrounding a long-established development paradigm founded on densification and single-use infrastructure have increased in response to climate challenges and the covid-19 health crisis. Some urban developments, such as La Défense, are dedicated to innovative experimentation, facilitating new ways of living, working, and designing urban spaces, with the ultimate goal of becoming post-carbon global business districts. This is because replicating the mechanisms of the past carries the ever-more tangible risk of ecological, social, and financial deadlocks. An evaluation of the carbon footprint of local activities highlighted the importance of climate issues for La Défense.

Despite a residential population of approximately 40,000 people, the emissions are comparable to those of the regional capital. Construction work is directly responsible for one-third of the carbon footprint, with building operations alone contributing to 15%. These different aspects show that to progressively reduce the world's reliance on fossil fuels, it is vital to address the issue of bioclimatic building design as well as the issue of lifestyles (travel, shopping, eating, etc.). New construction in La Défense is expected to be low-carbon, adhere to eco-conditionality standards, and operate in a manner that leaves a minimal environmental footprint. By 2030, the district's greenhouse gas (GHG) emissions are expected to halve, exceeding the national target of a 40% decrease in GHG emissions.[59]

New York City's New World Trade Center

The reconstruction of the World Trade Center in New York City has generated more scrutiny and higher expectations than any other urban megaproject (UMP) in the contemporary American experience. It took 15 years after the terrorist attack on September 11, 2001, for the completion of the main structures of this 16 acre (6.5 hectare) parcel of land in Lower Manhattan, known as Ground Zero. In the world of urban construction, in which the payment of interest on the enormous sums lent to finance large construction projects is decisive, such a long gestation period usually indicates that the best decisions were not made. In this case, deliberations were probably motivated by factors other than the technical and aesthetic optimization of the project by planners, developers, contractors, architects, and engineers, as well as economic, political, or bureaucratic problems that prevented rapid and profitable development. The area was redesigned and almost completely rebuilt with five new skyscrapers, a museum, a memorial to those who died during the attacks, and a transportation hub. The magnificent National 9/11 Memorial, the mesmerizing pair of ponds reflecting the names of the victims by architect Michael Arad, and the surrounding park by landscape architect Peter Walker, dedicated to the tenth anniversary of the disaster, serve as both the physical and emotional center of the new complex (Fig. 6.4).

Fig. 6.4 Aerial view of the National September 11 Memorial & Museum, the Perelman Performing Arts Center (top) and the Oculus (top right) in the New World Trade Center in Lower Manhattan. Source: iStock.com

In May 2014, the National 9/11 Museum opened, a far less successful design that resulted from a behind-closed-doors partnership between two companies with little in common: Snøhetta was in charge of the external design, and Davis Brody Bond was in charge of the interiors. This division of work resulted in a disconnected structure that unintentionally reflected ongoing disputes regarding how the 2001 attacks could be understood through contemporary urban design.

Santiago Calatrava's World Trade Center Transportation Hub (also known as the Oculus), the most ambitious architectural component of the complex, officially opened to the public in March 2016 but without much fanfare, probably to avoid focusing attention on its egregious financial waste. Although construction was initially forecasted to take 5 years, the project took 12 years to complete, and part of the astronomical USD 4 billion price tag (nearly double the already exorbitant initial estimate of 2.2 billion) resulted from higher shipping costs. The Grand Central Terminal handles 750,000 passengers daily, whereas the Transportation Hub serves only 40,000 passengers on average during the week.[60]

The primary skyscraper in the complex, the One World Trade Center, is the tallest in the Western Hemisphere (a 1776 ft. (541 m) monolith) and the work of David Childs (2013) of Skidmore, Owings & Merrill. It has cost USD 3.8 billion in public funds. It is the world's priciest tower, costing more than twice as much as the Burj Khalifa in Dubai, which is partly to blame for the increase in tolls on the city's bridges and tunnels as well as the lack of funding for infrastructure repairs.[61]

One World Trade Center was originally intended to resemble Daniel Libeskind's lopsided asymmetrical Jewish Museum in Berlin or Toronto's Royal Ontario Museum. With Childs, One WTC has become a modernist-resembling, hybrid steel-concrete structure with chamfered corners that are presented as a sequence of triangular walls. The entire tower is clad in glass, creating a surface that appears like a kaleidoscope—refracting the sunlight and changing weather.

According to the New York Police Department, the One World Trade Center is a solemn fortress that rises from the ground floor to the fifteenth floor and is enclosed in glass, but constructed of concrete. Visible and unseen inscriptions at the World Trade Center reflect the post-2001 security measures. The new buildings in the complex incorporate additional emergency stairs and bomb-resistant protections, while the streets are lined with bollards and heavily armed police officers patrol the memorial square.[62]

One World Trade Center is a bright spot in the updated New York skyline, yet most no longer associate the building with the site. When the Occupy Wall Street movement electrified downtown New York in the fall of 2011, neither protesters nor their opponents paid much attention to the giant tower, which rose a block away. The Port Authority agreed to take over the construction of One WTC in 2006 after George Pataki, the governor of New York State, and Michael Bloomberg, the mayor of New York City, mediated an agreement that gave Larry Silverstein, the developer, a significant rent discount for the remaining towers. The deal made New Yorkers perpetual owners of a commercially unviable Goliath, but this did not trigger any significant debate in the media or public organizations.

The museum, which opened in May 2014, has attracted mostly international visitors. The lower levels of the twin buildings were protected from water by a large wall on 9/11, and the museum's primary display areas perch on the bedrock facing the wall. The only spot where one can experience the pain, hope, and civic ethics that the new World Trade Center should inspire is atop the 20-meter-tall dank industrial wall, which is arguably the best reminder of the first World Trade Center still standing in the neighborhood. When the moat was finally cleared, the remains of the dead collected, and the cancer-causing debris pushed aside, New Yorkers were willing to bet on a thoughtful and ambitious urban design to bring about the city's rebirth around the idea of unity and civic pride. We believe this has not happened transparently, although it is unreasonable to call it a failure of urban design. The New World Trade Center is a symptom of great contradictions and urban problems that, even with the election of the leftist Bill de Blasio as the new mayor in 2013 (re-elected in 2017), showed no signs of abating.

In the early 2000s, Ground Zero represented collective trauma and pain. Today, it evokes other more recent problems, including massive income inequality, impunity in the financial sector, a federal congress without much to offer cities, and a state government mired in fiscal tightening. On the other hand, an unprecedented tourist flow, which is the spearhead of the contemporary gentrification of New York City, promises to increase with the recent construction of the Perelman Performing Arts Center next to the One World Trade Center. The coming residential space in the Five World Trade Center (1200 apartment units, 400 of which were subsidized) might open the door to turning the site into a recognizably mixed-use neighborhood.

During several recent visits, we found the site packed with people. Children run undisturbed by the security devices, and the noise volume is high; the impression is not of a place of reflection, memory, or meditation. The new World Trade Center teaches us that one cannot simply design an urban scheme with attractive or innovative aesthetic standards and world-class architects and expect positive results derived from the iconic centrality of a great city. Street politics, community sentiment, solidarity, and civic virtues are needed: ideals that flickered for a moment on the uncertain days of September 2001 but disappeared in the Bloomberg era and were almost completely replaced by privatization, oligarchy, and urban neoliberalism, which are the hallmarks of the 21st century, both in New York and in most big cities. The construction of the One World Trade Center was additionally beset by environmental problems.

Green Concrete

"Green concrete" was used to construct the One World Trade Center. This reduced carbon dioxide emissions by approximately 12 million pounds, energy consumption by approximately eight million kWh, and freshwater consumption by approximately 30,000 gallons compared to conventional concrete. However, the production of coal fly ash is also a part of the production of green concrete.[63] Green concrete consists of three additional materials added to the standard ingredients of concrete: natural pozzolan, zeolite, and coal fly ash from blast furnaces. Coal fly ash is produced when the coal is burned. The flue gas used in coal combustion drives small particles from the boiler. When coupled with bottom ash (ash that settles at the bottom of the boiler), fly ash is known as coal ash. Fly ash was collected using electrostatic precipitators or other particle filtration equipment.[64]

Although the generation of green concrete is more sustainable than traditional concrete, it poses several environmental risks. Fly ash may contain heavy elements such as arsenic, lead, mercury, cadmium, chromium, and selenium, depending on where the coal is extracted.[65] These heavy metals can poison aquatic life when they fall into the water. The United States Environmental Protection Agency also found that people living next to coal ash disposal sites had an increased risk of cancer and other diseases. These toxic substances can also cause substantial damage to the nervous system, cognitive deficits, developmental delays, and behavioral problems.

Doubts on LEED Certification

The Freedom Tower is one of the few buildings on its scale to obtain the United States Green Building Council's LEED gold certification. To secure gold certification, the Freedom Tower needed to run at least 20% more energy efficiently than New York's present construction requirements and achieve net- zero carbon dioxide emissions by acquiring green energy credits.[66] In addition, the structure has modern fuel cell installation, rain harvesting, and lighting that responds to sunlight. Instead of burning chemical energy, cells convert it into electricity through an energy-efficient electrochemical process. The tower was also heated and cooled using the heat and water produced as by-products. David Paterson, the governor of New York, complimented the structure's "space-age energy technology" in 2008.[67]

However, in 2012, superstorm Sandy destroyed the fuel cells of the building. An internal document obtained by Climate Desk from the Port Authority of New York and New Jersey claimed that "Superstorm Sandy damaged Freedom Tower's USD 10.6 million clean-power fuel cells. Two hundred million gallons of water flooded and destroyed all nine fuel cells." A third of the fuel cells could not be repaired and replaced, partly due to the need to uphold the terms of the billion-dollar contract with Condé Nast, the prominent anchor tenant and a major player in the publishing industry. Towers Three and Four currently have modern fuel cells, but due to financial constraints, the power supply of the main tower has never been replaced with fuel cells.[68]

Hotter Environment

The Freedom Tower is covered with 1368 feet of glass parapet (partly to pay homage to the fallen Twin Towers), giving the building a glassy exterior. The Freedom Tower has a relatively lower temperature than the neighboring infrastructure, significantly lower than structures built with bricks. Glassy skyscrapers are often thought to be more environmentally friendly because the glass reflects sunlight away from the building, thereby reducing cooling and air-conditioning costs. However, surrounding buildings made from bricks become the recipient of the reflected heat, thus making the environment hotter.[69] Experts do not think that the effects of the Freedom Tower on the urban heat island effect are significant. However, John Frederick, a professor of atmospheric science at the University of Chicago, stated the following:

> As we get more shiny, glassy metal buildings, reflected light will become a bigger deal [...] The high reflection is good if you're inside trying to keep intense sunlight out, but if you're outside, you'll be the recipient of that reflected light. However, the effect of one building on another in its vicinity has never been discussed.[70]

Glassy towers can be environmentally friendly when considered in isolation, but may raise environmental concerns in the context of surrounding buildings.

Green Authoritarianism

Environmental Challenges in China

Infrastructure and megaprojects have opened the door for socioeconomic development in China. Although socioeconomic results are indisputable, China's performance in environmentally sustainable development leaves room for improvement. According to the PRC Ministry of Ecology and Environment (PRCMEE), two-thirds of China's lakes suffer chemical deficits caused by pollution. The availability of drinkable water is inadequate in two-thirds of China's cities due to pollution and rising use. The air is severely polluted along the heavily industrialized east coast and the northern heavy industry belt from the Shanxi to Liaoning provinces. Before many polluted industrial sites can be reused, significant soil remediation is necessary.[71]

According to a *New York Times* report, China is responsible for 47% of the world's coal burning, more than all other countries combined. Respiratory illnesses directly linked to air pollution are the leading cause of death in China. The country has some of the worst air pollution in the world and many heavily polluted waterways. According to economists, more than 50% of China's surface water is unfit for human use, and 60% of the groundwater beneath Chinese cities is seriously polluted.[72]

Regarding energy, the amount of electricity produced in China doubles roughly every 10 years. China now produces 18% of the world's electricity, which is less than the United States. According to the World Bank, China continues to rely primarily on nuclear and hydroelectric power to meet its electrical needs (96% combined). The goal of producing 20% of the electricity from renewable sources by 2020 will be more challenging. Despite rapid expansion, wind and solar energy still account for less than 1% of China's total electricity production. The government provides subsidies for either capital investments or operations because solar power is still more expensive to produce than energy from coal-fired turbines, but neither subsidy is sufficient to achieve break-even in the current economic climate.[73]

In addition to renewable power generation, there is a growing market for energy service company (ESCO) projects, which could aid in lowering energy use and GHG emissions. ESCO projects often use anticipated savings on future fuel costs compared with purchasing new, energy-efficient equipment to finance the purchase. Many local investors are cautious because they believe the 5- to 10-year payback period is too long, despite the World Bank and other smaller "green funds" already joining this market. This presents possibilities for international energy-saving businesses to work with local partners.[74]

Regarding water and wastewater, future efforts to increase water sector performance should adopt a more integrated approach. Stormwater, wastewater, and drinking water are the three main components of metropolitan water systems. Several government agencies frequently manage these systems and may have conflicting goals. By utilizing integrated water resource management, it is possible to increase the cost-effectiveness of treatment, elevate the quality of discharged water to safe levels for the environment, and match the water quality to water users. Cutting-edge water treatment technologies with low-energy requirements and straightforward operations such as reverse osmosis and membranes may also impact China's water market.[75]

Transit-oriented development (TOD) is the key to developing Chinese cities. Historically, land-use and density factors have not been systematically considered during design. However, after building subways, many cities changed their zoning laws to encourage development around transit hubs. This allows for larger use of nonmotorized transportation and an appropriate land-use mix. As of 2017, 54 subway lines were operating in at least 13 Chinese cities, totaling 1700 km of track. Current developments will add 76 additional lines (1600 km). By 2020, 40 subway systems, totaling approximately 7000 km of line, were operational. TOD's current trajectory and scope are poised to powerfully impact the long-term sustainability of urban living.[76]

Desertification challenges remain important. Owing to poor agricultural methods, inadequate land management, and dramatic climate change, China is also coping with severe soil erosion and desertification, a type of land degradation in which rich land becomes parched. Desertification and soil erosion both result in blinding dust storms and river-clogging sludge that have wreaked havoc in Chinese cities near the Gobi Desert boundary. Approximately 30% of China's land mass was already affected by desertification. Since 1978, the Chinese have followed the rules of the Three-North Shelter Forest Program, also known as the Great Green Wall, which entails planting over 66 billion trees to block storms in the Gobi Desert. Despite this reforestation effort, the extension of the desert continues to have an impact on several nearby cities.[77]

Schwärzel argues that the Chinese government is compelled to evict impacted residents from lands that have been degraded as towns continue to be buried in sand because of frequent storms. More than 650,000 people who had previously resided in the Chinese province of Inner Mongolia were forced to relocate to neighboring cities between 2003 and 2008. Even more worrying is that these dunes grow at a rate of over 2 miles per year, barely 44 miles away from Beijing. According to Schwärzel, the Chinese government must look into novel and inventive ways to restore natural ecosystems to prevent the capital city from being buried in sand.[78]

Pudong's Waterfronts in Shanghai

Changes in centralized Chinese policy trends have influenced the development of Shanghai. Shanghai has served as a point of entry for Western influence since the late 19th century. However, after Mao Zedong's ascent in 1949, when the communist government in Beijing consciously worked to stifle the city's development, Shanghai was cut off from the outside world. More recently, Shanghai experienced phenomenal growth from 1980 due to Deng Xiaoping's market economy reforms, which helped the city reclaim its influence with Beijing's central government and turn a new page in its history. Owing to its location near the Yangtze River mouth, which flows directly through China, Shanghai's port was able to restore its worldwide standing and has since surpassed Singapore as the busiest container port in the world.

Shanghai has also become one of China's financial centers, competing with Hong Kong. Economic liberalization sparked a massive rural exodus that resulted in the concentration of more than 500 million people in China's cities. The population of Shanghai increased from six million in 1980 to 24 million in 2015.

In the early 2010s, Chinese authorities changed their policy focus from growth alone to sustainable development in response to severe urban pollution and management issues in the nation's cities. Shanghai was tasked with displaying this new "eco-friendly" approach as the host city for the 2010 World's Fair. Beginning in 1993, Shanghai implemented a high-rise urban development strategy to create new business areas in Pudong. As seen by the city's brightly lit, advertisement-covered skyscrapers and energy-intensive infrastructure, Pudong is a prime example of the thorny interplay between China's newly discovered need for energy efficiency and its long-standing pride in its prosperity.

China's rapid industrialization not only worsened Shanghai's pollution but also caused an increase in the number of automobiles as household income increased. Relocating polluting companies was the initial response to their environmental issues. Around Shanghai's Disney, 153 emission-intensive industrial facilities were shut down in the summer of 2016 in preparation for the park's inauguration. Simultaneously, the government initiated the development of eco- industrial parks like the Shanghai Chemical Industry Park (SCIP), approximately 40 km from the city center, and enacted stricter regulations. The Shanghai Metro was an important component of eco-friendly developments, which included a large expansion of public transit. Shanghai has also adopted electric and gas-powered buses in addition to the metro. A high-speed Maglev train, which uses magnetic levitation to move without touching the rails, has also been installed in the city in keeping with its role as a showcase. This train can travel 30 km from Pudong International Airport to the Lujiazui Business District in just seven minutes.

Shanghai's urban development paradigm is shifting toward one that stresses the densification and reuse of existing areas after three decades of rapid urban expansion. The newest master plan for Shanghai has the motto "Striving for an Excellent Global City." As de Hartog has documented, developing industry-specific clusters for creative industries, financial institutions, artificial intelligence (AI), technology, media, and telecommunications is one route to achieving this goal. These clusters are high-density investment initiatives created to promote and hasten Shanghai's transition to a service-based economy. The Shanghai experiments in innovative planning, which include demonstration zones and pilot projects, show many similarities with urban labs in theories of sustainability transitions.[79]

After losing its former port function, waterfronts are once again the primary force behind urban development. The 120 km of Shanghai's waterfront is being transformed to remove polluting enterprises, establish a continuous open public space, form new ecological links, reuse industrial legacy, and construct new landmarks. The new waterfronts of the Huangpu River also reflect state control. According to the Constitution of the Communist Party, three elements must be brought into harmony as we move from an industrial civilization to an "ecological civilization": environmental, economic, and social. General Secretary Xi is adamant about the importance of ecological civilization as a more sensible strategy for economic development.

According to this plan, the river serves three important purposes: as a spatial and functional carrier; as the city's public living room with a rich human connotation (referring to heritage and identity); and as an ecological corridor for a peaceful coexistence between humans and nature, or "ecological civilization." A new equilibrium between top-down and bottom-up governance strategies, as well as an investigation of public-private partnerships and novel kinds of participation, are necessary for sustainable civilization. The plan's final chapter contains numerous references to citizen participation.[80]

In order to revitalize the eastern banks of the Huangpu River, which span 22 kilometers from the Yang Pu Bridge in the inner ring in the north to the Xu Pu Bridge in the outer ring in the south, the Pudong District decided to start a competition. The master plan, which was awarded to Agence Ter in 2016, was founded on legally binding fundamentals to guarantee that the riverbank parks were created consistently while allowing for the unique character of each area. Planning challenges included severe limitations on construction of buildings and commercial use, and the fact that some land belongs to the city, others to the district, and the area directly along the river is also under state jurisdiction.

Footpaths, bike paths, and jogging tracks make up a typical feature on the so-called East Bund (in Pudong) as well as the western side of the river (a total of 45 kilometers of riverfront). They must repeatedly traverse the canals and tributaries that flow into the Huangpu via bridges, and occasionally they run together but separate as needed due to the varying speeds of movement. The master plan called for a planned, partially terraced landscape to cover the flood protection wall, distinct vegetation zones—low toward the river to maintain a clear view, higher toward the land on the opposite side—and, lastly, towers every kilometer that are illuminated at night to highlight the route. Old industries and harbor buildings have been decorated and new parks have been established.[81]

Waterfront Gentrification

With textiles, shipbuilding, pharmaceuticals, soap production, tobacco, machine manufacturing, public utilities, and other industries, Yangpu District's shoreline served as a pioneer for modern Chinese industry. This waterfront is part of the historic American Concession, a foreign settlement established following the Treaty of Nanjing (1842) as a foundation for global trade, along with the North Bund (Hongkou District). Today, we can see the construction of a dozen large thematic real estate clusters, each of which is around the size of the Hudson Yards in New York or Canary Wharf in London. The West Bund has developed its own appealing brand. Various institutions, galleries, and art events were invited to West Bund following the models of the Museum Mile in New York and the South Bank in London. Under the snappy motto "Art & AI as engine," the region assumes a more formal corporate image in an effort to compete for global investment. On the East Bund, recently completed projects include the Pudong Art Museum, the Shanghai Museum East, the Shanghai Library East, and the Pudong Urban Planning and Art Centre.[82]

Similar to green urban megaprojects worldwide, those along the Huangpu River combine private and public funding. However, the demonstration zones and pilot projects along Huangpu provide unique aspects compared to the typical urban planning approach in contemporary China, such as pedestrian and car-free surroundings, functional blending, and the creative use of underground areas. Another innovative feature was the abundance of open spaces and greenery. The idea of a leisure society was still unheard of in Shanghai and China less than ten years ago, and there were virtually no walking or cycling trails. Additionally, China and other countries have never seen pilot programs involving high-end art, AI industries, or the Internet. Leading international architects, such as David Chipperfield Architects, Foster + Partners, Heatherwick Studio, Kengo Kuma, KPF, OMA, Sanaa, and Sou Fujimoto, as well as many up-and-coming local talents, such as Atelier Deshaus and OPEN Architecture, have designed the majority of buildings and public spaces.

The well-known Shanghai Tower is a prime example (Fig. 6.5). The 632-meter tower was inaugurated in September 2015 and is the third-tallest structure in the world, providing a representation of both urban gigantism and ecological innovation. With one layer of glass covering the offices and retail areas and a second twisting layer forming the building's exterior, the skyscraper has a "double-skin" façade. The building is advertised as sustainable (it has a LEED Platinum rating); however, more than one-third of its floors have been vacant since 2016. Landscaped atriums serve as thermal buffers, drawing on warm air in the winter and cool air during the summer provided by the central air- conditioning system. The structure is capped with a spiral-shaped parapet that collects rainwater for use in an air-conditioning system and has dozens of vertically aligned wind turbines at 500 m or higher.[83]

The new waterfronts primarily serve the needs of tourists and the new white-collar (upper) middle class.[84] Perhaps unsurprisingly, the obvious tensions and contradictions in the green megaprojects along the Huangpu River between (1) supporting the desired image of a global metropolis and boosting the economy through investment and tourism and (2) the explicitly stated objective of fostering

Fig. 6.5 Pudong skyscrapers (foreground), the West Bund (top right), and the Huangpu River seen from the observation deck at Shanghai Tower, showing a portion of the riverbanks where cultural and green gentrification have become widespread after innovative regeneration and reuse. Source: iStock.com

an "ecological civilization" are never addressed by Shanghai planners, developers, and politicians.

China is using its current period of prosperity to invest in the future by building more square meters of office spaces, museum spaces, and other facilities. This has been accomplished after decades of socioeconomic deterioration. However, replacing inexpensive working-class neighborhoods with upscale (and mostly empty) real estate is not inclusive. Although the physical borders of these areas are permeable, adjacent areas are gentrified, and new socioeconomic restrictions are imposed. These initiatives emphasize pedestrian traffic, in addition to eliminating polluting enterprises and replacing them with low-carbon structures. However, only 50% of real estate clusters are close to metro stops. Every building complies with the National Green Building Standard; many even earn international certifications such as LEED. Although a project's primary objective is economic, it also aims to enhance a community's reputation and quality of life. Environmental issues are considered; however, they have not yet been given priority.[85] Affordable housing considerations have been overlooked, and issues related to connectivity with surrounding neighborhoods remain a major problem.

The initiatives in Shanghai being discussed are characterized by extensive functional programming without connection to its environment. They are constructed quickly without a defined timeframe or plans for (re)use. In fact, the initiatives are sanctuaries of "neoliberalism with Chinese characteristics" (Harvey, 2005) where disruptive innovation occurs. The government supports innovation clusters as sources of transformation and incubators for emerging industries. However, because of the need to catch up, socioeconomic experimentation in China frequently resembles improvising or creating something rapidly and then making changes while it is being implemented.

When it comes to innovation, urban planning and design typically include gathering best practices for inspiration and then enhancing or modifying them as necessary. In Shanghai's governance procedures (as is presumably the case elsewhere in China), there is a complicated interaction between several groups or departments, as well as between top-down and bottom-up authorities. However, they all function as a single body with the authority to veto nongovernmental stakeholders. All urban land is under the municipal government's control and ownership. Consequently, the projects mentioned do not have the same open- ended goals as urban labs in the international (mostly European-centered) debate. System innovation is a managed and controlled process in Shanghai, China. This means that end-user feedback and other social learning variables are typically disregarded.[86]

Riyadh's Quality-of-Life Megaprojects

Rapid economic development in the past ten years has led to physical changes in Gulf capitals and cities. Dubai took the lead in constructing urban megaprojects that changed the skyline of what was once a fishing village. Other Gulf cities, including Abu Dhabi, Doha, and more recently Riyadh, have followed Dubai's lead and implemented their own urban megaprojects. Cities, particularly those in developing nations, use megaprojects as change agents to restructure their metropolitan areas in accordance with standardized livability indices and improve how they are perceived in the international marketplace.

These initiatives have publicly stated goals of luring international retirees, professional elites, and tourists, as well as expanding the domestic market while promoting economic diversification toward a more knowledge-based economy.[87] Narratives of sustainable development are now being incorporated into the national policy objectives of the Arab Gulf countries.

Governments in the Gulf are vying with one another to build the "greenest," "eco-friendliest," or "most sustainable" city in the area, which has resulted in the emergence of multiple megaprojects and opulent master plans that have all been carefully marketed as sustainable.

Saudi Arabia's government program Vision 2030, launched in 2016, aims to diversify the nation's economy, society, and culture. This effort includes numerous massive projects, which we will briefly describe here before focusing on Riyadh. The NEOM ecosystem, which is based on technology, innovation, and sustainability, seeks to become a global benchmark for tourism and logistics. It is situated in the province of Tabuk, in the northwest of the nation. It is anticipated to cost USD 500 million and extend approximately 460 km. It was first announced in 2017, with an initial plan to open a portion in 2025. The concept outline began in January 2021 with the smart-city called Line. It is 170 km long, stretching from the Red Sea's coastlines to the valleys in the country's northwest, and has a population cap of one million people.

Transportation systems, base services, and infrastructure are currently under development. The complex's 5G network was installed by the telecom provider STC. The company has also constructed an innovation hub. NEOM has moved its headquarters from Riyadh to the project's location in the Tabuk region, where there are now residences, offices, and supporting structures.[88]

On the outskirts of Riyadh, there is a huge entertainment complex called Qiddiya, which spans 334 km^2. It will have water and amusement parks, including Six Flags, as well as places to support arts and culture, such as pavilions, concert halls, theaters, and movie theaters. A complex with a stadium, golf course, aquatic facility, schools, and other facilities aims to encourage sports. Additionally, it features a track, go-kart, and rally driving experience. Because of the building of an equestrian center and a safari park, one part will be dedicated to nature and the environment, and the other will be used to organize exploration and adventure activities.

The Red Sea Project, another megaproject, involves the creation of more than 90 islands between the villages of Umluj and Al Wajh. The primary center, Shurayrah, has 11 resorts: an airport, boat marinas, and recreational amenities will be added to its five islands, two coastline complexes, and 16 hotels, with a combined capacity of 3000 rooms. Completion of the project is anticipated by 2030, with an additional 8000 rooms. Another megaproject, Amaala, is a Red Sea luxury tourist development situated between the NEOM and Red Sea projects on the northwest coast of Saudi Arabia. It has three zones: Coastal Development (culture center, golf course, seven hotels, 200 villas, and condominiums), Triple Bay (16 hotels with 1325 beds, 435 villas), and The Island (artistic center, botanical garden, 200 villas and condominiums, seven hotels with 400 rooms).

Additionally, the Al Ula megaproject aims to create five thematic zones connected to nearby historical sites of interest: Old Town, Ancient City of Dadan, Jabal Ikmah, Nabataean Horizon, and the Ancient City of Hegra, each of which has up to

15 different attractions, including galleries, museums, gardens, markets, and research institutes such as the Kingdom Institute. By 2035, it hopes to attract two million visitors annually, create 38,000 new jobs, and contribute USD 32 billion to Saudi GDP.[89] Although commercial redevelopment has produced many positive results, the adaptive reuse of the existing urban fabric and the preservation of vacant or underused buildings and their diverse communities, milieus, and activities—many of which continue enduring traditions—are not considered.[90]

In 2019, King Salman bin Abdulaziz of Saudi Arabia announced the establishment of four significant projects in Riyadh to address Saudi Arabia's environmental deficit, with a combined budget of USD 23,000 million.

According to Vision 2030, these four projects seek to enhance Saudis' quality of life by giving them access to green spaces and encouraging them to engage in the city's cultural and sporting activities.

- *King Salman Park* promises to be the largest urban park in the world at 13.4 km^2, constructed on an abandoned airfield (Fig. 6.6). When completed (2026) it will be four times as large as Central Park in New York and five times as large as London's Hyde Park. The park contains several amenities, including a national theater, an opera house, and an art academy, in addition to its enormous green spaces and various leisure, environmental, and cultural activities.
- As part of *Green Riyadh*, one of the largest urban planning initiatives in the world, 7.5 million trees will be planted throughout the capital's parks, gardens, mosques, schools, and health facilities. New irrigation networks will be created utilizing recycled water, and tree species will be planted that can thrive in Riyadh's arid environment, requiring little water.
- *Riyadh Art*: There will be 1000 pieces of art that will be visible across Riyadh's streets, parks, and transportation hubs.
- The region's first of its kind, *Sports Boulevard,* will have a 135-km multisport circuit for jogging, cycling, and horseback riding. Instead of prioritizing vehicles, neglecting those who prefer walking, cycling, or running, the concept seeks to create a "human place that will bring back human life to Riyadh."[91]

As Klingmann has shown, despite each project focusing on one or more distinct lifestyle domains, all projects together meet the requirements for each lifestyle category listed in Saudi Arabia's Quality of Life Program: design, environment, infrastructure and transport, social involvement, and safety. Every project also provides a variety of economic and educational opportunities for various populations, as well as steps to increase livability in urban areas.

Regarding the criteria for sustainable development, Klingmann's study showed that all case studies offer sufficient steps to improve Riyadh's mobility by constructing greenways for bicyclists and pedestrians that connect to public transportation. Additionally, the projects showed the potential to create urban synergies between various lifestyle domains when strategically combined into a series of urban layers. This potential is especially apparent when extended through an inclusive participatory planning framework, which in turn could significantly improve the quality of life for a wide socioeconomic demographic.[92]

Fig. 6.6 A rendering of King Salman Park in Riyadh, Saudi Arabia. The Park is one of the four Riyadh Vision 2030 megaprojects. Source: MrWallpaper.com

On the planning front, establishing a sizable public infrastructure that promotes walkability and inhabitants' active involvement in leisure, cultural, entertainment, and sports activities is part of creating a new comprehensive lifestyle setup for residents of Riyadh. However, Klingmann's analysis has also shown that, while the projects include a significant number of public facilities, they do not adhere to a mixed-income project model, which would enable low-income families to live in a market rate setting. Additionally, one initiative called for the relocation of migrants and low-income groups. Although the government has a separate initiative to provide affordable housing in other parts of the city, these model destinations mainly cater to wealthy Saudis and luxury visitors, which could further exacerbate the socio-spatial divisions already present in the city. The megaprojects thus show Saudi Arabia's conflicted response to entrepreneurial neoliberalism: on the one hand, progressive attempts to promote an egalitarian approach to urban livability; on the other hand, strategic efforts to use megaprojects as spectacular showcases in the international marketplace.[93]

Yasser Elsheshtawy, an expert on Arab cities and the author of a recent book on Riyadh, believes that the capital of Saudi Arabia can provide a distinctive urban model that contrasts with the more speculative trends prevalent in the area. The emphasis on quality-of-life issues in announcements signals a paradigm shift in urban planning in the Gulf region. Projects have typically tended to concentrate on exceptional features, emphasizing branding and the extent to which they would attract investment, thus fostering real estate speculation and commercial development. More recently, the emphasis is clearly on people, the provision of green spaces, and locations for exercise and enjoyment. The paradigm shift is not only noticeable in connection with the larger Gulf Arab setting, but it also signals a change in planning strategy for Riyadh, which was based on car traffic and real estate speculation, inevitably leading to the expansion of the city and urban sprawl.[94]

Elsheshtawy argues that the fact that it is difficult to navigate Riyadh's current built shape on foot is one of its remarkable features. The humanization program attempted to reduce this by constructing walking routes and redesigning the streets to accommodate people. The success of these municipal initiatives demonstrates an environment that is more open to change, and residents of Riyadh are eager to interact in public areas and participate in a variety of outdoor leisure activities. This is evident in the removal of social barriers to mixed-gender activities, which is one of the reforms associated with Vision 2030. By removing service roads and limiting the number of car lanes, traffic patterns can be significantly altered to focus less on cars, encouraging more people to use public transportation. Development will include a mix of residential and office spaces in addition to these open venues and public areas, providing possibilities for private investment to make the project financially self-sufficient. The King Salman Park and Sports Boulevard projects stand out for their extreme scope and ambition. If Riyadh can give in to the pull of Dubaization, it will demonstrate its capacity to present a distinct urban model that can counterweigh the more speculative trends prominent in the Gulf area. Modern urban development places people first, and the extent to which Riyadh is doing so may serve as an example for the rest of the region.[95]

Green Capitalism

Cities play a significant role in climate change because they produce substantial amounts of greenhouse gases from human activities. According to estimates, cities produce 75% of the world's CO2 emissions, with transportation and construction comprising two of the largest sources. Simultaneously, urban living has been significantly affected by climate change. Increasing sea levels, the frequency of extreme weather events such as floods, droughts, and storms, and the spread of tropical illnesses are all effects of rising global temperatures. All of these have expensive effects on essential services, housing, livelihoods, and the health of those living in cities.[96] The 14 coastal cities among the top 17 largest in the world are particularly susceptible to climate-related extreme weather conditions. In order to prevent and lessen catastrophic health impacts, the world's most vulnerable cities require improved preparedness and mitigation measures, just like they would after an earthquake or hurricane. Early warning systems, infrastructure mitigation, education, and heat wave management strategies can help communities get ready for urban heat islands and the effects they bring. It is critical to guarantee the safety of city citizens, particularly the most vulnerable, such as the elderly, the impoverished, and people living in risky areas.[97]

The fight to lessen and mitigate the effects of climate change places cities at the forefront of conflict. The type of urban expansion within cities is also a key factor determining the vulnerability of urban residents to environmental stress. The effects of urban growth on the environment extend well beyond metropolitan regions. Agriculture expands to new places, rapidly intensifying undeveloped land, urbanizing areas, and placing pressure on available land resources. Urbanization also has an impact on climate. Approximately 5% of total emissions from tropical deforestation and land-use changes is expected to come directly from areas with a high likelihood of urban growth. There is still much to learn regarding the size and extent of these effects. The impact of urbanization on CO2 emissions and heat budgets has been the subject of several studies, and new research has documented how urbanization affects the movement of water, aerosols, and nitrogen in climate systems.[98]

One of the main areas to work on is building construction and operation regulations as well as planning rules that support sustainable transportation. Less than 1% of building projects currently consider the lifecycle carbon implications. Research indicates that the industry must reduce emissions by 50% by 2030 to attain net-zero by 2050, as the built environment accounts for about 40% of global emissions. Because the construction sector typically focuses on lowering operational energy usage, most buildings do not undertake carbon footprint assessments throughout their life cycle. Embodied building sources, such as the production of materials and construction methods, are responsible for up to 50% of their lifecycle carbon emissions.[99] A city equivalent to the size of Paris is anticipated to develop each week for the next 40 years. Given that the manufacture of concrete accounts for 8% of all global emissions, our approach to designing and constructing new buildings must shift drastically to meet our climate targets. A reduction in the "whole-life carbon" impact of new developments is made possible by advances in design, cutting-edge materials, modern construction methods, and increased use of technology. The use of digital twins to control energy consumption in buildings is promising. The ability of digital twins to diagnose and correct such inefficiency significantly contributes to reducing emissions, given that typical buildings use 3.8 times more energy than is intended for them.

Enhancing the efficiency of the vast array of existing structures poses a significant challenge. The practical obstacles to modernizing facilities without interfering with corporate operations, as well as the more nebulous problem of cost allocation and financial feasibility, are presented with retrofitting occupied buildings. Early adopters are setting examples for others to follow in both public and private sectors. However, government regulations of building standards in the U.K. and EU have already hastened obsolescence to tip the scales of cost-benefit analysis. This is likely to be followed by other jurisdictions. Although real estate is geographically fixed, every area on Earth is subject to a distinct set of legal and environmental concerns. Parts of the world and segments of societies that are least equipped to defend themselves or recover from climate disasters are affected disproportionately by climate change. Millions of people face an increasing wave of social and economic effects due to climate change, ranging from the loss of housing or livelihood to the loss of life. Climate change is more than an environmental catastrophe; it is a societal crisis, as the World Bank has eloquently stated.

From the perspective of megaprojects, debate on the preferred modes of governance (liberal or authoritarian) regarding environmentalism is largely irrelevant, although there are sharp differences between the two. A major example stemming from authoritarianism is labor rights and abuses for workers, as seen in Abu Dhabi. This issue may become more prominent as authoritarian governments increasingly invest in megaprojects. Authoritarian environmentalism describes a highly centralized system of environmental governance in which environmental policies are designed and promoted by a small number of elite bureaucratic institutions, frequently at the central level.

Shanghai and Riyadh are examples of megaprojects in authoritarian countries. On the other hand, environmental policies in liberal democracies are characterized, as we know, by more or less inclusive processes of public decision-making, where various stakeholders often develop conflicting strategies and tactics, and where obstacles must be removed to bring about transformative change.

However, as we have seen in this chapter, both modes of governance converge around the goal of global competitiveness, for which megaprojects are used as an urban "sustainable" strategy. Both modes of governance function as varieties of "green capitalism." Sustainability appears to depend on the dimensions of entrepreneurship, innovation, and competitiveness of economies. Indeed, the goal presented to us in the majority view on sustainability is "green capitalism," that is, not a sustainable global society with clearly established limits to growth but rather the sustainability of the information and knowledge economy, to which reformers and planners add more or less ambitious commitments to the environment.[100]

As countries and cities transition from urban property investment and finances to science, knowledge, and innodistricts, the limitations of many sustainability strategies persist. Therefore, one of the key questions regarding "sustainable" megaprojects is whether and to what extent a capitalist framework pursuant of competitive development (be it liberal or authoritarian) is an effective strategy for tackling environmental challenges and enabling meaningful sustainability practices.

Environmental quality is becoming crucial to ensure sustainable economic growth as cities and nations move up the industrial value chain and broaden their service industries to meet the rising domestic demand. Residents of cities in more affluent economies are already paying significantly more for an excellent urban environment (i.e., ecological or "sustainable").

Cities should step up their game further if they want to draw in the right labor pool.[101] Thus, in global cities such as New York, London, and Sydney, certain strategies aimed at sustainability can have the perverse effect of gentrification, which starkly reveals the problems of anthropocentric sustainability.[102] From this perspective, the planet continues to be considered exclusively a resource for human use.[103] We return to this question in Chapter 8.

Toward Complex Sustainability

As we have seen in our analysis of megaprojects, these megastructures have significant environmental impacts, and thus attaining acceptable levels of environmental sustainability needs to become a priority for planners, developers, and other stakeholders. However, attaining environmental sustainability does not ensure megaproject sustainability—a goal that must be pursued holistically. One approach is to use the notion of "key or multiple success factors."[104] This notion is not new in the field of project management and is a highly debated topic among specialists. It is increasingly important "to evaluate projects and their impacts at different times based on multiple criteria to fully evaluate their performance. Success is often driven by political- and power-related factors."[105] Due to the strong political nature of stakeholders throughout the supply chain and their different underlying objectives, success factors are often deemed inadequate. This configuration requires innovative governance solutions that align stakeholders' interests in a complex environment with a large number of key actors.[106]

Following the notion of "multiple success factors," we contend that several requirements should be met to achieve sustainable megaprojects: environmental sustainability (sustainable infrastructure delivery and sustainable development zones), sustainability in design and planning, sustainability in megaproject management, institutional sustainability, and socioeconomic sustainability. We consider sustainability as an organic process, including the environment, economy, and community: form and efficiency (environmental factors in design, architecture, engineering, and construction) as well as policy (urban plans and practices that explicitly aim to maintain and improve the social and economic well-being of citizens).

Sustainable Infrastructure Delivery

Cities can raise the bar for urban infrastructure delivery to safeguard the natural environment and open doors to the next phase of socioeconomic development. Environmental quality will become crucial to ensuring sustainable economic growth as cities and nations move up the industrial value chain and broaden their service industries to meet rising domestic demand. Urban dwellers in affluent markets are willing to pay significantly more in high-quality urban settings. Cities must intensify their efforts to attract the ideal workforce.[107] The backbone of the next phase of infrastructure development should follow a "one-system" philosophy. Infrastructure planners must consider the growth of a complete city's infrastructure system, including its subsystems for energy, transportation, land use, and water use. It will take technology for real-time information, conservation pricing, and demand management to realize the potential synergies between subsystems.[108] Foreign infrastructure providers and seasoned experts should be poised to significantly contribute to achieving higher infrastructure performance standards in the coming years. Government funding and stimulus programs are essential. Central government can strengthen its commitment to environmental sustainability by maintaining ambitious goals for resource conservation and economic productivity, while still providing financing for infrastructure projects.[109]

Sustainable Development Zones

In early 2019, the Chinese government approved three sustainable development zones—Shenzhen, Guilin, and Taiyuan—which formed the leading axis in Chinese innovation.[110] These zones are implementing the 2030 United Nations Sustainable Development Goals. Shenzhen is China's innovation engine: it incorporates technologies for sewage treatment, waste utilization, ecological restoration, and artificial intelligence to address problems ranging from resource management to pollution. Guilin concentrates on desertification-related technologies and develop remedies for areas at risk from advancing deserts. Taiyuan promotes creative solutions to air and water pollution that other resource-extraction-dependent regions may adopt.[111] Shenzhen, Guilin, and Taiyuan function as large sustainable innovation districts (or innodistricts). In newly developed urban areas, achieving sustainability usually implies achieving environmental sustainability. This goal by itself, however, does not guarantee the sustainability of megaprojects and territories in the knowledge economy.[112]

Sustainable Planning and Design

Megaproject planning should be oriented toward socially progressive goals and ensure sustainable development rather than focusing exclusively on growth and competitiveness. Megaproject financial planning must avoid strategic misrepresentation of costs and benefits and include clauses for equitable community benefit distribution. In the megaproject design process, contextual elements such as history and local culture are important factors in interpreting the architecture and assigning a specific meaning (local, regional, national, or global) to the architectural practices used to build megaprojects and make them visible.[113]

Sustainable Management

Planners and managers of megaprojects tend to see their projects as unique. This "exclusivity bias" prevents them from learning from other projects. An early "block" or "capture" caused by an overwhelming commitment to a specific project concept prevents the exploration of alternatives and forces ad hoc commitments. The completion and delivery of the megaproject is a high-risk stochastic activity with disproportionate exposure to "black swans or extraordinary occurrences that have profound detrimental effects." Instead of handling projects as though they live in a deterministic cause-and-effect Newtonian world, managers should bear this reality in mind. Complexity and unanticipated events must be considered; they must be effectively addressed by budgets and time contingencies.[114]

Institutional Sustainability

When studying megaprojects in connection with urban growth coalitions, we can ask ourselves if the form of development represented by these large projects simply legitimizes growth machines and commercial interests or if this phenomenon can be analyzed from the perspective of the role that actors and state agencies play in urban restructuring. Besides growth coalitions, the governance of megaprojects needs to consider additional stakeholders to ensure institutional sustainability, as follows: (1) the role of civil society; (2) the role of local context, history, and culture; (3) the importance of maintaining urban variety and diversity; and (4) the importance of megaproject local embeddedness, (5) the role of public spaces, and (6) the role of urban design professionals.

Socioeconomic Sustainability

The embeddedness of multiple scales of socioeconomic action has paradoxically emerged with a tendency toward providing megaprojects with their own strategic spatial planning tools that function independently of state and urban regulations. Thus, aligning the megaprojects' goals with regional and national policies is urgently needed. New institutional arrangements between administrative levels vital for implementing strategic policies must emerge because separating the objectives of megaprojects from regional and national policy goals yields negative results. Therefore, megaprojects, which usually function as catalysts for urban development, should be used at the national level as tools for advancing sustainability policies. Accordingly, complex sustainability is an organic process that includes the environment, economy, and community; form and efficiency (environmental factors in design, architecture, engineering, and construction); and policies (plans and urban practices that aim explicitly to maintain and improve the social and economic well-being of citizens). Thus, a development project can be defined as sustainable if it is planned and implemented to consider the capacity, adaptability, resilience, diversity, and balance of the ecosystem in which it is located. This constitutes a symbiotic element.[115] The result can be an optimization of sustainable policy outcomes owing to synergistic and multiplier effects. From this perspective, the key question for future research is not how to plan and build urban megaprojects that simultaneously foster both sustainability and competitiveness; rather, urban growth machines must change so that cities are more likely to build megaprojects with public value.

Conclusion

This chapter presented the evidence of the impacts of some megaprojects on the surrounding environment. We analyzed some strategies to develop allegedly "sustainable" megaprojects to counter environmental damage. Owing to the considerable environmental effects of megaprojects, planners, developers, and other stakeholders are prioritizing the achievement of acceptable levels of environmental sustainability. However, achieving environmental sustainability does not guarantee megaproject sustainability, because this objective must be approached holistically. Megaprojects often follow a sustainability logic based on strategic disinvesting in issues that appear secondary to the competitiveness of cities, while investing in issues crucial for forming urban regions as strategic scales for capital accumulation. This logic neglects to address the socioeconomic sustainability issues. We proposed some essential elements to understand megaproject sustainability as a complex and multidimensional concept. Research on complex sustainability related to megaprojects aims to explore the actual significance and effects of decision-making when there are overlapping webs of interdependence between different groups of stakeholders (political parties, urban officials, technical experts, advocates, and civil society groups). Research has focused on strategies and techniques that have been successful in enacting transformative sustainable change, such as coalition-building, issue framing, benefit-sharing, shifting the location of decision-making, disaggregation, privatization, and public-private partnerships.[116]

We must pay attention to the complexity that can be observed in shifting interactor networks of decision-making; the development of preferences, perceptions, coalition-building activities, and choices by different stakeholder groups; and the structural and contextual contexts (regional and urban factors) in which they operate that have an impact on the outcomes. This is a networked and embedded agency approach. It effectively depicts sets of intersecting transactions between actors and groups of actors in particular settings where emergent processes and outcomes play out. It also explores relationships and co-variations between processes and outcomes, and longer-term institutional capacities (relationships, commitments, mandates, and responsibilities among actors). This network-based relational approach is strengthened by considering contextual and structural factors (such as how problems are framed, time horizons, spatial scales, boundaries, components, drivers, what is known, what is uncertain, what might change, and what outcomes emerge from processes) and proposing pertinent comparisons of appropriately grouped sets of cases.

We shall devote Chapter 8 to the question of complex sustainability. Before undertaking this task, we explore, in Chapter 7, the notion of disruptive complexity as an analytical inference from the focus on megaproject impacts in the previous chapters.

Part III
Inferences and Conjectures

Chapter 7
Disruptive Impacts

Introduction

Megaprojects represent a socio-spatial manifestation of disorderly, disruptive, and contentious complexity in urban development and socioeconomic planning, as found in Chapters 3 to 6. As complex and disruptive urban innovations, megaprojects require a relational analytical perspective that describes interrelations and interdependencies, networks, assemblages, and flows, and avoids reductionist perspectives. This complexity approach, introduced in Chapter 2, is advantageous for understanding the nature and implications of megaprojects as situated (embedded) within multiscalar political, economic, and sociocultural systems and contexts. As an *inferences* chapter, Chapter 7 summarizes the evidence in Chapters 3–6 and then fleshes out the notion of *disruptive complexity*: the complex web of interrelated and interdependent impacts that emerge after megaproject construction and affect the urban living of many citizens, particularly those more exposed and vulnerable. This Chapter emphasizes the megaproject impacts imperative: the importance of identifying impacts as a prerequisite for critically assessing whether megaprojects should be built.

When discussing disruptive impacts we should mention the notion of risk. *Risk* can be defined as the probability of an impact. *Impacts* can be defined as probable consequences or materializations of risks. Some materialized risks (impacts) are never identified in the risk phase and are not included in any sustainability planning. Thus, uncertainty—the reason for conceiving a risk—is also present when risks materialize into impacts. This is expected because the urban impacts of megaprojects are a function of megaproject embeddedness in the polity, economy, and society. In Chapter 2, we saw how embeddedness begets complexity, which in turn, is inherent to uncertainty. Therefore, impacts can sometimes, but not always, be predicted by examining the risks. In other words, megaproject impacts are *emergent properties* that cannot always be predicted in advance in any meaningful or

G. del Cerro Santamaría, *Contesting Megaprojects*,
https://doi.org/10.1007/978-3-031-98382-5_7

comprehensive manner. The uncertainty of emergent impacts manifests in three main ways: (i) impacts cannot always be predicted, (ii) impacts evolve over time, and (iii) impacts are interrelated or interdependent. Building megaprojects entails a double manifestation of uncertainty, which increases their disruptive potential: the uncertainty involved in megaproject risks and the uncertainty involved in megaproject impacts.

Many cities (not only in developing countries) are caught in a perfect storm of population growth, escalating adaptation needs, substantial development deficits created by a shortage of human and financial resources, increasing levels of informality, poor governance, environmental degradation, biodiversity loss, poverty, and growing inequality. In most cases, building megaprojects has not improved this situation. In fact, megaproject disruptive complexity has contributed to fostering and worsening it. Abundant evidence exists of cost overruns, negative environmental impacts, gentrification risks (including high tech, green and tourism gentrification), accentuation of socioeconomic inequality, drawbacks of top-down cultural engineering, neglect of local cultural identities, uncertain economic success, population displacements, spoiling of cities´ visual and structural uniqueness, rise of congestion and overpopulation, political gridlock, threats to ecosystems, and others.[1]

As we know, costs and benefits from projects are differentially distributed. Typically, vulnerable and underrepresented groups and lower classes experience a disproportionate share of the considerable negative impacts of megaprojects.[2] This includes the potential human rights infringements created by megaprojects.[3]

We characterize megaproject impacts as disruptive complexities by using the notion of *entropic impacts* (defined as disruptive, negative impacts). What we find is that disruptive impacts occur even in ideal situations where the best combination of planning and decision-making strategies has been deployed. This is a widespread pattern in a context where worldwide urban planning practices and cultures have become more complex since the mid-1990s due to the rapid expansion of information and communication technologies.[4] Three interrelated developments further support adopting complexity research strategies in urbanism: social learning, collective intelligence, and the convergence of epistemic cultures.[5] Complexity and relationalism lead us to transdisciplinarity, understood as both a process of innovation in knowledge creation and application and a procedure of integrating apparently divergent perspectives into a "rhizomatic research culture."[6]

Disruptive Impacts: Interpreting the Evidence

The question of impacts can be approached from the notion of *unanticipated consequences*, a term popularized by Robert K. Merton and crucial to understanding the impacts and consequences of human action.[7] Most environmental problems, from chemical pollution to global warming, are unanticipated or unintended consequences of the application of modern technologies. This does not imply dismissing the notion of environmental justice and the concept of "dumping"—particularly on underrepresented communities. It rather points to the fact that impacts can be considered emergent properties in a complex system. Merton classified unintended consequences into three groups: unexpected benefits (serendipity or windfall), unexpected drawbacks (detrimental to desired effects), and perverse results (contrary to intentions). However, a more relevant classification for our purposes is a twofold complexity, both disruptive and disorderly, generating perverse results. Megaproject complexity triggers a substantial degree of disruptive capacity in various respects and dimensions.[8]

Disruption and disorder are the perverse results of megaprojects. The causes vary from case to case: autocratic rule [e.g., Istanbul's urban megaprojects (UMPs)], erosion of intent (e.g., Sydney's Barangaroo), or misrepresentation of targets (e.g., Hong Kong's West Kowloon). Better management of megaprojects, based on avoiding human cognitive bias among planners, developers, and managers (as proposed by project management authors), could improve megaproject performance. However, such a strategy is unlikely to cancel the disruptive and disorderly impacts that emerge after megaproject implementation. Any application of optimal rationality models that disregard contextual conditions as well as potential ecological, socioeconomic, and equity impacts would not prevent a megaproject from negatively disrupting affected communities in major ways, particularly when communities and citizens are excluded from key stakeholder deliberations during the planning process.

Recent megaprojects are marketed as "sustainable" under the burgeoning "green capitalism" regime. Nevertheless, it is still unknown under what circumstances these allegedly sustainable megaprojects may promote fairness, livability, and shared wealth while minimizing environmental harm. Evidence demonstrates that they merely symbolize the lofty ideals of political and economic leaders in their attempt to create a globally competitive society. Will megaprojects bring economic benefits not only to the entrenched interests of megaproject players but also to the host cities and their residents? According to mounting research, the benefits might be limited, as with the Olympic mega-events.

New paradigms, such as "smart cities," have gained prominence in discussions about urban revitalization strategies in the West. This is due to growing skepticism about spectacular and large-scale urban development, as well as the numerous effects of the Great Recession of 2008 and the coronavirus pandemic. While advocates of the "smart-city" concept correctly emphasize the importance of digitalization and technology for managing urban areas, they seem to overlook that improving governance models and practices is just as crucial as technological tools.

Disruptive Landscapes

The evidence in three disparate cities (Istanbul, Sydney, and Hong Kong) presented and discussed in Chapter 3 demonstrates that megaprojects are disorderly and contentious regardless of context and location. They have significant economic, sociopolitical, and environmental impacts that are often difficult to resolve for the benefit of urban communities. These impacts are incompatible with any definition of sustainability and work against any sustainable conception of the urban. This is especially true when multiple simultaneous effects have cumulative and cascading consequences on the affected urban regions and populations.

Istanbul's megaprojects faced local opposition, political clashes, and the realignment of political alternatives to end up with an ostensibly eco-friendly Vision 2023. However, the megaprojects still caused significant environmental damage. Similarly, Kanal Istanbul, aiming to control Bosphorus Strait maritime traffic, is under construction with substantial financial and environmental costs. In Sydney, Barangaroo development strives to strengthen its global position, but political controversies focus on citizens' rights and potential urban inequality. Additionally, Hong Kong's WKCD, initially promoted as an art "hot spot," in reality favors wealthy developers over local artists through top-down governance.

Our investigation into how UMPs affect Istanbul, Sydney, and Hong Kong clearly reveals that bioeconomy and biodiversity could be at risk. Enhancing urban resilience and livability would require pursuing the following goals: (1) mediate between the interests of the public and private sectors; (2) encourage greater local and regional government involvement in projects carried out in their communities; (3) actively involve citizens in consultation and direct participation before, during, and after UMP development; (4) maintain local heritage as a key element of any UMP agreement; and (5) protect the environment across air, water, and land. These are the primary issues facing urban decision-makers and leaders committed to promoting the economy, environment, and equity in urban governance. These disruptions exemplify the dominance of diverse political and economic interests at the expense of urban dwellers. Unsurprisingly, the global adoption of the megaproject strategy by urban elites in recent decades has prompted the study of controversies in megaproject development. Our analysis identifies key obstacles for cities seeking global exposure and investment and aims to signpost the way to more sustainable urban policy.

Impacts of Emblems

Centrality and Visual Power

Megaprojects follow an agglomeration logic in urban and regional development characterized by both the concentration (clustering) and dispersion of economic activities, whereby a central location can be interpreted as the spatial counterpart of economic and political power.[9] According to the ideology of growth that has shaped urban fortunes during the neoliberal era, centrality is typically assessed by considering flows and links in the global city network. In this context, it is relevant to mention the notion of "iconicity" and its "visual power." Iconicity has become a crucial idea for comprehending the dynamics of urban megaprojects and how places are created and transformed in the age of globalization. We view "iconicity" as a mechanism by which cities rise in the global hierarchies of centrality. Both the abstract and physical realms of globalization have placed emphasis on becoming visible. Image transformation for a city can have numerous beneficial externalities, and iconic buildings and emblematic megaprojects are typically excellent strategies to achieve this. One of the tools of choice for an urban growth machine when attempting to overcome structural barriers that hinder a city from "globalizing" or "becoming global" is the visual power of iconicity.[10] This is how megaprojects act as growth catalysts, even if they do not usually function as primary sources of growth.

In addition to helping cities globalize, iconic buildings are a powerful tool for transmitting consumerist values and practices that sustain capitalist globalization. If iconicity is a visual manifestation of power, it follows that the drivers of iconic megaprojects will seek ways to establish both hegemony and legitimacy. UMPs are used to establish a "hegemony of vision" that secures legitimacy for the new coalitions of their public and private stakeholders through the spectacularization of development. This frequently removes the focus from the substantive, on-the-ground transformations of the urban-regional socioeconomic fabric, including the drawbacks.

On the one hand, UMP architecture needs to respond to the urban environment and find a method to stand out from nearby structures and locale-specific architectural styles and materials. On the other hand, architects, planners, and politicians in charge of project design must solve the dilemma of how to stand out compared with global role models and similar projects in rival cities.

The argument over the iconicity of UMP architecture is heavily focused on the concepts of place specificity versus global uniformity, and its connections to the development of urbanity. However, the local environment, its planning history, trajectories of path dependency, and the political process heavily influence which architectural features are regarded as place-specific or global outstanding.

Exposure and Vulnerability

Centrality and the visual power of iconicity lead to exposure to global networks and local vulnerability. The need for exposure, recognition, and differentiation is driven by cities' expansion and a variety of resources and roles, as well as by an intense rivalry. Globalizing cities need to improve their business climate, talent pool, and connections to compete in global markets. City leaders typically prioritize the development of global identity. Standard urban policies in digital times dictate that a city may stand out from its crowd and make its local assets more noticeable by developing a strong global brand. Megaprojects thus become "digital monuments."[11] Urban global identities are not expected to be static, as they develop cyclically over time. A metropolis' flaws, shortcomings, and frictions are extensively reported once they become well known. A metro area with a distinct global identity must manage its reputation more actively and prepare for opportunities and problems. Developing a global identity is hardly the result of chance or accident but requires deliberate policies establishing a worldwide identity anchored in urban emblems. An urban global identity may become an even more crucial instrument for positioning cities at the forefront of international commerce in light of isolationist or populist headwinds in an increasing number of countries.

However, the "overexposed city" (Paul Virilio)[12] is also vulnerable to shocks and developmental contractions overseas, both in the material economy and digital worlds. "Vulnerability" is a concept recently associated with discussions on resilience and risk, climate conditions, conflict, violence, and sustainability.[13] In our analysis, urban vulnerability emerges as a possible outcome of the exposure and visibility perceived as requirements to play the global game of competitiveness in neoliberal times. Global visibility via emblematic megaprojects can become a conduit for urban fragmentation, inequality increases, and the expulsions reported by Saskia Sassen about the evolution of the global economy: soaring income inequality and unemployment, expanding populations of the displaced, and accelerating destruction of land and water bodies.[14] Here, we consider vulnerability to result from the interactions of many risks that could impact the urban system as well as the interconnections between them. Cities are complex, dynamic systems that create entropy as they develop. The energy lost during the system's evolution correlates to some inefficiency-related factors affecting the entire system. Consequently, this energy loss can be considered "systemic entropy" resulting in an integrated territorial impact triggered by megaprojects.

High-Tech Disruption

Despite the incremental nature of smart-city implementation, its disruptive character is undeniable. Disruptive innovation develops within the framework of capitalist creative destruction. Such a disruptive thrust adds to the process of megaproject development as an essentially disruptive and contentious enterprise. A disruptive AI urban regime effectively responds to biased strategic interests among entrepreneurs, technology stakeholders, and urban growth machines. Consequently, there is a lack of research on alternative venues for urban development based on other models, values, priorities, and regulatory frameworks. This scientifically planned blue ocean strategy, which knocks down all alternatives and turns AI into a necessary and unavoidable component in urban development (potentially leading to monopolization), is a major threat to cities' chaotic and haphazard character.[15] It is also a risk to the centrality of public spaces and the human civic friction lying at the core of the urban as self-organized complexity.[16]

Further, relying on a single, encompassing, and interrelated set of strategies for urban development delivered by a single industry—the tech industry—presents serious risks of dependency on capital mobility. Since the "spatial fix" process is decided fully by tech corporations, private interests can determine the success and failure of urban policy and urban development. A case in point is the failed negotiations of the City of New York in 2019 to try to have Amazon build a new corporate headquarters in the borough of Queens. The mode of production promoted by Facebook, Google, and Amazon claims to be run on "artificial intelligence." However, tens of millions of anonymous workers hold it together "tucked away in warehouses, data centers, content moderation mills, electronic sweatshops, lithium mines, industrial farms, meat-processing plants, and prisons, where they are left unprotected from disease and hyperexploitation."[17] As a top-down strategy, the nascent AI regime will likely foster a race for biased valuations in actions, processes, devices, and outcomes, tilting urban economies toward a heightened focus on networked land and intellectual property rights. Tech gentrification (interrelated with green gentrification) is an emerging feature of spaces adjacent to innodistricts and other urban areas. Consequently, large segments of urban populations (those living in poverty, without access to basic services, or with accessibility problems owing to disabilities) can potentially be displaced. These are some shortcomings in the political economy of the nascent AI urban regime. However, smart-city strategies trigger many other risks, disruptions, and ethical dilemmas:

- Control, privacy, and security in a situation of "surveillance capitalism derived from the high level of big data collection, analytics, scanning, and identification for predictive policing."[18]
- Cognitive risks, such as "intellectual automation" and "automation complacency."[19]
- Deep-learning operation on a black-box model that is not always understandable to humans.[20]
- The hacking of AI enabling bad actors to interfere with energy, transportation, early warnings, or other crucial systems.[21]
- Unpredictable outcomes from AI systems interacting autonomously; uncontrolled AI could potentially represent an existential danger to humans.[22]
- AI-based automation leading to massive job losses in any industry, including megaconstruction.[23]

Disruptive Impacts of Data Centers

While industry leaders such as Microsoft's Mustafa Suleyman are well aware of the risks of AI, his company (and others) keep building AI-enabled data centers to train AI models and deploy AI and cloud-based applications around the world at an increasing speed. Data Centers cause highly problematic impacts, due in part to their energy demands.[24] We must also consider the environmental damage they cause (carbon emissions, water footprints, noise pollution, toxic electronic waste), as well as the precarious labor conditions of many of the workers they employ, particularly in countries in the Global South. As more data centers are built in cities, their impacts will affect an increasing number of people.[25] When approving a data center, authorities usually neglect to consider the facility's whole life cycle. The supply chain begins with the extraction of natural resources, moves on to the fabrication of chips, continues with the data center industry and artificial intelligence, and ends with the disposal of e-waste. Even in cases where projects are sustainable, the particular industrial environment that includes mining, data centers, and green energy projects has a real and significant impact on their locations.[26]

Carbon Emissions

The carbon footprint of the Cloud now surpasses that of the aircraft sector. The amount of electricity used by one data center could power 50,000 households. The high energy use is due to the air-conditioning systems that cool servers and other equipment. Paradoxically, the Cloud cools by burning carbon. Cooling can account for up to 50% of the electricity used in any data center. Hyperscale data centers, such as those run by Google, Facebook, and Amazon, have committed to using carbon offsetting and investments in renewable energy infrastructure, such as solar and wind power, to make their locations carbon-neutral. Nevertheless, most data centers lack the funding and resources necessary to undertake comparable sustainability projects.[27]

Water Footprints

Since liquid is a better convective agent than air, large amounts of chilled water are often routed through the server rack latticework in data centers to cool the space more effectively. Although there is a cost involved, switching from cooling air to cooling water is an attempt to lessen carbon impact. Communities in the western United States are facing rising water resource constraints as they endure record droughts and heat domes. Several lawmakers have publicly opposed the building of data centers, arguing that the water usage of these facilities is unnecessary and reckless in light of the limited resources available.[28]

Noise Pollution

The quiet whirring of air chillers, pounding of diesel generators, tiny vibrations of hard drives, and mechanical fan spinning are all examples of the acoustic waste that data centers release into the atmosphere. This is a cause of physical and emotional suffering in addition to being an irritation. Industrial noise pollution has been shown to cause hearing loss, higher stress hormones such as cortisol, hypertension, and sleeplessness, among other acute and long-term physiological impacts. Some locals claimed that because human ears are sensitive enough to detect these frequencies above others, the monotonous drone—a frequency that lingers in the range of human speech—is especially unsettling. Because the data centers are powered by diesel generators, their noise levels may go beyond allowable decibel limits.[29]

Toxic Electronic Waste

Data center components that contain hazardous polychlorinated biphenyls (PCBs) include cables, batteries, transformers, air conditioners (CRACs and CRAHs), power distribution units (PDUs), and uninterruptible power supplies (UPSs). These parts should be disposed of rather than reused. The scenario is getting worse as AI grows more quickly. Semiconductors are made using hundreds of chemicals, some of which are extremely hazardous and belong to the PFAS (per- and poly-fluoroalkyl substances) family of chemicals, which is composed of over 12,000 compounds that have a tens of thousands of years half-life in the environment. PFAS have been connected to cancer and other significant illnesses. Furthermore, the list of minerals required to produce semiconductors has expanded from 11 to over 60, including gallium and germanium, for which there are no federal quality standards in the United States or anywhere else; the United States Environmental Protection Agency states that the available data on their effects on human health and the environment is "limited."[30]This combination of chemicals has the potential to pollute water supply in Latin America or anywhere else in the Global South, with governments there having few resources for major cleanups. Emerging data storage infrastructures currently researched—molecular (synthetic DNA) and 5D optical ceramic (digital cuneiform)—can be alternatives to silicon-based storage at a time when artificial intelligence (AI) intensifies storage and computing demands in the Anthropocene.[31]

Worker Exploitation

Big Tech frequently sets up shop in nations with lax regulations and low wages for workers. This is partly due to a little-known fact about AI: producing these ostensibly intelligent, automated, and magical cyberproducts takes an army of human labor. Humans must classify, clean, and detoxify AI text, photos, and videos in accordance with American cultural norms, which include rules for hate speech, pornography, and extreme violence. Those who perform this emotionally taxing, monotonous, and frequently dull task are usually foreign nationals. This work was mostly completed in Bangladesh, Pakistan, and India. The Philippines has also been linked to reports of worker exploitation and abuse, as well as sweatshops for AI data annotation. Latin America has become a hotspot in more recent times. Workers may be immediately fired if the algorithm determines their productivity is insufficient because they are remote, nonunionized, and constantly monitored. Tech companies in North America and Europe have become experts at "outsourcing damage." It has been demonstrated that this kind of demeaning AI employment is destroying the mental health of workers everywhere, proving the algorithmic harms associated with extractivism, dispossession, and the destructive potential of capitalism.[32]

Impacts of Smart Megaprojects as Platform Economies

Megaconstruction companies expect that executives can control and lessen concerns affecting the project from other domains, such as economics, finance, or geopolitics, with the aid of artificial intelligence (AI) monitoring systems. Risk management procedures can be aided by information judgments based on foresight intelligence and account for many risk perspectives, including accurate and reliable forecasts, at an early stage. Executive decision-makers are interested in learning to use monitoring technology for risk management because they are still determining how to conduct themselves in a complicated and unpredictable global context. Furthermore, the accessibility of big data increases the amount of noise and complexity, which could lead to executive irrationality and failure to recognize long-term hazards. Designing a flexible risk management framework is difficult because of the complexity of domain knowledge, computational constraints, knowledge fusion, and the interdependencies among domains. Consequently, current developments in artificial intelligence (AI) and machine learning (ML) are profoundly changing corporate processes and how leaders make decisions. However, many executives are unsure of the optimum way to integrate these technologies to support decision-making. Calculating the probability of a rare (or undetected) occurrence is difficult. Although most economic forecasts, including for oil prices and GDP growth, have been inaccurate, people continue to rely on them. Research methodologies, findings, and implementations may either increase or decrease the system's redundancy, in a risky field such as technology.[33]

Paralleling the race among countries over which one can create superior techniques for AI safety and reliability, there is increasing competition among megaconstruction companies to gain competitive advantage as megaprojects become intelligent environments built via on-site techniques of modularity; that is, by prefabricating parts, sections, or components that are then assembled.[34] Theoretically, a project could aim to achieve a competitive edge by focusing on safety issues alone and not discussing their solutions. It is unlikely that this technique will be more effective than the open, global process of collaborative scientific investigation, because it is ill-suited to the fundamental, broadly applicable concerns that underlie AI trustworthiness. Furthermore, it is unclear how to address the brittleness, opacity, and bias of deep-learning systems. Deep neural networks have complex statistical internal workings that are difficult to understand using well-proven computer science approaches to assess the safety of prior generations of hand-programmed automated systems, such as airplane autopilot. Many machine learning researchers anticipate that AI systems will continue to be unpredictable and incomprehensible in the foreseeable future.

Megaproject developers and businesses that approach AI cautiously and thoughtfully could view the indications that the competition is implementing deep learning in its systems as an exciting opportunity to exploit the known vulnerabilities of such systems. At the same time, it would be in everyone's best interest to consider how the growing usage of AI and automation can affect strategic stability, for example, by causing unintentional escalation or other unfavorable dynamics.[35]

To the extent that new megaprojects are designed as intelligent environments, they share the problems and controversies associated with platform economies. Mazzucato discusses some of these issues: (1) whether platforms have special types of rent to collect more value than their retailers make from the supply of goods and services, (2) whether a platform's market strength means they can extract rents, (3) how the current platform economy overextracts profit from data, and (4) whether a giant's position to extract rents originates more from the price to retailers than from the price to customers by influencing how both users and retailers understand how to navigate shifting digital opportunities. Platforms do not adapt to the market; rather, industry titans define norms for the market that digital enterprises must follow to create sustainable advantages.[36] The problem is not necessarily with data production per se, but rather with the kinds of market incentives built into the existing paradigm of data collection and value that gave rise to platform capitalism, as we know it today.

Platform capitalism did not emerge naturally. Rather, it is part of the fundamental political economy of capitalist evolution, including the types of institutions for governance; options for using data to create value; and methods for incorporating privacy, transparency, and agency into the design process. An innovative and healthy service ecosystem (i.e., an intelligent megaproject) is created through interactions between the platform, its users, and its merchants. This raises some important questions: How transparent are the processes used to distribute value among customers, vendors, advertisers, and the platform itself? What ecosystem and institutional capabilities can be developed to spur investment, standards, enforcement, and coordination capacities to transition toward privacy and public value, empowering the Internet and economy? Who benefits from the creation and evaluation of data, and what other types of governance institutions could mediate? Changing the value of data means changing the fundamental structure and direction of the digital economy, changing the types of data worth collecting, and whether to collect them at all.[37] Beginning with the premise that megaprojects are results, and that a combination of the public, private, and civic sectors create their value, we can question the "public value" of megaprojects, and to what extent public value is a measure of megaproject sustainability. We explore this question in Chapter 8.

Feeble Sustainability

In Chapter 6, we described some of the environmental challenges encountered in four cities (Tokyo, London, Paris, and New York) and the green sustainability strategies of Shanghai and Riyadh. In Tokyo, we discussed the urban heat island effect and problems related to liquefaction of reclaimed land. In London, we addressed some controversies and impacts related to the 2012 Olympic Games (cooperation with the Rio Tinto Mine, loss of biodiversity, mismanagement of nuclear waste, loss of allotments and greenspace). In Paris, we explored the *La Défense* megaproject and its associated environmental challenges (disturbance of the skyline, criticism of the Grande Arche, and unequal distribution of natural light). Regarding the new World Trade Center in New York, we discussed using green concrete in construction, persistent doubts regarding LEED certification, and creating a hotter environment. In both Shanghai (Pudong Waterfronts) and Riyadh (four *Quality of Life* projects), green sustainability plan development and presentation appear to be interdependent on entrepreneurship, innovation, and economic competitiveness. The goal of sustainability is actually "green capitalism," which refers to the preservation of the information and knowledge economy with more or less ambitious environmental obligations added to by reformers and planners rather than a sustainable global society with clearly defined development boundaries.

The growing importance of megaprojects in the sustainable development agenda warrants careful analysis to assess how urban leaders foster sustainable urban practices and whether these practices could be improved, in a context where promoters and developers pressure officials and others to expand building plans. We found that several requirements need to be met to achieve an acceptable level of sustainability in megaprojects, such as sustainability in the environment; in design, planning, and management; institutionally; and socioeconomically. Urban development centered around megaprojects is paying close attention to the interplay between sustainability and competitiveness and to building megaprojects that meet both goals, rather than prioritizing a strong notion of sustainability over growth and global scope. The covariance in practices related to these two concepts and practices might be at the center of future research and analysis.

As indicated in Chapter 6, the main lesson of our analysis is that environmental challenges are gradually becoming key factors affecting the capacity, performance, and impact of megaprojects in urban settings. As countries transition from urban property investment and finance to science, knowledge, and innovation districts, sustainability strategies—including environmental sustainability—become a central element in market positioning. Despite this systematic and holistic approach, sustainability strategies have limitations. Environmental quality is becoming crucial to ensure sustainable economic growth as cities and nations move up the industrial value chain and broaden their service industries to meet the rising domestic demand. Cities must step up their games further if they want to attract the right labor pool. Urban residents in more affluent economies are already paying significantly more for an ecological or "sustainable" urban environment. The perverse effect of gentrification starkly reveals the problems of anthropocentric sustainability, which considers the planet as a resource for human use.

Within this framework, people are portrayed as consumers and "eco-entrepreneurs" who actively participate in corporate marketing initiatives and look to engage in "green consumerism." In this scenario, the expectation and the goal is to combat climate change without stopping the current rate of global consumption growth.[38] This viewpoint holds that there is no conflict between economic growth and environmental sustainability. The accommodative idea of "green capitalism," which asserts that corporate activities may effectively address environmental problems by turning sustainability into a profit-making activity, furthermore meshes nicely with the prevailing political and socioeconomic paradigm. The power of global capitalism is cultivated through the construction of the controlled discourse of sustainability in politics and industry and is enhanced by the advertising of every new "green" product.[39] The dominance of corporate capitalism, and their blatant disregard for the recommendations by science, is contributing to *feeble sustainability,* whereby several profit-making eco-friendly innovations and strategies (innovative design and construction materials, renewable energy integration, resource-efficient systems, biodiversity promotion, high tech and smart devices and infrastructures, digital enhancement, AI integration, etc.), widely used in recent megaproject construction, coexist with the steady deterioration of our habitable climate while socioeconomic disparities and inequalities persist.

A Typology of Disruptive Impacts

The findings from our empirical analysis are summarized below for easy reference and to underscore the wide-ranging and numerous disruptive impacts of megaprojects.

Socioeconomic Impacts

- **High-Tech Gentrification**—Influx and project-induced in-migration is a huge issue for communities that host megaprojects. Two of the main consequences of this influx are gentrification and displacement. Our gentrification typology includes high-tech, green, cultural, tourist, and commercial gentrification. High-tech gentrification occurs in spaces adjacent to innodistricts and other high-tech megaprojects (Barcelona, London, Boston). Meeting the global standards (sustainable, technologically advanced) of an innovative megaproject—e.g., Barangaroo—does not mean an improvement in the well-being of urbanites. For detractors, innodistricts are useless ecologically, uncertain economically, and counterproductive in terms of real estate values; they harbor false promises in an already uncertain and precarious era. As a top-down strategy, high-tech megaprojects and the nascent AI regime are likely to foster a race for biased valuations in actions, processes, devices, and outcomes, tilting urban economies toward a heightened focus on networked land and intellectual property rights. Displacement of some segments of urban populations (those living in poverty, without access to basic services, or with accessibility problems owing to disabilities) is common.
- **Green Gentrification**—Greening megaprojects helps to attract the right labor pool, with the perverse effect of gentrification, which starkly reveals the problems of anthropocentric sustainability. Residents of cities in more affluent economies are already paying significantly more for an ecological urban environment. Barangaroo's commitment to environmental sustainability (carbon-neutral materials, supply of recycled water, rooftop solar energy) entails the significant displacement of disadvantaged populations due to "green gentrification,"

whereby these benefits are only available to a privileged few. The Olympic Park in London is yet another example that greening megaprojects for sustainability leads to gentrification, displacements, and exclusion.

- **Cultural Gentrification**—Urban leaders in Denver tried to transform the city's civic center from a place of homeless people into a place where the upper classes spend time. The Hamilton Building was integral to Denver's attempts to transform its downtown, and particularly the Golden Triangle, into habitable "urban villages" intended for upper-income residents: gentrification by design.
- **Tourism Gentrification**—In Shanghai, replacing inexpensive working-class waterfront neighborhoods with upscale (and mostly empty) real estate is not an inclusive process; it gentrifies the area and imposes socioeconomic restrictions. Challenges, both exogenous and endogenous, related to the WKCD and Hong Kong's tourism development, include an aging population, risks of gentrification and displacements, natural disasters, climatic risks, and disease. Tourism associated with megaprojects can lead to a rise in congestion and overpopulation.
- **Commercial Gentrification**—Megaprojects harm retail business and the fabric of cities, thus contributing to the "retail apocalypse" perpetrated by e-commerce while intensifying the transmission of consumerist values. In Barangaroo, for example, only business behemoths such as KPMG and Westpac can afford the high commercial rents. Consequently, retail stores, restaurants, and other entertainment-oriented services will primarily serve affluent people. It will become increasingly difficult for the less fortunate to patronize the site's residential, retail, or commercial facilities.
- **Displacement—**The residents of Istanbul are generally unaware that many will lose their homes as lands are cleared for the Kanal construction. Because major renewal projects normally spearhead gentrification and dislodge low-income populations, they cause tremendous social and economic upheavals and tend to be highly controversial. Displacement is closely associated with issues of land acquisition, expropriation or resettlement, and the consequences of this (i.e., how the land is acquired and what happens to the relocated residents).
- **Worker Exploitation**—The high-tech component of megaprojects claims to be run on "artificial intelligence." However, tens of millions of anonymous workers hold it together tucked away in warehouses, data centers, content moderation mills, electronic sweatshops, lithium mines, industrial farms, meat-processing plants, and prisons, where they are left unprotected from disease and hyperexploitation.
- **Unemployment**—AI-based automation leads to massive job losses in any industry, including megaconstruction and manufacturing megaprojects; lights-out manufacturing boosts productivity while requiring less investment in labor, thus increasing cost-effectiveness. Ironically, some in the corporate world still believe that people are "the most programmable, efficient machinery for most assembly activities."
- **Rent Overextraction**—To the extent that new megaprojects are designed as intelligent environments, they share the problems and controversies associated with platform economies: high rent extraction, overextraction of profit from data,

and issues relating to the relative contribution to rent extraction by retailers and customers.
- **Land Use and Labor Issues**—In the case of high-tech megaprojects, local industries and social communities want protection from expanding tech companies, the real estate developers who build for them, and the conflicts of interest over land use and labor issues that can be generated. City council members, state legislators, and community organizations in many cities oppose the city government's attempts to satisfy big-tech companies.
- **Drawbacks of Growth and Competitiveness**—Megaprojects epitomize globalized urbanization and its discontents: accentuation of socioeconomic inequality (because of gentrification and displacement); dependency on capital mobility; private interests determining the success and failure of urban policy and urban development. Megaprojects of all kinds often develop alongside the pursuit of growth and sustainability targets, leading to benefit shortfalls due to unrealistic expectations. For example, Riyadh's megaprojects show a conflicted response to entrepreneurial neoliberalism: on the one hand, progressive attempts to promote an egalitarian approach to urban livability; on the other hand, strategic efforts to use megaprojects as spectacular showcases in the international marketplace.

Environmental Impacts

- **Threat to Ecosystems**—Urban development in Istanbul has reduced its air quality, the amount of green space, and the health of its flora and fauna while increasing waste, risks of deforestation, anthropogenic climate change, and other ecosystem problems. The key issue for constructing Istanbul's Kanal—a shipping channel project—is the danger of losing green areas for business development. The forests in the north contain rich flora and fauna, water basins, and natural resources, representing protected areas.
- **Harm to Water Supplies**—The Kanal and the recently built third airport could jeopardize the city's water supply. The projected route would damage marine and freshwater ecosystems, as well as lakes and rivers supplying Istanbul's water. The Kanal will be built on the European side of Istanbul, where 40 percent of the city's water originates.
- **Harm to Hydrological Balance**—The new Kanal is anticipated to upset the natural balance of currents and countercurrents between the Black Sea and the Sea of Marmara. Some scientists assert that the human-made canal will work as a siphon, drawing contaminated Black Sea waters into the Mediterranean Sea. Irreversible environmental harm is expected, triggered by the likely inversion of the hydrological balance between the cold and fresh waters of the Black Sea and the warm and salty waters of the Mediterranean Sea.
- **Waste Treatment Problems**—if the Kanal is built, it will create problems with waste treatment. The current system uses Bosphorus crosscurrents, which have a denser flow moving north at the bottom than the surface flow from the south.

Wastewater and leftover waste are currently being discharged at the confluence of the Bosphorus with the Marmara Sea. The Kanal would upend the delicate balance of aquatic life and disrupt the city's distinctive waste system, which was designed after years of modeling and scientific research. The city might begin to smell of hydrogen sulfide because of the shift in salinity, which might also result in an anoxic state in the water.

- **Release of Toxins**—Barangaroo South is built in an area where many enterprises have contaminated the soil over the decades. There was fear that toxins might be released into the environment in the project's development phase.
- **Energy and Water Waste**—Servers and other equipment in data centers are kept cool through the use of air-conditioning. Paradoxically, the Cloud cools by burning carbon. In any data center, cooling can use up to 50% of the electricity used. Large amounts of chilled water are often routed through the server rack latticework in data centers to more effectively cool the space. This accentuates water constraints and shortages in some areas.
- **Toxic E-Waste**—Data center components and semiconductors contain hazardous combinations of chemicals that have the potential to pollute water supply in Latin America or anywhere else in the Global South, with governments there having few resources for major cleanups.
- **Soil and Air Pollution**—Because of the presence of hazardous materials in some areas of Barangaroo, pollution has become a major public concern. Critics and many residents remain worried about the health and environmental consequences of developing contaminated land. Poor air quality during the construction and operational phases of the WKCD was also a problem in Hong Kong.
- **Noise Pollution**—Data centers are powered by diesel generators, producing noise levels that may go beyond allowable decibel limits.
- **Water Pollution**—Dredging was required in Hong Kong to remove underwater obstacles and replace large volumes of sand, exacerbating water pollution. This negatively affected the marine ecosystem, including plants, sea organisms, and dolphins.
- **Impacts of Reclamation**—In Hong Kong, reclamation entailed construction site runoff, pumped groundwater discharge, drainage diversion, sewage effluent, and accidental spillage that contains high levels of suspended solids and chemicals such as oils, solvents, and cement-derived materials. In Saadiyat Island, the location of the Guggenheim Abu Dhabi, new lands and breakwaters have had to be created. Breakwaters are known to stagnate the flow of ocean waters and can lead to environmental problems. Multiple flooding events have also occurred during these reclamation projects. In Tokyo, most of the Bay Area is reclaimed land. When an earthquake occurs, the reclaimed soil acts as a liquid: the sand and water from underground rise, and structures on the surface sink into the water.
- **Impacts on Biodiversity**—In Abu Dhabi, sensitive marine communities, including mangroves, sea grasses, and coral communities, were adversely affected by dredging. Impervious surfaces, such as roadways, increase runoff rates and carry water mixtures containing pollutants; these waters are eventually discharged into coastal waters, negatively impacting the coral. In London, the Olympic Park was constructed in an area with immense biodiversity that was severely affected.

London's "green lung, the lower Lea Valley," is also impacted by the megaproject, including the habitat and migration path of many birds. In Barangaroo, the proposed five-star hotel, built on a pier projecting into the port, raised concerns that aquatic life would suffer irreparable harm.

- **Urban Heat Island Effect**—In Tokyo Bay, a temperature increase was attributed to the Urban Heat Island effect. Because concrete and asphalt are used to construct buildings and skyscrapers, the heat energy from the sun and other structures is retained, turning the city into a gigantic absorber and then an emitter of convected heat. Glassy skyscrapers are often thought to be more environmentally friendly because the glass reflects sunlight away from the building, thereby reducing cooling and air-conditioning costs. However, the exterior environment, which is made from bricks, becomes the recipient of heat, thus making the environment hotter.
- **Mismanagement of Nuclear Waste**—Radioactive nuclear waste was buried under the London Olympic Park. Over 7000 tons of waste were buried in a radioactive storage bunker beneath the Stratford International Station, where Olympic attendees traveled.
- **Loss of Allotment and Greenspace**—Public spaces were lost during construction for the London Olympics, for example, the compulsory purchase of 100-year-old allotments with waiting lists.
- **Excess of Emissions**—An evaluation of the carbon footprint of local activities in La Défense highlighted that despite a residential population of approximately 40,000 people, the emissions are comparable to those of the regional capital. Construction work is directly responsible for one-third of the carbon footprint, with building operations alone contributing 15%. To progressively reduce the world's reliance on fossil fuels, it is vital to address the issue of bioclimatic building design as well as lifestyles (travel, shopping, eating, etc.)
- **Green Concrete Drawbacks**—Although the generation of green concrete is more sustainable than that of traditional concrete, it poses several environmental risks. Fly ash may contain heavy metals that can poison aquatic life when they get in the water. This was a problem in New York's NWTC.
- **Fuel Cells Damage**—The NWTC has advanced systems, including modern fuel cells, rain harvesting, and sunlight-responsive lighting. The fuel cells efficiently convert chemical energy into electricity and use byproducts for heating and cooling. However, water from Superstorm Sandy damaged Freedom Tower's $10.6 million fuel cells. Owing to financial constraints, the main tower's fuel cells were never replaced, although Towers Three and Four still have modern fuel cells.

Spatial/Territorial/Visual Impacts

- **Residential Segregation**—After construction forced the displacement of local residents, only 13,000 homes have been constructed on and around the London Olympic Park, 11 years after the 2012 opening ceremony. Only 11% of these are affordable to people with average wages. Generally speaking, when new housing projects are built that are only financially available to a small portion of the population, social boundaries are formed: these places essentially become enclaves for a specific socioeconomic group. These patterns are exacerbated through housing financialization and the increased control of housing markets, including megaprojects, by hedge funds as global landlords (in Kings Cross, Masdar City, WKCD, Toronto Quayside, NWTC, and others).

- **Perverse Exposure**—Centrality and the visual power of iconicity (hegemony of vision) in megaprojects as catalysts of growth and competitiveness lead to exposure to global networks, impacts coming from developmental contractions overseas, destructive creation, and local vulnerability.
- **Iconography of Order and Control**—High-tech megaprojects, intended as a blue ocean strategy, knock down all alternatives and turn AI into a necessary and unavoidable component in urban development (potentially leading to monopolization). This is a major threat to cities' chaotic and haphazard character. It is also a risk to the centrality of public spaces and the human civic friction lying at the core of the urban as self-organized complexity. Toronto residents reacted forcefully against Alphabet's dystopic plan for Quayside.
- **Terraforming**—Kanal Istanbul will bisect Istanbul and create an island between Europe and Asia. It will chop off the city and fundamentally damage its vital systems.
- **Contention over Right to Public Space**—As in most other iconic megaprojects that are at the core of urban redevelopment schemes, political contention in Barangaroo centered on one key issue: the right of citizens to urban public space versus the private interests of developers. The prominent political ambitions of local officials and the pressures from globalization (as perceived locally) drive cities to make extraordinary efforts to align with international trends, even at the cost of jeopardizing many local citizens' standards of living and exacerbating urban inequality.
- **Loss of Aesthetic Appeal**—In the WKCD, the construction of the metro station, Elements Mall, international commercial centers, and large apartment buildings will dominate the area. The future construction of skyscrapers and other enormous complexes will further reduce the beauty and access to the views of Victoria Harbor in Hong Kong. The skyscrapers in La Défense have been criticized for being disruptive to the sky view of certain monuments, including the Arc de Triomphe. The public criticized the "forest of towers," in which buildings came to disturb the view of the Arc de Triomphe, as seen from the Étoile. The skyscrapers were also criticized as generic buildings—towers considered a postmodernist style that could be built anywhere. Thus, megaprojects can spoil cities´ visual and structural uniqueness.
- **Unequal Distribution of Natural Light**—In Paris, green activists demanded "shadow studies," where "new designs are computer tested in a virtual landscape to see where shadows will fall at different times of the day and year," so skyscrapers do not ruin the skyline's general harmony.

Political and Geopolitical Impacts

- **Increased Political Complexity**—In Sydney, An "erosion of intent" regarding Barangaroo's initial aims and goals has generated multiple controversies and led to a highly controversial, complex, and gridlock-bound political landscape. Public management in the context of a large economic infrastructural development, especially when the government is attempting to position a city globally, is far more complex and political than considered rationality in new public management theory suggests.
- **Opaque Governance**—Like most megaprojects, Barangaroo has exhibited a significant lack of transparency and accountability in governance. For example, as a series of adjustments were made to increase development profitability while diminishing public benefit, it was seen as "perverse" that such a large enterprise should be subject to less scrutiny rather than more. The original master plan had a generic nature that could have been easily exploited to allow an "anything goes" development in the future.
- **Disregard for Citizen Needs**—Megaprojects frequently prioritize the requirements of the property and leisure industries over the needs of citizens. In Helsinki, The Next Helsinki initiative was launched as an alternative to the controversial Guggenheim Helsinki project, addressing shortfalls in citizen needs. The organizers called upon architects, urbanists, artists, and environmentalists to imagine how Helsinki and the South Harbor site allotted to the proposed museum could be transformed for the maximum benefit of the city's residents and visitors.
- **Privatization of Governance**—Because of its connections to the large-scale retail industry and its commercial and residential development interests, the WKCD was criticized as being primarily aimed at developing commercial property, rather than cultural projects. Government funding in Hong Kong primarily supports shopping, dining, and entertainment facilities, with an expectation of increased revenue, which benefits wealthy developers. Not surprisingly, property prices began to increase in residential areas in 2010 and continue rising today.
- **Religious and Political Censorship**—In Saadiyat Island—the location of the new Guggenheim Abu Dhabi—only purely ornamental art is permitted, even though modern art to be exhibited in the Guggenheim Museum in other sites is generally regarded as "largely liberal, tolerant, and exuberant." This contrast between local culture and modern art creates a societal clash where cultural biases threaten to divide the project philosophically, while practical differences suggest that the users and providers also disagree about the eminence of economic or aesthetic interests.
- **Postpolitical Agendas**—Big-tech companies, in general, and discourses on smart cities, in particular, are prone to pursuing a postpolitical, expertise-led, technological agenda. Quayside in Toronto is a notable example of a large-scale, digital firm entering the field of urban planning and attempting to control growth over a specific area to further its own commercial objectives. According to postpolitical theorists, formal state-instituted processes and venues for participation

increasingly do not provide a forum for meaningful public discourse or appropriate spaces for contest. However, urbanites created new opportunities to participate in politics as Alphabet Inc. became a major player in Toronto's urban planning and Canadian politics.

- **Iconic Power**—If iconicity is a visual manifestation of power, it follows that the drivers of iconic megaprojects will seek ways to establish both hegemony and legitimacy. UMPs are used to establish a "hegemony of vision" that secures legitimacy for the new coalitions of their public and private stakeholders through the spectacularization of development.
- **Geopolitical Impacts**—With Istanbul Kanal, Türkiye will have the capacity to strategically set regulations on maritime traffic at its discretion and hence increase its own geopolitical influence in the area. The major powers have expressed a lack of support for the project. Russia is opposed to it because it would potentially decrease its strategic power by giving NATO ships a more direct path to the Black Sea. The United States has not yet publicly rejected the proposal but does not seem to support the project. Kanal's development has both benefits for the European Union, by limiting Russian influence, and drawbacks—increasing Ankara's power and role on the world stage.

Financial Impacts

- **Uncertain Economic Success**—In Hong Kong, the WKCD proposal was a typical package of urban entrepreneurialism, with an emphasis on chasing after megaprojects, iconic buildings, and media visibility. It avoided questions concerning the substance of the entire project, consensus from below, and a vision of cultural development. Its failure shows that without addressing these basic questions, city competition by means of developing global architecture, megaprojects, and fabricated urban culture is inevitably futile.
- **Benefits Shortfall**—Poor planning and the inherent complexity of megaproject construction and management mean that building an extraordinary megaproject is insufficient to ensure success. As a perhaps extreme example, the Chinese city of Ordos has sprung up quickly and is relatively wealthy, thanks to discoveries of oil and gas. However, the new museum there has no collections and few plans for exhibitions, and consequently has few visitors.
- **Uncertain Financial Prospects**—The accuracy of estimations regarding the growth in shipping traffic has significant financial implications for Kanal Istanbul. If that estimated figure is not achieved, the government must pay companies an amount established in their contracts. Further, currency fluctuations (such as that experienced by the Turkish lira in recent years) sharply raise the fiscal implications of state guarantees for the Kanal and other megaprojects.
- **Threats to Economic Security**—The Kanal financing included the issuance of bonds, a mechanism used when there is reduced global liquidity, making capital markets less appealing. Experts have noted that the success of project bond

financing is far from guaranteed, and Türkiye's debt is already sizable. How future governments and citizens will pay the debts incurred daily by the country's megaprojects is a difficult challenge.

- **Possibility of Public Profits**—Megaprojects may thus be a tool for "financialized municipal entrepreneurialism": they certainly create "anchor capital" by remaking the built environment to the benefit of investors and developers, but they also use public land speculatively to generate substantial financial returns for state governments. Barangaroo development shows that urban development need not primarily seek to benefit private developers. By acting as a quasipublic firm driven by speculative ambitions, public authorities may turn the goal of "public benefits" into a strategy for increasing public revenue.
- **Cutting Cultural Budgets**—A major reason for Helsinki rejecting the Guggenheim was the widely shared perception that public funds should not go to support a private organization. Everything would have been financed by the city—the land, the construction, the upkeep of the building, the salaries, the license fee. This would mean big cuts would have to be made elsewhere in Finland's culture budget.

Cultural Impacts

- **Neglect of Local Culture**—Because the strategy focused on "the bundle sale of the development of creative industries," the development proposals for WKCD would not benefit local culture. Civic organizations such as Project Hong Kong and Citizen Envisioning a Harbor feared the project would solely benefit real estate developers and urban elites, who clearly prefer Western culture. Local artists were concerned by the deep involvement of Western cultural institutions; they feared that WKCD would prioritize the works of international artists to attract new art buyers and generate income for the city at the expense of local artists and cultural organizations. In fact, the development of tourism was not designed to advance culture but rather to strengthen Hong Kong as a "traditional economic base." Top-down decision-making structures and a lack of voice from cultural practitioners have undermined cultural citizenship and civil society in Hong Kong.
- **Fading Away of Bilbao Effect**—The Bilbao Effect has proven difficult to replicate in most places, even in the case of other designs by Frank Gehry. Further, iconic architecture is no longer a hegemonic visual discourse in urban revitalization. In addition to the usual gentrification effect caused by cultural megaprojects, researchers show that supply may have outstripped demand, and cities have been left with the responsibility to maintain or even pay for cultural centers they do not entirely need.
- **Hegemony of the Global**—In Helsinki, a set of critiques stressed that the international competition designs did not incorporate any elements from the local context, culture, or heritage. Critics pointed out that architecture needs to be

responsive and considered as a form of "critical regionalism." In Abu Dhabi, this debate about architecture as a domain and creative reflection of local culture was proposed as a vehicle to interact with the global appetite for knowing "the other." The critique of architectural products such as the Guggenheim focused on them being exemplars of a globally crafted strategy. Also, the franchise model imposed by Guggenheim means that local officials have no autonomy to make major decisions on matters from exhibition calendars to budgets and investments.

Systemic Impacts (Interrelated and Cascading)

- **Vulnerability**—"Vulnerability" is a concept recently associated with discussions on resilience and risk, climate conditions, conflict, violence, and sustainability. In our analysis, urban vulnerability emerges as a possible outcome of competitiveness requirements: the exposure and visibility needed to play the global game in neoliberal times. Megaprojects are intended to expose cities to global networks, and those "digital monuments" travel in cyberspace as urban global identities. Centrality and the visual power of iconicity are intended to enhance local competitiveness but they also increase local vulnerability, which results from the action at a distance of many interrelated impacts on the urban system. Vulnerability is thus a systemic impact of megaprojects. The "overexposed city" (Paul Virilio) is thus vulnerable to shocks and developmental contractions overseas, both in the material economy and digital worlds. Global visibility via megaprojects can become a conduit for urban fragmentation, inequality increases, and the expulsions reported by Saskia Sassen about the evolution of the global economy: soaring income inequality and unemployment, expanding populations of the displaced, and accelerating destruction of land and water bodies.
- **Strategic Instability**—The growing usage of AI and automation in megaprojects can affect strategic stability, for example, by causing unintentional escalation or other unfavorable dynamics. Such a disruptive thrust represents a risk for the centrality of public spaces and civic friction between humans that should occur, according to Jacobs, in the core of the urban as self-organized complexity. People are drawn to chaos, captivating interactions, and fortuitous events among a vastly different mix of residents. Instead, proponents of smart cities have embraced the notion that a city is something that can be measured and managed. The paradox of AI urban environments lies in the promise of attaining a certain type of sustainability at the risk of further jeopardizing urbanity, equality, and livability. Smart-city technology should be able to reduce carbon emissions, accelerate the construction of affordable housing, improve the effectiveness of public transportation, and shorten commuting times. Rather than focusing on what it ought to accomplish, supporters of high-tech and AI

systems emphasize on what it can do. Paradoxically, this may be a pathway to urban instability.

- **Network Dissemination**—High-tech megaprojects are part of the smart-city complex and evolve toward a more comprehensive and integrative approach, potentially enhancing the impact that high-tech can have on urban environments. Processes have gone from connecting various threads of public infrastructure and knitting the data collected through them to a more integrated network aimed at bringing together government, universities, industry, and citizens—evolutionary smart-city processes that promote an integrative approach on three fronts: strategy, entrepreneurship, and organization. Integrated smart-city processes aim to connect the physical spaces of cities with the economic and social spheres. In doing so, they foster increased coordination of entrepreneurial actors involved in further efficiency. This scheme intensifies as the nascent AI urban regime grows. As a result, the diffusion of smart megaprojects is already impacting urban systems systemically, through interrelated, interdependent, and cascading effects. Pervasive high tech is increasingly contested and confronted by citizens' opposition and resistance. For example, Toronto's citizens did not object to Sidewalk's plan due to concerns of architectural preservation, building heights, densities, or construction methods—often grounds for protests. It was the project's tech-first approach that irritated many people. The main reason it failed was undoubtedly its apparent disregard for the Torontonians' privacy concerns. It is then questionable whether better megaproject performance via high-tech enhancements can by itself contribute to a situation of greater urban sustainability. Urban sustainability *for all* would need to take precedence over high-tech innovation by supply and AI schemes because it ultimately aims to preserve life on Earth.
- **Feeble Sustainability**—The dominance of corporate capitalism, and their blatant disregard for the scientific recommendations, is contributing to *feeble sustainability*. In feeble sustainability, several profit-making eco-friendly innovations and strategies widely used in recent megaproject construction coexist with the steady deterioration of our habitable climate while socioeconomic disparities and inequalities persist. The goal of sustainability is actually "green capitalism," which refers to the preservation of the information and knowledge economy with more or less ambitious environmental obligations added to by reformers and planners rather than a sustainable global society with clearly defined development boundaries. Like any other type of megaproject, green megaprojects epitomize the growth machine process of investing in issues crucial for forming urban regions as strategic scales for capital accumulation. In this scenario, the expectation and goal is to combat climate change without stopping the current rate of global consumption growth. The question is: to what extent is a capitalist framework pursuant of competitive development an effective strategy to tackle environmental challenges and enable meaningful sustainability practices?

Impact Diffusion: Transcalar Complexity

Disruptive complexity travels and spreads with the megaproject model of urban development, and its manifestations recombine and interact with one another in different contexts to appreciate key differences and to offer insight into policy challenges.[40] This process underpins a mode of global urban development whereby transnational policy circuits foster spatially unbounded collaborations and implicit partnerships between policy creators, receptors, and adopters. Transnational urban policy researchers analyze the factors enabling and constraining the formation of transnational circuits of policy adoption and adaptation as well as the social organization and consequences of the complex interconnectivity of cross-border networks. Urban policy mobility can thus be said to underline "the socio-spatial processes by which social actors and their networks forge the *translocal* connections and create translocalities that increasingly sustain new modes of being-in-the-world."[41] This complex transnational interconnectivity is multidimensional, encompassing social, economic, and political relations as well as cultural and interpersonal networks and technological linkages. This complex process also engenders misplaced expectations and failures. We briefly illustrate urban policy mobility using three examples: the Bilbao Effect, Dubaization, and Manhattanization.

The Bilbao Effect

As discussed in Chapter 4, the Bilbao Effect is an attempt by urban elites worldwide to implement a policy of building icons. This effort was largely motivated by a superficial and media-based understanding of the Bilbao case, which led many people to believe that iconic architecture could revitalize a city facing economic difficulties. These urban policy thought leaders disregarded the diverse patterns of iconic megaproject development and the factors that influence how successful or unsuccessful such megaprojects are in various locations, including the institutional circumstances, policy tools, and territorial-based social dynamics. The difficulties that the diffusion of this type of urban growth must overcome are illustrated by three of the numerous cities that have implemented icon building: Denver, Helsinki, and Abu Dhabi. The Helsinki Project was shelved and will never be implemented due to government and general public opposition. In Denver, urban planning changes in the years since the opening of the Bilbao-designed museum extension have pushed the Denver Art Museum (DAM) to the periphery of regional initiatives for urban redevelopment. If Abu Dhabi's new Guggenheim mirrors that of Masdar City's eco-experiment on the city's outskirts, the museum's success is far from guaranteed. The new Guggenheim operates in a complex environment in the context of expanding the diversification of the local economy.

The "Bilbao Effect" has faded, due to the constraints of current political reasoning and decision-making processes in periods when globalization pressures metropolitan leaders to redevelop and become internationally recognized. Additionally, outsiders fail to grasp the context and fundamental drivers of Bilbao's urban revitalization success. This has little to do with iconic buildings and more with the financially self-sufficient Basque government's sound economic policy and a well-designed, all-encompassing revitalization plan, of which Guggenheim was merely a minor and ad hoc component. Thus, the persistence of different cultural settings in various locations poses a significant barrier to urban policy travel and circuit construction. Each city has a distinct political makeup that affects local decision-making, as well as a local history and spatial constraints. Every city exhibits unique characteristics that help to explain its decline, and every city requires tailored redevelopment plans. Relying solely on adopting such plans to duplicate a city's success is frequently doomed to failure.[42]

Dubaization

Most Arab world cities compete to imitate Dubai in its unprecedented effort to build the tallest and largest-ever-built architectural and urban icons, a planning approach known as "Dubaization." Dubaization is qualitatively similar to the "Bilbao Effect" and has even spread to cities outside the Gulf area, such as Istanbul and Vancouver. The term "Dubaization" has been popularized in numerous talks, publications, and forums. For instance, the government of Barbados is criticized over the creation of offshore islands in a recent piece about the "Dubaification of Barbados." Similarly, a Panamanian urban forum focused on the overabundance of these structures in Panama City described Dubaization as "the dense construction of high-rise buildings." Indian energy specialists criticize the "Dubaization" of their cities, equated with building what they call "glass towers." While in these instances, the word refers to locally driven developments, at other times, there is direct involvement from Dubai-based developers. Locals in Istanbul describe the development of opulent towers in the Levent financial district—among them, Dubai Towers—as "the Dubaification of Istanbul." In another example of Dubai's economic reach, Asad Yawar discusses the extent to which Dubai-based companies purchase football stadiums in the U.K. and support British teams. He notes that Dubai's economic vitality "has made such a powerful impression globally that some are beginning to talk of a 'Dubai model' that poor countries may rocket themselves into first-world levels of luxury."[43]

Dubaization triggers crucial questions: What are the consequences of this urbanization strategy on the future of Arab cities? What kind of social life has emerged from this development? Is this an elite-driven process of constructing, reconstructing, and deconstructing the identities and territorial outlook of Arab cities? Is there any future for sustainability in the development strategies of Arab and Middle Eastern cities? Dubai, as a model of urban development, is primarily based on images and icons rather than sustainable concepts and processes. Major conflicts result from this, including failure to adopt sustainability, limited interpretation of globalization, and degradation of locality. To address these challenges, Arab cities must consider a holistic approach to sustainable strategic development. Architecture as a domain and creative reflection of local culture can be used to maintain local culture and interact with the global appetite for knowing "the other."[44] The main condition for these architectural products to be exposed to the other is that they should come from a deep and original local vision rather than being exemplars of a globally crafted strategy. The multiple controversies and disruptions associated with the Guggenheim Abu Dhabi indicate errors and failures in planning, policy, and implementation. The complexities of diffusion, adoption, and adaptation require further intercultural research, analysis, and practice.

Manhattanization

The phrase "Manhattanization" refers to the transformation of large cities brought on by the rapid growth of skyscrapers (vertical megaprojects) in their downtown areas. Critics of the 1960s era high-rise structures in San Francisco coined the phrase to describe their disdain for their downtown's changing landscape, as the ocean and hill views were obscured. For critics, Manhattanization means transforming a city into a playground for its wealthiest citizens, while disregarding the needs of its most vulnerable citizens, and making downtown more expensive and less accessible to residents in the middle and lower classes.

Miami experienced a real estate boom from 2003 to 2008, which resulted in the construction of more than 50 high-rise structures across the city. This has been referred to as the "Manhattanization" of Miami. From 2012 to 2015, Miami experienced a second-home market boom that led to the addition of approximately ten thousand housing units. Downtown was renovated, and the popularity of walking and public transportation increased, similar to Manhattan. Another reason Miami is known as the "southern Manhattan" is because of its sizable financial center.[45]

Although the term Manhattanization initially had negative overtones, over time the phenomenon gained acceptance, and the phrase became popular in Las Vegas, Dubai, and Los Angeles as a catchphrase for high-density development. Further examples are the rapid growth of skyscrapers in Hong Kong and Tokyo since the 1990s, and in Toronto after 2007, a city with more skyscrapers than New York. Many smaller United States cities have used the word to describe their downtown high-rise rental buildings, which have increased significantly since the turn of the 21st century.

Frankfurt's skyline, particularly that of its Bankenviertel central business sector, is referred to as "Mainhattan." The word is a mashup of Main, the river on which Frankfurt is located, and Manhattan, a nod to the aesthetically stunning high rises in the inner city, a unique feature for a European metropolis. The New York transformation has also inspired the term "Manhattan transfer": the trend for outstanding skyscraping monuments in the Asia-Pacific area in the 1990s.

Systemic Obstacles

The structuring of cities around borders and citadels, virtual and symbolic or cultural walls, and ghettos, as well as the challenges to translation, adoption, and adaptation of urban policies across distinct local planning cultures, are obstacles for the transfer of urban knowledge around the world, and thus for the possibilities of global urban development around megaprojects. Issues of decentralization and devolution of powers, building trust, achieving fair representation, enabling resources and support systems, and building transparency through engagement platforms represent potential limitations to this approach. The fading away of the Bilbao Effect and the limited impact of Dubaization illustrate several drawbacks in the materialization of circuits and best-practice adoption. In addition, the sheer complexity of alliance formation, circuit efficacy, and the predominance of different epistemic cultures (with distinct conceptual sets) among participants in urban politics, analysis, and practice present substantial challenges for effective intercultural communication. Different values and cultural contexts complicate interpretation and fair judgment among the parties involved in intercultural research practices.

Research has shown that the processes of urban policy travel and the establishment of transnational circuits present challenges and can run into various structural barriers due to the Manhattanization of the world, the Bilbao Effect, or Dubaization. Recessionary economic conditions are just one of the numerous challenges urban travel policies must overcome. Another is political in nature, specifically the lack of powerful metropolitan governments equipped with the tools needed to launch initiatives that can reshape the urban fabric and image. This is the situation with Mumbai, which is keen to "Shanghaize" itself despite encountering significant obstacles. As a significant illustration of urban policy evolution, megaproject development additionally faces significant organizational challenges.[46] An urban megaproject is typically developed over the course of several phases, which leads to numerous reorganizations, corrections, additions, and errors, not to mention the developers' typical inability to prevent cost overruns. The growing activity of civil society, organized to oppose the objectives of the political and economic elites, can be an additional challenge to developers. In turn, this can result in a lack of transparency, and the projects become increasingly difficult to sustain. It is important to acknowledge that UMPs can develop in conflict situations. Organized resistance against megaprojects may reach a level where the state and project promoters face significant implementation challenges and project failure.[47] An example of this occurred with Mexico City's proposed international airport project, halted by divisions between and within the political class and citizens, which were triggered by the progressive democratization, decentralization, and globalization of the country.[48]

Understanding Disruption

The Logic of Interdependent and Interrelated Impacts

Because they involve interrelated disruption and damage processes, the impacts of megaprojects are complex. To precisely analyze the intricate relationships and influence of these various, sometimes conflicting, elements, the focus should be on understanding the interaction of natural and anthropogenic processes and their impacts. We have seen how megaproject impacts may interact and manifest in increasing urban inhabitants' vulnerability. Megaprojects with converging environmental, political, social, and economic repercussions (such as those discussed in this book) negatively impact several urban population groups and cascade effects on cities' organized complexity—its ability to function.

One example of the complex logic of disruptive impacts is the interdependence between the greening of megaprojects and gentrification. While the greening of megaprojects is purportedly done to save the environment, it also causes green gentrification in the surrounding neighborhoods, driving away working-class residents and drawing in wealthier immigrants. Megaproject greening, a subset of urban greening, is a concerning illustration of the detrimental societal effects of green growth coalition initiatives as it transforms the metropolis for the sustainability class. Global cities experience negative redistribution and a new constellation of cumulative inequalities from urban greening if equity-oriented public policy action is not implemented. Green discontent (an outcome of advancing environmental sustainability at the expense of socioeconomic sustainability) also has a geographic dimension, as the green transition contributes to leaving already left-behind regions and metropolitan areas further behind.[49] As a result, disadvantaged people and many citizens in stagnant regions and cities see the green transition and the high tech digital transition as politically motivated strategies benefitting the elites. More equitable sustainability results can be achieved through early public policy initiatives targeted at neighborhood stabilization. Taxing citizen behavior to incentivize green transitions needs to be progressive (discriminating among pollution levels produced) and use redistributive mechanisms to minimize impacts among low-income populations.

In addition, systemic approaches are required to help people, families, neighborhoods, and systems within cities deal with the disruptive impacts of megaprojects. Such approaches contribute to better understanding how manifested risks (impacts) that cause vulnerability and harm can develop rapidly when megaprojects are under construction or more gradually once they are up and running. An integrated strategy that addresses: (1) risk and protective factors, such as avoidance, coping, mitigation, (2) adaptation strategies, and (3) assets that interact to enhance potential protective responses and alternatives is necessary to gain a better understanding and response to these interconnected impacts.[50] The Sustainable Development Goals (SDGs), the New Urban Agenda, the Sendai Framework for Disaster Risk Reduction, and the Paris Agreement on Climate Change are just a few examples of global frameworks that reflect the structures resembling a systems approach to megaproject impacts.[51]

A systemic analysis of urban processes is vital, as urban residents heavily rely on complex infrastructure for supplying electricity, food, water, transportation, and communication. All of these are susceptible to unexpected consequences from various shocks and pressures at various scales, such as individuals, households, neighborhoods, cities, and nations. It is vital to have a comprehensive grasp of such processes to provide the appropriate evidence to oppose the development of megaprojects in metropolitan areas.[52] Systems respond with a positive or negative feedback loop when they are perturbed by bottom-up or top-down chains of causation or by internal and external shocks and pressures, for instance, a decrease in energy consumption per capita due to urbanization. The amount and caliber of the data required are crucial, and it is not always simple to identify the nature and veracity of the underlying correlations between many components.[53] The ultimate objective is to create a roadmap supported by empirical evidence that offers insights and suggestions on how the complex political economy at the root of megaproject impacts in these cities can be reconfigured to allow city residents, especially those who are the most vulnerable, to take advantage of protective factors that improve their ability to challenge megaproject implementation.

Disruptive Complexity as Entropy

Urban systems are complex adaptive innovation ecosystems—networks of people in close proximity exchange information and opinions, create new knowledge, and interact, in actor networks, with matter as well as other forms of human and nonhuman life. Urban self-organization, which encourages innovation and economic viability, is a dynamic process. In many cases, dynamic urban models function at the regional level, replicating broad-scale phenomena such as land use, population dynamics, and economics, or examining patterns that occur from different conditions between urban players or the dynamic states of an urban system. Urban dynamics at the regional level are impacted by specific local-scale defined and self-organizing zones that continually emerge in the city and have significant potential implications for the city's cultural and economic life. It is crucial to look into ways to promote these areas' self-organization because they sustain the cultural and economic survival of the entire city on a complex, resilient trajectory.[54]

Disruptive complexity (the complex web of interrelated and interdependent megaproject impacts) undermines urban organized complexity and generates disorder; disruptive impacts can thus be referred to as *entropic impacts*. Entropy, which is the degree of disorder in a system, is a measure of the likelihood that energy and matter are arranged in a particular state. Entropy is inversely related to energy because the higher the entropy of a system, the less energy is available in the system to perform work.[55] It was first introduced in physics to relate the velocities of particles (microscopic world) to the temperature (macroscopic property). This concept helps to describe and analyze the large complex systems defined by decision-makers' macroscopic concerns that are related to the microscopic dynamics of individual elements in the system. In a closed system, entropy will always increase, whereas open systems, including all environmental and social systems (also megaprojects), can manage the rate of entropy generation to some degree by maintaining a network structure.[56]

Other real-life complex, entropic systems are networked. For example, food webs are networks of species that feed on each other. Supply chains are networks of firms supplying intermediate goods. In systems where the number of network elements is large and the connections are subject to many factors that can be seen as random to some degree, the concept of (information) entropy becomes applicable. In an analysis of global commodity trade (showing a complex network structure that arises from bilateral and multilateral trade agreements) using information entropy, Kharrazi et al. found that trade agreements can make commodity trade networks more efficient and lead to more rapid growth in trade volume.[57] The research shows that gains take a toll on resilience levels, particularly in economic shocks, a prominent example of which is the 2008 financial crisis. It also shows that adaptive strategies in some domains may reinforce structural problems that create risks in others.[58] Results also demonstrate that networks with greater redundancies do not have to sacrifice growth, an example of compensatory mechanisms between risks, impacts, and resilience.

Available evidence shows that the dynamics of information entropy (stakeholder coordination) and compensatory mechanisms (strengths and opportunities) in megaprojects does not offset the disruptive complexity of their impacts. Take for example megaprojects' sheer scale and complexity. It not only poses challenges to implementation and management but it also causes significant environmental impacts (land use and carbon emissions associated with construction, transportation, and energy consumption) even in recent, green and high-tech megadevelopments. Or consider how public infrastructure—including utilities, transit systems, and public services—becomes under stress. This pressure may result in traffic jams, a rise in the need for resources, and difficulties meeting the requirements of both current and prospective residents. Disruptive complexity (entropy increase) becomes evident, unavoidable, and very hard to manage when we consider that megaprojects are embedded in society, the polity, and the economy: recessions and geopolitical instability can impact megaproject financial stability and political support; supply chain disruptions may affect resource availability; community resistance and protests can stop construction and even lead to abandon projects; climate change and extreme weather can physically damage infrastructure.

Thurner et al. applied the notion of entropy to systems that are both networked and path-dependent: complex human-earth systems. The researchers focused on situations with winner-take-all dynamics, where simple path-dependent systems can be studied using generalized entropy. Winner-takes-all dynamics appear in many socioeconomic and environmental contexts, which show strong reinforcement and hence fat-tailed distributions. Population displacements, a common disruptive impact of megaprojects, exemplify the winner-takes-all dynamics triggered by generalized entropy in complex systems. Such dynamics imply that catastrophic events with high impacts occur more often than commonly suggested.[59] This leads us to consider megaprojects as failed systems to the extent that disruptive or entropic interdependent and interrelated impacts are an intrinsic component of megaprojects and thus not preventable.[60]

Disaster-Prone Disruption as Failure?

Given the wide occurrence of disruptive impacts, megaprojects can be considered failed systems. If we apply Cook's ideas about the failure of complex systems, we realize that megaprojects, as complex systems, are virtually impossible to manage in ways that prevent disruptive impacts. Cook contends that complex systems are intrinsically and inescapably hazardous. Although it is occasionally possible to alter the frequency of hazard exposure, the processes constituting a complex system are inherently and irreversibly dangerous.

Systems develop hazard-specific defenses. Owing to the serious repercussions of failure, several levels of defense build up over time. The apparent technical and human components of these defenses are backup systems and "safety" aspects of equipment, for example, as well as organizational, institutional, and regulatory defenses (e.g., policies and procedures, certification, work rules, and team training). These precautions create several shields that typically direct operations away from hazards. Paradoxically, the cost of eliminating all latent defects serves as a major barrier to their removal; the challenge lies in predicting how these defects might contribute to an accident before it actually occurs.[61]

According to Cook, the implication is that complex systems function as broken systems. The system continues to work because of all its redundant parts and because people can still work despite its many shortcomings. Nearly invariably, postaccident evaluations mention that the system has a history of earlier "proto-accidents" that almost always result in catastrophe. Arguments based on naïve assumptions about system performance typically claim that these degraded conditions should have been identified before an overt accident.

System operations are dynamic, with organizational, human, and technical components constantly breaking down and requiring replacement. Owing to evolving technology, modifications in how work is organized and initiatives to eliminate errors and failures have changed continuously.

There is a possibility of catastrophic failure in complex systems. Disasters can occur at any moment and almost anywhere, and human practitioners are frequently nearby, both physically and temporally, to these potential failures. The possibility of such a disastrous collapse cannot be eliminated because it is inherent in the nature of the system.[62] Overt catastrophic failure occurs when several seemingly unimportant, minor problems combine to give rise to the chance of a major accident through the logic of disruptive complexity as an emergent property of megaprojects. Although each small error is not enough to trigger a catastrophe, their combination is sufficient to allow failure. In other words, there is far greater potential for failure than there are obvious system failures.[63] Failure entails much more than cost overruns, poor management, and their consequences. Failure is a function of the uncertainty involved in megaproject impacts (materialized risks), which are interrelated, change over time and cannot always be predicted or avoided.

Furthermore, innovations can induce disruptions and failures, as saw in our discussion of recent high tech and green megaprojects (Chapters 5 and 6). The low rate of overt accidents in complex systems may promote modifications, particularly the adoption of new technologies, to reduce the frequency and number of failures with little impact. The trade-off is that these modifications may trigger brand-new, infrequent, and severe breakdowns. New technologies frequently open new paths to massive, catastrophic failures when utilized to eliminate well-known system flaws or achieve high-precision performance. Because the purportedly advantageous aspects of modifications are typically the focus of attention, it might be challenging to anticipate these new types of failures. Identifying how innovative technology contributes to the failure is problematic because these new, high-consequence incidents rarely occur and because several system changes may take place before an accident.[64]

Given the fragility of complex systems and their probability of failure—the massive disruptions caused by megaprojects—it is legitimate to ask whether mega-urbanization ought to be conceptualized as a global risk leading to catastrophe, similar to other global risks such as technological disruption, nuclear proliferation, or climate change. This begets the question—what would be an efficient and adequate response of urban systems to the disruptive impacts of megaprojects? One obvious response is to improve megaproject performance. This response might effectively reduce exposure to external impacts affecting megaprojects but is feeble in reducing or avoiding megaprojects' impacts in the urban system. Two other potential responses for consideration are resilience and contestation.

Resilience and Contestation

Self-organization is a characteristic of life (both human and nonhuman) in complex systems driven by anti-entropic effort facilitating adaptation and survival. Such an anti-entropic effort can be viewed as resilience or systemic energy: the ability of any urban system to maintain continuity after shocks or catastrophes while contributing positively to adaptation and transformation. Thus, urban resilience is one of the forces that work against entropic risks caused by urban megaprojects. This adaptation process can unfold together with a transformative component of contestation against megaprojects.

The tensions, shortcomings, and hazards inherent in the complex unfolding of megaprojects' path-dependent trajectories produce excessive entropy that affects the functioning of the urban system. When there are disruptive complex processes inside the system that cause entropic impacts on the surrounding environment, the inevitable development of entropy in the system's evolution is a component of systemic vulnerability. A properly timed and planned response to such malfunctions can reduce the generation of systemic entropy, thereby lowering the overall susceptibility of the urban system and increasing its resilience.[65] System resilience can be defined as the system's capacity to respond to risky events in one of two ways: either by becoming highly adaptable to the circumstances that arise after an event that may disrupt the course of its evolution or by identifying preventive measures to lessen the system's vulnerabilities and weaknesses.[66]

Given that, by and large, the benefits of megaprojects accrue to a small portion of the population in urban areas where they are built, most residents need to develop adaptation strategies relating to the disruptions and disorders of large urban projects. These can be considered as resilience strategies.

According to Davis, to the extent that urban, social, economic, and environmental ecologies are interconnected both at the local level and through territories linking cities with regions, mega-regions, and other (economic and governance) units, any resilience strategy must be based on the appreciation of the entire urban ecosystem and its properties as an integrated system in a larger ecology. This appreciation must start from considering systemic risks that determine the adaptive processes of ecosystems.[67] Such an analysis would lead us to examine the interrelationships and compensatory mechanisms between forms and patterns of risk and resilience, as we consider megaprojects *embedded* in urban systems, not only between different residents or locations in the same city, but also in terms of immediate versus long-term gains in livability. Thus, adaptive strategies in some domains (e.g., the environment) may reinforce structural problems that create risks in other domains (e.g., inequality). Davis argues that we need to consider the interdependencies between elements of the system and disaggregate adaptation strategies to understand when adaptive responses to vulnerabilities or crises will establish a path toward a better future. In other words, under what specific conditions will adaptations, whether made voluntarily by citizens, imposed by government authorities, or crafted by planners and designers, decrease rather than increase vulnerability?[68]

However, a traditional complex systems approach to urban systems excludes the potential for megaproject contestation. While resilience consists mainly of adaptation, contestation involves resistance and transformative change leading to sustainability. Grassroots efforts in community revitalization can reshape public processes and institutional frameworks involving the design and development of public spaces. This requires the inclusion of stakeholder engagement in the planning phase. Hou and Rios suggest that the current participatory model involving professionals and users' needs to be expanded to consider three sets of factors: (1) mobilization structure (people's engagement in collective action), (2) political opportunity (likelihood of access to power), and (3) cultural framing (shared meanings and definitions).[69] Some researchers argue that consensus building creates three types of shared capital among participants: social, intellectual and political.[70] Social capital facilitates discussions, intellectual capital enables the transformation of information into actions, and political capital makes it possible to transform urban reality.

Successful grassroots-based political action can modify the intentions and plans of urban growth machines. These machines routinely favor megaproject construction as an allegedly beneficial strategy for urban prosperity, even though megaprojects trigger processes of gentrification, displacement, exclusion, and expulsion.[71] Besides grassroots political action, growth machines need to be ideologically confronted in a forceful manner with notions of degrowth based on ideas from ecological economics, political ecology, and environmental justice. By adopting degrowth theories we emphasize the need to reduce global production and consumption and put into question development and megaprojects as privileged particles in the development process.

From Complexity to Transdisciplinarity

Our relational examination of disruptive complexity leads us to transdisciplinarity as a promising analytical approach to tackle relational webs among complex adaptive systems. In Chapter 2, we conceptualized complexity as a logical derivative of megaproject embeddedness in the polity, economy, and society. Complex knowledge also influences the centrality of the knowing subject in analytical endeavors, the uncertainty of the knowledge enterprise itself, and the incompleteness and undecidable nature of *the homo complexus´*s human action. Through complex knowledge, the rhizomatic nature of capitalist urbanization naturally leads to a transdisciplinary conception of theory-building and practice development. The Macy and Alpbach conferences held in the 1950s and the 1960s called for unification and the dismantling of silos. More recently, in Edgar Morin's understanding, complex thought, education, and knowledge is the way forward in order to grasp contextual, global, and multidimensional factors and devise strategies conducive to more fruitful action. To present a different way of engaging scholarship that draws on the fundamental insights of systems, cybernetics, and information theories, as well as philosophy, Morin most notably followed this original inspiration and developed his epistemology of complexity (not to be confused with complexity theory, associated with the Santa Fe Institute).[72] Researchers have increasingly published on complexity in a wide range of journals.[73]

Complexity begets transdisciplinarity, which involves the notion that the world is interconnected, interdependent, and creative, and also that research should reflect this reality by acknowledging and incorporating systems theory and complex thought—which is inquiry-based rather than discipline-centered. Transdisciplinarity is also paradigmatic; it acknowledges various approaches to a subject and that the literature on a particular subject may already employ one or more theoretical frameworks. Furthermore, transdisciplinarity develops integrative complexity and epistemic cognition, metacognition, and the ability to cultivate creativity, which is a way of integrating the inquirer, as a corollary to the principles of second-order cybernetics.[74]

Three interrelated developments further support adopting transdisciplinary research strategies in urbanism: social learning, collective intelligence, and the convergence of epistemic cultures. One factor that underpins the potential of transdisciplinary research is the idea that the way cities learn is essentially a social and collective process. Learning and research are cultural practices bounded by values reflected in practice. They are fundamentally collective. According to Vygotsky, we learn through interactions and communication with others. Interactions between students and learners with their peers, teachers, and other experts are conducive to enhanced learning. Therefore, it is possible to create appropriate learning environments that maximize the chances of participants learning through discussions, collaboration, and feedback. We also learn and construct knowledge within the boundaries of our cultural frameworks, rules, skills, and abilities. Thus, culture is the single most significant factor for learning, research, and knowledge creation.[75] For Vygotsky, "language is the main tool that promotes thinking, develops reasoning, and supports cultural activities like reading and writing."[76]

From the concept of "social learning," we can logically move toward the idea of "collective intelligence." To the extent that cities learn socially and collectively, collective intelligence can reasonably be applied to cities as imagined communities. Most discussions of collective intelligence or the "wisdom of crowds" refer to the Social Web. Here, the idea is that the individual contributions of web participants create value for anyone simply by being available for reading and usage. However, new knowledge emerges from the juxtaposition of a large amount of data and information sources. Gruber argues that emergent knowledge occurs when "the system enables computation and inference over the collected information, leading to answers, discoveries, or other results that are not found in the human contributions."[77] "Emergent knowledge" is a potential outcome of transdisciplinary research. Knowledge emergence will materialize if research participants can "bootstrap" their collective intelligence. Transdisciplinary research can be conceived as an information repository. It can also be conceived as an active framework in which participants actively interact to express their knowledge interests and needs, request specific knowledge, and apply transdisciplinary knowledge to problem-solving and, in general, to address their collective needs.

Another factor or precondition that works in favor of transdisciplinary research is the idea of converging epistemic cultures. Scholars in the field of social studies of science and technology have provided examples that indicate a certain degree of convergence of research fields under a new techno-scientific paradigm. These discussions mostly refer to the macroscale and adopt a broad understanding of the convergence. Kastenhofer introduces a focus on epistemic cultures and raises the question of what convergence might imply at the micro-level in everyday research practices. She distinguishes three forms of scientific change over time (convergence, divergence, and emergence) and three modes of convergence (cooperation, integration, and assimilation).[78] Furthermore, as Knorr Cetina argues, "the concepts of knowledge culture and epistemic culture are used here against the background of contemporary transformations in global financial architecture."[79]

Several attempts have been made toward less reductive approaches to space and design: approaches that no longer *choose* between theory and practice as the ideal locus for critique, but instead allow critique to be processed in ways that are more complex and entangled, and approaches that advocate hybrid modes of inquiry and research. For example, the hybridization of nature and technology, engineering and the social, facts and values, human and nonhuman, and the explicit attention to agency in science and technology studies (STS) and actor-network theory (ANT).[80] Such approaches suggest viewing urban issues not according to predefined ideologies or (critical) theories but to study them as a problem of the outside—as situated, complex gatherings of all sorts of agencies, where the notion of transdisciplinarity can be meaningfully applied in three crucial dimensions (theoretical, phenomenological, and experimental), as we have shown elsewhere.[81]

Transdisciplinarity implies the development of a *mentally mobile attitude* informed by flexibility and creativity that can translate symbolic codes across fields of endeavor and practice. In urbanism, such a disposition to translate can be expressed as "practicing places," and the notions that best capture such a mobile disposition are "assemblages" and "hybridization."[82] Both notions connect spatial concerns with cultural politics to provide multiple identities, challenging all binaries that are part of the homogenization and universalization of human existence with singular analytical categories.

Transdisciplinary research in urbanism can potentially overcome the barriers to alliances and circuits through an ontology and epistemology developing around assemblages or "rhizomatic research cultures."[83] As a result, urbanists and urbanites can be thought of as sets of actors who have the potential to cross over binary opposition and overcome the challenges of complexity.

Cities are fractal and trans-scalar, suggesting inherent linkages across sizes, and highlighting the significance of small-scale events. A smaller target scale may facilitate an examination of local interaction features based on informal information exchange. Qualitatively, lower-scale nodes, such as economic or cultural concentrations, can be significant on regional, national, or even global scales. In terms of city planning, this would entail not only first acknowledging the uncertainty inherent in complexity thinking but also realizing that bottom-up urban processes, such as the dynamics that determine where activities are located, require guidelines rather than controls to be implemented. More adaptable planning and complexity planning tools (such as measurements based on fractality, scaling, or computation) can contribute to advancing transdisciplinary planning cultures aiming to tackle wicked problems such as sustainable development.

Conclusion

This chapter has emphasized the megaproject impacts imperative: the importance of identifying impacts as a prerequisite for critically assessing whether megaprojects should be built. The chapter has defined megaproject impacts as *emergent properties of megaprojects* that cannot always be predicted in advance in any meaningful or comprehensive manner. We defined *disruptive complexity* as the complex web of interrelated and interdependent impacts that emerge after megaproject construction and affect the urban living of many citizens, particularly those who are more exposed and vulnerable. Thus, megaproject impacts are complex (interrelated), disruptive (entropic), and emergent (they cannot always be predicted from risk). Both urban systems and megaprojects are complex adaptive innovation ecosystems; that is, networks of people in close proximity exchange information and opinions, creating new knowledge and interacting, in actor networks, with matter as well as other forms of human and nonhuman life.

Megaprojects can be characterized as complex systems (organizationally and scale-wise) embedded in urban, political, and socioeconomic systems. They are path-dependent and open systems, where entropic risks (as in fat-tailed distributions) occur often, particularly when we look at one of their emergent properties: their disruptive and disorderly impacts (materialized risks) in urban areas. A characteristic of life (both human and nonhuman) in these complex systems is that they are driven by an anti-entropic effort that facilitates adaptation and survival. Such an anti-entropic effort can be viewed as resilience or systemic energy: the ability of any urban system to maintain continuity after shocks or catastrophes while contributing positively to adaptation and transformation. Urban resilience is tied to sustainability because it is one of the forces working against the entropic impacts caused by urban megaprojects.

Similarly, contesting urban megaprojects would work toward a sustainability goal by integrating communities into the megaprojects' planning process. The three pillars of sustainable development—society/community, economy, and environment—can be equally promoted through the concepts of intergenerational and intragenerational equity.[84] The first is concerned with maintaining the quality of natural ecological systems and their services over time, while the second is based on promoting equitable access to resources within current generations and providing human populations with basic needs.

Based on this anthropocentric conception of sustainability, human settlements can be defined as sustainable if planned and executed to consider the surrounding ecosystem's capacity, suitability, resistance, diversity, and balance. We consider sustainability an organic process that includes the environment, economy, and community. This includes form and efficiency—environmental factors in design, architecture, engineering, and construction, as well as policies—urban plans and practices that explicitly aim to maintain and improve the economic well-being of citizens. Although environmentalism is receiving significant attention from the perspective of urbanism, it is just one of the components of any comprehensive strategy for sustainability. Green and high-tech capitalism claim to contribute to environmental sustainability while neglecting the socioeconomic dimension of sustainability. To plan and build sustainable megaprojects, we need a multidimensional systems approach based on transdisciplinarity and a new knowledge generation agenda vis-à-vis the urgency to understand the challenges and opportunities in a rapidly urbanizing world. This effort needs to factor in the idea that both "cities" and "nature" belong to the realm of organized complexity, and thus the notion of an "urban ecology" would deal with the emergence and self-organizing power of complex adaptive systems. This, in turn, entails developing a nonanthropocentric notion of complex sustainability. We address these matters in Chapter 8.

Chapter 8
Complex Sustainability

Introduction

In Chapter 7, we made several inferences from the empirical analysis (Chapters 3 to 6), which point to the multifarious manifestation of megaproject disruptive impacts as sets of interdependent and interrelated webs. We developed an analytical explanation of such evidence around the notions of *disruptive complexity* and *entropic impacts*. Their disruptive complexity turns megaprojects into inherently unsustainable endeavors, and their construction compromises any attempt (even modest) to achieve urban sustainability. Two possible responses to disruptive complexity are resilience and contestation, which would need to be framed around a renewed and robust conception of sustainability as a transcalar normative framework for the common good that also encompasses the dignity of individual urbanites.[1] Reframing and renewing the idea and practice of urban sustainability is necessary in order to build a policy scenario that sets incentives away from the megaproject solution to urban development. To this end, we need frameworks with a key strategic and evolutionary component aiming at containing systemic risks as well as learning, inventing, and creating.

Thus, Chapter 8 addresses urban sustainability—in the form of a set of conjectures—and lays out foundations for sustainable policy frameworks, rather than specific policy proposals, that provide a new direction for urban governance in a period of global policy experimentation. Such policy frameworks or principles, subject to adjustment resulting from urban learning processes,[2] coalesce into the notion of *complex sustainability* and intend to fundamentally critique and renew the concept and the practices that have been associated to the hegemonic regime of green capitalism, where "sustainability" overlooks social justice and conflicts of interest.[3]

In this chapter, we present complex sustainability as an assemblage of two main components, socioeconomic sustainability and deep sustainability, which are intended to address the urban question and the Earth system question in the face of

G. del Cerro Santamaría, *Contesting Megaprojects*,
https://doi.org/10.1007/978-3-031-98382-5_8

relentless megaproject urbanization. Our discussion on socioeconomic sustainability includes fundamental vectors or directions for urban sustainability that dispute existing strategic public policy and private priorities.[4] Such vectors aim at addressing priorities of urbanites away from the "megaproject solution" to urban development. Our discussion on deep sustainability highlights some key elements intended for a wider framework where megaprojects and capitalist urbanization align with the boundaries and limits of the Earth system. The combined elements in socioeconomic and deep sustainability produce a dynamic equilibrium that prioritizes protecting vulnerable human and nonhuman life and fostering planetary preservation. In summary, this chapter fleshes out the notion of complex sustainability, which includes two different yet complementary and interrelated responses (socioeconomic sustainability and deep sustainability) to the disruptive complexity generated by megaprojects.

Socioeconomic Sustainability

A focus on socioeconomic sustainability is warranted for two main reasons: (1) sociopolitical and socioeconomic events have historically impacted cities to a greater extent than physical events (pandemics, droughts, and floods)[5]; (2) the socioeconomic dimension of sustainability has been largely neglected in recent transitions to "green" and "high-tech" forms of urban projects. In the neoliberal era, megaprojects became a policy of choice among urban, regional, and national elites concerned with the relative status of cities, regions, and other geographic locations in newly conceived world hierarchies and networks. Urban policy shifted toward prestigious flagship projects and foreign investment, frequently at the expense of locally based bottom-up initiatives.

Policymakers have become increasingly concerned with building venues that will suit new types of "global investment" in cities. Directly addressing local communities' needs and priorities has received less attention or at least little research attention. The elites assert that populations would instead indirectly gain from the higher growth of new types of "sustainable" economic expansion, such as high tech and green projects, while overlooking socioeconomic approaches to achieving more cohesive and less unequal societies. Cities have been—and still are—eager to highlight the quantity of their sporting and cultural institutions as well as the breadth of their linkages to international networks.

Prevailing ideas about the facilities and assets expected in a city of a specific status and the perception of engagement in place competition as symbols of inclusion in global capitalism have greatly influenced government agendas and the implementation of policy objectives. These problems raise questions for further study. For instance, what form will urban competitiveness take in the future, and how will imagined hierarchies change? Can place competition, a major factor in spatial and urban policy, change to become a more progressive force? In what follows we suggest four areas for utopian pragmatist policy experimentation that can help address and improve "socioeconomic sustainability" in a context of declining but damaging neoliberalism in which high-tech and green initiatives and strategies still adopt an "urban trickle-down" model that overlooks socioeconomic challenges. Such areas are: (1) the need to go beyond green capitalism (or decoupling sustainability from profitability); (2) sustainable innovation in the new artificial intelligence (AI) urban regime (or framing innovation around sustainability goals); (3) cohesion versus competitiveness (or finding ways to design equity policies with a local focus); and (4) public value (or developing bottom-up market shaping and value-co-creating eco-strategies).

Beyond Green Capitalism

The 2015 Paris Agreement is regarded as humanity's final opportunity to address climate change.[6] A global agreement was reached, binding the world to considerable reductions in greenhouse gas emissions. However, the main problem may be that "the much-advertised solutions" of emissions reduction goals and market mechanism control appear ineffective for mitigating or limiting the worldwide repercussions of climate change.[7] Furthermore, the major producers of fossil fuels and coal were significant funding sources for the Paris meeting. Although not particularly shocking, this fact suggests that we are dealing with a more serious issue: corporate capitalism's global economic system cannot decarbonize to the extent required to avert the likely catastrophic repercussions of climate change.[8]

The concept of "creative self-destruction" has become increasingly prevalent. Companies are meshed in a cycle of resource exploitation that is becoming increasingly irreversible. Oil-drilling firms rapidly move forward with projects such as oil-sand processing, new coal mega-mines, "fracking" (which currently appears to be in decline), shales, and coal seam gas. These instances demonstrate the ingenuity and dynamism of corporate capitalism, as well as governments' and industries' neglect of the environmental damage they cause.[9]

Despite scientific warnings, large corporations continue to abuse the environment by purposefully hiding the connection between unending economic growth and global ecosystem deterioration. They are frequently successful in refuting critical views of climate problems in civil society. Instead of framing the issue as a social, economic, and political one that has implications for the future of humanity, they often approach it as a subject of party discussion. Additionally, through the lens of "green capitalism," markets and corporations are considered the most effective way to address climate risks. This corporate image claims that "green" goods and services, increased eco-efficiency, and creative and technological corporate entrepreneurship will prevent us from facing a grim future for the world.[10]

Investors and shareholders naturally want to get the most out of their investments, which requires increasing sales and customers' resource consumption. Corporate board chairpersons have attempted to address this issue in two different ways, both of which include evading the issue and carrying on as usual. Using one approach, the most ideologically militant corporations, such as ExxonMobil, financed climate change deniers. Exxon is exploring oil in the Russian Arctic, as they deny global warming.[11] In a different approach, companies promote "green" products, including organic foods, less hazardous cleaning supplies, light bulbs, and energy-efficient appliances. These companies manage production in "green" buildings with solar panels. However, the core corporate priority of expansion and profits above the environment has not truly changed despite "green" production.[12]

In one of many such examples of "greenwashing," the Ford Motor Company makes hybrid vehicles and covers the roofs of its auto factories with solar panels, yet it continues to make automobiles, which are millions of global heaters. According to Naomi Klein, "in order to preserve mankind, what we need is not electric automobiles, but a lot less cars."[13] We may need and want phones, but if they are disposable, they are not sustainable. However, if Apple stops producing throwaway goods, how will it continue to operate? To protect Earth, we probably need more extreme solutions. For instance, we should consider changing our tax laws so that they tax damaging, nonecological behavior instead of just income. Business ought to be serious about including among their stakeholders the public interest, not just investors and company shareholders. Urban growth machines would need to radically reorient their priorities. While the debate on these issues is ongoing, most governments have yet to make fundamental decisions. Can green capitalism protect the environment? The maximization of profit and the preservation of the environment are intrinsically at odds. As long as environmentalism boosts earnings, the business community will support its goals. However, to protect Earth, the pursuit of wealth would have to be subject to further ecological considerations. Despite what is at stake, this transition seems difficult to implement.

Sustainability in the AI Urban Regime

As technology, particularly AI, becomes "the new currency of power"[14] and the United States Congress is recommended "to establish and fund a Manhattan Project-like program dedicated to racing to and acquiring an artificial general intelligence (AGI) capability,"[15] high-tech innovation and "smart-city" strategies, including decision-making algorithms, are surreptitiously becoming a normative framework for urban environments. In fact, the focus on high-tech capabilities and big data—in particular, AI technologies, automation, machine learning, deep learning, and the autonomous decision-making capabilities of bots—is diverting attention from the key problems (welfare, equity, and social justice) of urban governance. It also prevents a careful reconsideration of what constitutes a restructured "growth machine" capable of fostering livability in cities.[16] Legislators, policymakers, and society—rather than algorithms—could prioritize among the various objectives of the smart-city, in which innovation is framed in values imposed on the community. Regulatory frameworks must ensure trust without hindering innovation or development. Decision-making around AI would have to effectively integrate different smart-city domains by optimizing the communication between systems.[17]

Crucially, the high-tech's allegedly benign transformative capacity lies at the root of promoting (and even strategically misrepresenting) innovation ecosystems, intelligent environments, smart cities, and the new AI regime as key elements for achieving resilience and sustainability in the urban realm. The nascent urban AI regime focuses on developing solutions to enhance efficiency using bots that make autonomous decisions. However, there is a debate on how sustainability ought to impact the larger urban context or, conversely, how the larger urban socioeconomic context ought to be considered to frame the overall innovation and AI effort.

Smart cities and intelligent environments have been proposed as key approaches toward urban sustainability, under the assumption that the more data we have and the better we analyze them, the better decisions we can make. This outcome is not guaranteed because good, sustainable decision-making is not something we can infer from the data themselves, even if we use a robust data-driven process.[18] Good decision-making is prone to bias and is also a socially constructed and politically contested goal. We need to ask who benefits from good decision-making and according to what criteria. As a result of what deliberations among diverse interests are decisions made? Once the specific objectives of a megaproject have been set and specifically articulated, good decision-making is a moving target subject to context. It involves several other elements concerning relevance, approach, methodology, behavioral biases, and philosophical outlook.[19]

To this end, it is increasingly important "to evaluate projects and their impacts at different times, based on multiple criteria, to fully evaluate their performance. Success is often driven by political- and power-related factors." This is due to the strong political nature of stakeholders throughout the supply chain and their different underlying objectives; the success factors that are usually considered no longer seem sufficient (iron law, iron triangle). Innovative governance solutions are required to "align the interests of the different stakeholders in a complex environment with a large number of key actors."[20] Thus, the sustainability factors described in Chapter 6 can also be applied to AI megaprojects.[21] Accordingly, in complex adaptive innovation ecosystems, such as cities, an AI urban environment can be defined as sustainable if it is planned and executed to account for the capacity, fitness, resilience, diversity, and balance of its urban ecosystem, including the environment, economy, and community.

Enhancing data efficiency by optimizing big data management is not a substitute for good policies. The quest for urban sustainability would benefit from (1) a comprehensive sustainability policy in the growth machine encompassing the various components and targets of sustainability and (2) an ex ante collective deliberation of the outcomes to be achieved by implementing AI strategies. Thus, rather than conforming to the kind of sustainability potentially resulting from AI megaprojects, we need to revisit our vision for the design of a sustainable urban future, and then establish its main components and conditions.

Framing AI Innovation

As Dunn reminds us, to promote urban sustainability, we must evaluate the objectives that AI decision tools try to serve.[22] The linkage between artificial intelligence and sustainability requires a wider policy framework. Such a framework specifies priorities and expected outcomes so that we can precisely understand the conditions under which AI innovation leads to sustainable urban scenarios and outcomes. The nascent AI regime also needs to be scrutinized alongside the fundamental questions of (1) what are the benefits that megaprojects bring to cities, (2) which megaprojects ought to be built and which should be discarded, and (3) how can the wider socioeconomic impacts of large-scale development projects be mitigated?[23]

Class reference experimentation and test results regarding safety, explainability, transparency, and validity would contribute to improving the record of smart cities and AI disruptions in megaproject planning, implementation, and success.[24] Besides concerns about improved megaproject management via high-tech innovation and AI, researchers should develop analytical blueprints to account for megaprojects as socioeconomic particles that produce negative externalities and hugely negative societal impacts.[25]

Ultimately, the values and objectives of human development and sustainability must determine the necessary technological innovations. In other words, technological innovation and AI alone do not necessarily foster urban sustainability, even if specific innovations (technological or otherwise) do.

Further, in the case of innovations and AI technologies that foster sustainability, we need to factor in both their benefits and risks, and thus seriously consider matters of resilience or even "antifragility."[26] The key question, then, is how to craft and build a framework for sustainability that can discriminate and tell which AI innovations can contribute to the sustainability outcomes at hand, rather than hindering sustainability postinnovation. As in the case of the idea of "quality of life," the concept of sustainability probably needs to be redefined, articulated, and specified. In this process of conceptual and practical clarification, both common global trends and specific local conditions play an important role. Sustainability imagery comprises a complex set of assemblages, which suggests multiple interrelated components and scales for action in a baroque rhizomatic structure, inviting analysis via both complexity and transdisciplinarity.[27]

Cohesion vs. Competitiveness

One of the factors contributing to the complex nature of sustainability is that in conceiving and presenting the goals of preserving sustainable strategies, sustainability appears to be interdependent on the dimensions of entrepreneurship, innovation, and competitiveness of economies. Indeed, the goal presented to us in the majority view on sustainability is "green capitalism," that is, not a sustainable global society with clearly established limits to growth, but rather the sustainability of the information and knowledge economy, to which reformers and planners add more or less specific commitments to the environment.[28]

As an example of green capitalism, the three Chinese sustainable development zones—Shenzhen, Guilin, and Taiyuan—are presented as the leading axis in Chinese innovation, and implement the 2030 United Nations Sustainable Development Goals.[29] As discussed in Chapter 6, these zones will develop technologies for sewage treatment, waste utilization, ecological restoration, artificial intelligence, desertification-related technologies, and solutions for air and water pollution.[30] The three zones function as large sustainable innovation districts (innodistricts). In newly developed urban areas, achieving sustainability usually implies achieving environmental sustainability.

However, this does not guarantee the sustainability of innodistrict development in the knowledge economy.[31] This objective must be pursued comprehensively and holistically (United Nations SDG 2030), such that it is integrated into innodistricts, infrastructure projects, industrial corridors for advanced manufacturing, factories of the future, and other development projects. However, is this effort sufficiently efficient in the quest for sustainable megaprojects? Available evidence suggests that this may not be the case, which questions prevailing policies on urban socioeconomic sustainability and the primacy of the "megaproject solution."

One way toward this goal would be to rethink the competitiveness-cohesion binary at the time of policy experimentation under the perception that we are witnessing a shift away from the golden years of neoliberal practices. One shift in neoliberalism has been that, rather than being a source of competitiveness, social cohesion is seen as predominantly an effect and treated as a by-product outside of the concern of public policies. This has been portrayed as a widespread evolution in urban and national political economies, even though, to handle shifting policy issues and opportunities, various national welfare systems, political cultures, and socioeconomic realities combine and recombine. The focus of "competitiveness and cohesiveness" programs is on competitiveness. The cohesion component acts more as a consequence and is justified by its purportedly positive effects on competitiveness. Urban megaprojects (UMPs), as emblems of global capitalism, have become the main spatial manifestation of these dynamics. The adoption of neoliberal policies from the center to the periphery (and research on urban policy mobilities) has been made possible by this dominant discourse, which is imposed on various urban and regional realities as a decontextualized universal analysis and policy direction. The global nature of megaprojects also minimizes the significance of addressing contextual specificity in tackling local challenges. Consequently, there is a glaring lack of meaningful alternative agendas beyond place competition, knowledge economies, and cluster-based growth strategies.

Further, even if research highlights the significance of internal forces, disputes, and resolutions in urban communities, as well as the variety of responses to global concerns and interpretations of shared policies, we also know that the still prevailing neoliberal narrative affirms that more competitive cities have greater potential to create tools for social cohesiveness and address social issues. Urban entrepreneurship is more likely to trigger win-win scenarios, not only because these cities are likely to create better (and more) policy solutions, but also because the social dynamics in these cities have been (and are) directly impacted by the economic development they experience. However, increased economic competitiveness does not always translate into improved forms of social cohesion unless certain conditions are met. For example, we must develop new urban governance models, implement structural changes to promote competitiveness without increasing exclusion, stimulate low-skilled labor without marginalization, and organize new socialization and social control mechanisms.[32]

Despite the alleged propensity to balance their two components, competitiveness-cohesion discourses represent significant regression in equity policies and discourses. Cohesion discourses and policies prioritize protecting the most vulnerable from significant marginalization rather than promoting social justice. In Istanbul, one of the cases included in this book, global trade and export activities led to a significant increase in job opportunities, particularly for disadvantaged groups, such as women from low-income households. However, most of these positions had low wages and were unprotected. Residential segregation has decreased due to reforms in the educational system, increased opportunities for upward mobility, and rising female labor market involvement, even though the economic gap between lower- and higher-income groups in society remains wide. Households leave their neighborhoods as possibilities grow, and the growth of mixed-character neighborhoods is compelled by new opportunities offered by the urban land market. Evidence in this instance demonstrates that oversimplifying the relationship between social cohesiveness and competitiveness may be insufficient or even deceptive in comprehending the dynamics of development in various cities.

According to Krueger and Savage, there is a growing understanding that the availability of top-notch social infrastructure and natural resources is necessary to sustain economic prosperity.[33] For example, housing and transportation are essential for ongoing development. These more widely acknowledged connections appear to produce promising grounds for additional investigation into what causes economic growth and how larger agendas might be reshaped to promote subsequent development cycles.[34] Despite their almost utopian appeal, the results have been largely disappointing at the public policy level. It is this realm—that of public policy—that we see as holding the potential for utopian pragmatist policy experimentation in the quest for reconstruction, following neoliberalism's widespread damage to urban societies.

Many unanswered problems remain regarding the definitions of sustainability used worldwide and how it has been interpreted and used as a policy discourse. Reconfiguration of social cohesion via welfare statistics would not suffice as an effective strategy for urban sustainability without simultaneously refocusing on preventing further damage by neoliberal individualism and breakdowns in community structures. Strong communities serve as a barrier against social exclusion and alienation, and as a foundation for promoting social cohesiveness and stability. This is Etzioni's communitarian argument: that much social conduct is supported and influenced by the informal web of social ties and by the moral voices of the community. If people are social beings, powerful communities will provide desirable goods and prioritize societal and civic values over individual values, while acknowledging the interconnectedness and embeddedness of human life.[35]

Communitarianism positions itself as a complement to social democracy and would engender the promotion and strengthening of communal social ties.[36] There is a pressing need to get past the focus on traditional liberal-individualist ideas of society, including the identity debates, and discuss the practical issues of gendered violence, poverty, and unemployment, as well as the need to fight for safe working conditions, quality healthcare, and inclusive public education in urban areas. We must rediscover a language of universality, democratic rights, and secularist politics to do this, which leads to conceiving an active space for urbanites as both citizens of the polis and transdisciplinary co-creators of sustainable urban futures, as shown below in the section "Urban Experience."

Megaprojects and Public Value

Understanding "value" is a key component to improve urban sustainability and shift the focus away from building megaprojects. Value should be decoupled from, and go beyond, the market and the private sector. In fact, the issue of how value should be distributed arises when we acknowledge that it is created collaboratively by public, commercial, third-sector, and social movement organizations.[37] High-tech megaprojects must include fresh ideas for data ownership and governance that consider human intelligence as well as technological independence. Public investments in the underlying technology have directly benefited digital companies, and the data they monetize come from citizens. To maintain sustainable capital and resources for ongoing innovation, the benefits of innovation should also be shared, because value is created jointly via the participation of public and private actors. The issue of inclusive growth must also include ways to ensure that the production and distribution of values are constantly communicated. For this reason, narratives and stories about how and where value is created are important. If the story is that value creation and risk-taking only occur in the commercial world, value can, at best, be redistributed from risk-takers to others, with the public sector merely serving to reduce risk in the private sector. Instead, the question of how to ensure that not only risks but also rewards are socialized arises if we have a true community understanding of value, and the market itself is regarded as a result of how public and private spheres connect. Mazzucato argues:

> There are many publics in "the public." In a public value framework, contestation of actual value production and evaluation systems is a critical success factor. Involving civil society organizations to frame public policy goals (missions) is a central part of the co-creation process. Producing public value requires collaboration and co-creation; public value cannot be created using a top-down approach. Missions present an opportunity to put citizen participation at the heart of innovation policy and to directly connect (R+D) spending and broader policy measures to issues that matter to people [...] This is a "market-shaping" and "market-creating" role rather than a "market-fixing" role [...].[38]

The argument could be extended to all types of megaprojects and lead us to interrogate the concepts of action, ethics, responsibility, distributed morality, and the public. Planning tends to be haunted by a fraught relationship between plans, and the consequences meant to follow. Therefore, it is necessary to consider shifts and alterations in intentions, knowledge, actions, and consequences in a planner's work. In describing the formation of publics (alliances, assemblages), Robert Beauregard's new materialism is not interested in "facts" or "measures" but rather in "matters of concern." While matters of fact are a combination of models and measures, matters of concern also include actor networks that help entangle ideas and reality into the world, thus becoming more difficult to oppose. Thus, planners would need to connect knowledge to action "by translating between possibilities (truths) and the material manifestations of those possibilities (realities)."[39]

Does the public sector have resources to learn from and try out such things? Innovative thinking, flexibility, and adaptability are abilities that managers of private businesses teach. They are at the core of MBA programs globally in strategic management, organizational behavior, and decision sciences. The marginalization of the public sector in urban planning may reflect a lack of skills and knowledge. Over the last few decades, many public organizations have concentrated on improving the marginal efficiency of their operations, which has hurt their capacity to develop strategic goals (within the bounds of political and legal constraints) and align organizational resources to achieve these goals. Governments are becoming more aware of the need for dynamic capabilities; however, they do not have a theory to support this transformation. Policymakers and scholars recognize that relatively little is known about what makes governments and particular public organizations dynamic and capable of responding to shifting societal demands and requirements.[40]

Co-creative approaches see individuals as active participants in their communities rather than passive recipients of services, prioritize cultural change over immediate problems or successes, and involve the entire community rather than just a subset. Citizen-centered and co-creative approaches assist people in forming and promoting their own decisions, creating new stakeholder maps, building capacities for self-government, and developing open-ended civic processes rather than asking them to "plug into" already existing, predetermined programs, initiatives, or campaigns. Co-creation has the potential to be used as a tool to include citizens and demand openness and cooperation from the government. Co-creation has been used for some time in the private sector by businesses understanding the benefits of client feedback. However, it is also becoming more widely acknowledged as a useful tool for the public sector. A true community understanding of value enables the public and private sectors to work together on urban environments: co-creation could be applied to public-private partnerships (PPPs) in megaprojects,[41] thus contributing to reforming urban growth machines from the bottom up.

Deep Sustainability

Our conjectures on socioeconomic sustainability aimed at addressing priorities of urbanites away from the "megaproject solution" to urban development. We proposed some ideas that dispute existing strategic public policies and priorities. The following discussion on deep sustainability highlights some key elements intended for a wider framework where megaprojects and capitalist urbanization align with the boundaries and limits of the Earth system. Such elements are: (1) urban ecology and nature (as self-organizing complex systems); (2) low entropy (a desired sustainability state would be one of minimum entropy production); (3) materialities (contributing to the nonanthropocentric thrust of complex sustainability); (4) urban experience (on the role of agency in shaping human understanding and inquiry through civic frictions), and (5) transdisciplinary sustainability (the integrative analytical approach for complex sustainability).

Urban Ecology and Nature

Deep sustainability relies in that both urban ecology and nature—the urban and the natural—are self-organizing complex systems and can become intertwined beyond the nature-culture divide. In an increasingly urbanized society, "nature" is not "the other," but rather a coding of the harmonious coexistence of life with the natural world. People and the environment are shaped together, and this interplay is most visible in cities.[42] The word "natural" does not necessarily imply "nature." Natural concerns innovation, emergence, and capacity of complex adaptive systems for self-organization. The concept of the "ecology of reconciliation," in which the built environment is organized to suit the requirements of various species in addition to our own, can inspire inventive ecological designs. Innovative examples can be found everywhere: for example, in New York, the High Line Park was built on an abandoned rail line in Manhattan and encompasses nature, art, and design. These human-scale initiatives are easily accessible and frequently supported and maintained by the local community.

The romanticized view of nature and the artistic idealization of exurban areas have terribly affected the planet's ecological stability. The image of a "wild" or "natural" life encourages urbanites to tolerate many of the city's issues on a daily basis and positions cities as deserving of aesthetic scorn. Human conscience is imprisoned in a romantic ecologically consuming mode that turns the forest into a museum and permits the surroundings of a museum to be perceived as a temple of nature.[43] However, the concepts of cities and nature are intertwined. While cities are organized complexity, nature may be perceived as a self-regulating system of autonomous, harmonious interactions that, in the absence of interference from humans or other living things, always return to balance.

Yet, while this image is not inaccurate, it is incomplete. Species have always benefited from throwing things out of balance throughout their natural history, as seen in the emergence of eukaryotes billions of years ago.[44] Moreover, let us remember that Darwin did not praise the harmony of nature, but rather how geographic drift and climate change can cause small differences to suddenly become significant, and how various interspecies relationships can create new evolutionary vectors that can evolve in completely unexpected directions.[45] This scenario is comparable to the organized complexity of urban ecology. Anthropocentrism, which underlies the prevalent ecological vision, treats nature as an ideal that must be preserved (so that we might save ourselves) rather than as a community to which we belong. In our view, this is a fundamental ideological roadblock to deep, transformative, and complex sustainability. The nonanthropocentric thrust of deep sustainability needs to be taken into account as we respond to the disruptive complexity of megaproject impacts and strive for the viability of life, matter, urban ecology and nature in the Earth system.[46]

Low Entropy and Complex Sustainability

Deep sustainability also requires low urban entropy. Cities are extremely complex systems composed of several interconnected systems that are structured to utilize the environment for human growth. Thus, it is possible to consider a complex system as consisting of dynamic, interconnected, and ordered hierarchies with nonlinear results. Urban areas are filled with complex systems, including enterprises, educational institutions, religious institutions, municipal administrations, and governmental bodies. Cities can also be seen as a collection of different types of flows (people, money, energy, matter, information, etc.) structured to support the socio-economic growth of humans. Furthermore, cities are spatial entities with location, connections, and spatial variation that are interconnected with their surrounding ecosystems and landscapes as open thermodynamic systems exchanging matter and energy with the environment.

However, as complex systems, cities are expected to strive for a steady state characterized by nonequilibrium thermodynamics and minimal entropy creation.[47] To the extent that entropy plays a unifying role in the sustainability of the processes of development—economic, social, and environmental—we can hypothesize that a desired sustainability state would involve minimizing entropy production. Although thermodynamic considerations appear to favor systems that maximize entropy production, the system that ultimately develops may not be sustainable. Furthermore, understanding system feedback or, more generally, couplings between system components is crucial for comprehending sustainability in socio-technical–environmental urban systems. Multiple scales and dimensions—transcalar complexity, discussed in Chapter 7—contribute to this insight (local to global; ecological, social, and economic).

Another method for identifying feedback loops is *transfer entropy*. It gauges the degree of knowledge (reduction in uncertainty) that one variable's knowledge relates to another. A feedback analysis of interrelated megaproject impacts would need to be evaluated against the idea that a sustainability requirement should be a minimum production of entropy, along with the analytical arguments mentioned above, which run contrary to the disruptive complexity of entropic impacts by megaprojects, as discussed in Chapter 7. This sustainability requirement entails a viable and desirable collection of flows (people, money, energy, matter, information, etc.) that can be maintained despite internal and environmental changes. The capacity of these flows to be sustained and increased in the near future while responding to changes in population and/or population expectations is referred to as sustainability. Therefore, sustainability implies that an urban system is effective with low entropy, and redundancies and losses do not undermine the system's ability to continue performing its functions in the future.

Understanding resilience (how systems adapt or reorganize themselves in response to change) is another crucial aspect of sustainability in urban systems. After a structural change, the former less efficient paths may be easily taken to maintain the necessary flows because the urban system's ability to continue operating in the face of a significant disturbance seems to imply some level of redundancy or entropy.[48] Entropy can be considered a sign of the resilience of organizational structures in urban settings. Organizational systems are better prepared for intergenerational survival by minimizing their entropic tendencies when sustainability is strong or complex; responsive to changing economic, social, and environmental situations (as previously defined); and open to these changes.

Thus, complexity science, based on the notion of entropy, can help understand the interrelations between resilience and sustainability in urban settings. It is important to understand a system's susceptibility to change or its robustness, for instance, by measuring the brittleness or ductility of feedback. The role entropy and information must play in this comprehension is still to be determined; however, given the complexity of feedback traits and behaviors, assessing their impact on the sustainability goal-oriented trajectories of urban systems that integrates social and environmental sustainability with the process of economic growth calls for a comprehensive or holistic approach that complexity theory can offer.[49]

When we shift our attention from risks to actual impacts, megaprojects are "entropy generators," impacting on other urban subsystems. Our research shows that a single entropic subsystem (megaprojects) tends to develop entropy in other subsystems (the city as a whole). Ultimately, megaprojects undermine the sustainability of urban systems as a whole. Further, given the scope and consequences of the interrelated entropic impacts produced by megaprojects in urban areas the notion of a "sustainable megaproject" is meaningless from a complexity perspective. Some megaprojects may suit a "feeble sustainability" profile (discussed in Chapter 7), which implies that natural and man-made capital are freely substitutable. Weak sustainability holds that man-made capital is more significant than natural capital and easily replaceable. By contrast, strong (complex) sustainability acknowledges that natural capital cannot always be replaced by man-made capital. Building on our discussion of socioeconomic sustainability in this chapter, we believe that shifting from feeble to complex, strong sustainability can be facilitated through public value co-creation, sustainable innovation, and cohesion-building in network governance processes.

Sustainability, Mind, and Matter

Deep sustainability replaces the prevailing, anthropocentric idea of sustainability that evokes environmentalism without an environment, and ecology devoid of living creatures, except for human. A typical standard definition of sustainability is found in the 1987 Brundtland report: development that "meets the needs of the present without compromising the ability of future generations to meet their own needs."[50] "Generations" refer here to human generations, and the animate world is reduced to that which can satisfy human needs. This view is fundamentally opposed to that of Earth systems science, based on Lovelock's and Margulis' idea that the oceans and the atmosphere are completely intertwined with the living biosphere, and can be understood as a coupled system; thus, life in the biosphere is an agent of its own survival.[51] Transposing Earth system science and the mind-matter assemblage to urbanism is an interesting proposition because it can help set the foundation for understanding urban systems as complex assemblages of socio-technical–environmental vectors or dimensions that cannot be analyzed separately in any meaningful way.

Faced with a prevailing anthropocentric attitude, deep sustainability embraces new materialism's contention that the open gaze toward a radically transformed world and the observant attitude must prevail over the existing conceptions, visions, plans, analyses, or solutions based on the schemes that created the problems we need to solve.[52] In urbanism, materiality and assemblage thinking have found friendly ground.[53] The built environment may be comprehended from the outside by "the observation of concrete materials, not from the working of the isolated mind," as Sennett reminds us. Buildings, streets, and communities operate as dynamic organisms that change in response to how people interact.[54] A material conception of sustainability affects how we conceptualize space, place, scale, and context as "places" and environments that significantly interact with development planning. *Place* is to be seen as a relational space, a notion originating with Leibniz.[55] Thus, we can conclude that place does not exist until it becomes interactive in actor networks and subject to the network's dynamics.

Network relationalism is here metaphor conveying the complexity of trying to capture the multiple and changing relational dimensions of always-mobile assemblages and connections among norms, technologies, and life-worlds through networks of human associations, technologies, natural ecologies and places.[56] This strategy questions the central role of nonhuman elements in transdisciplinary research within urbanism and urban planning[57] in order to overcome what Alfred

North Whitehead named "the bifurcation of nature" expressed in the secular dichotomy nature-culture.[58] A focus on materialities is therefore a key component of non-anthropocentric, deep sustainability as presented in this book.

New materialism invites us to rethink life and matter, and recognize the planet as interdependent and interrelated.[59] This perspective emphasizes the importance of objects, things, and matter over the mind and ideas, a viewpoint with both supporters and detractors.[60] We must not know by defining the objects of knowledge but by responding to the immanence of vibrating matter, its influences, results, and consequences. In this sense, French sinologist François Jullien has stated that "a wise man does not have ideas" that are independent of matter.[61] By focusing on matter, we can move away from the secular attitude of placing humans at the center of reality and experience and instead appreciate the power of the "forgotten masses," that is, the artifacts that populate the world.[62] New materialities contribute to building a nonanthropocentric conception of complex, deep sustainability, showing the promise of meaningfully tying the political-economy analysis of capitalist urbanization and megaprojects with Earth system science at a time when the wicked problem of a "green transition" manifests itself as complex, systemic, interconnected, and urgent.

The Urban Experience

Alongside a new focus on matter, deep sustainability must be sympathetic to inclusive epistemologies that affirm ontological, critical realism while giving room for the shaping role of the knowing subject via perception, imagination, memory, and affects. This is important because the pretensions of pure objectivism in some interpretations of materiality—rejecting or downplaying the crucial role of the mind in shaping human understanding and inquiry—are difficult to defend. It is crucial not to misrepresent the causal capacities of nonhuman objects while considering the significance of the capacities of human beings. Human attributes, such as intuition, affect, and emotion, are pulses of sociomateriality.[63]

Such human shaping of socio-materiality needs to remain aware of the importance of civic friction zones and invoke a certain urban imagination that has to do with the practice of everyday life in the city, a practice disrupted by megaprojects. Urban megaprojects are neutralized spaces that wall off the differences between people and reflect our fear of exposure to others. They are spaces that remove the often-perceived threat of social contact: street walls faced in sheets of plate glass, highways that cut off neighborhoods and people from the rest of the city. Producing boundaries is clearly a political strategy, an often aggressive one, a strategy of power to control, obtain legitimacy, and neutralize or eliminate resistance in order to create hegemony in space.

By bordering, megaprojects can produce and enhance urban fragmentation. This is obvious for gated megaprojects, which are instances of socio-spatial segregation, as in citadels, enclaves, or ghettos. However, it is also present in other megaprojects that claim to be "open" spaces in the city. The debate by political economists on the "dual city" and later the idea of a "quartered city" echoed the analysis of fragmented societies.[64] It analyzed the dynamics of power produced by the workings of the global economy, which are reflected in the spatial dynamics of power at work through bordering during the planning and implementation of megaprojects. The negative experience of walls, borders, isolation, and detachment—all of this contrary to the ideal of inclusiveness—may be rooted in the centuries-old tendency to privilege the inner space of consciousness over the external space of the physical world.[65]

The origin of the blandness and neutrality of urban space, which we see exacerbated in urban megaprojects, can be traced back in history to the belief that the outside world of things is not as deserving as the inner world of the mind; this is equivalent to privileging the perspective of epistemology over ontology, over the conscience of the eye, and over the value of observations. Megaprojects often preclude experiencing the positive and unsettling value of mingling by encountering something alien. Sennett reminds us that we need to expose ourselves to the diversity of city life and develop an empathetic eye that perceives, affirms, and celebrates differences. Uncertainty, exposure, and discovery come from observing physical objects rather than from isolated operations of the mind and run contrary to the iconography of order, control, and border exerted by megaprojects. In Sennett's words,

> Like the 19th-century romantics, we discover that unity can bring isolation and withdrawal. Finally, sympathy is outlawed by wholeness. This insufficiency causes us to send out shoots to others whom we need to trust, a trust whose signs are only found in shared moments and spaces. Urbanites, including researchers, need to develop a creative art of exposure. We need to expose ourselves to the diversity of city life and develop an empathetic eye that perceives, affirms, and celebrates differences. Uncertainty, exposure, and discovery result from a focus on the observation of concrete materials, and not on the workings of the mind in isolation. This outward turning involves the renunciation of certain impulses to wholeness and completion. This is what the city needs to invite an ethical vision.[66]

The deep sustainability of an urban experience where citizens co-create public value through the *ethos* of exposure to difference also requires a solid sense of territorial identity tied to local distinctiveness and universal values: *topophilia*.[67] Instead of emphasizing the axes of difference or social power relations, universal human experiences are captured under the premise that relationships between culture and the environment and between topophilia and the environment help to understand the formation of human values. Yi-Fu Tuan believes that, since humans are one of the "dominant" factors controlling the environment (an idea that is fully in force today in the "Anthropocene" concept), consideration of environmental problems must take into account the degree of affinity of humanity toward that environment. Therefore, topophilia provides a comprehensive and comparative view of all affective ties of a human being with the material environment.

Cultural differences affect attitudes and values. They also affect how those attitudes and values are reflected in art, literature, landscape architecture, in the dreams of an "ideal world" and in aspirations toward "contentment and joy," as well as in our "landscapes of fear."[68] The enriching aesthetic experience is linked to a balance between the cosmos and home (hearth) as two spaces that anchor the human being in happiness.[69] "Hearth" is our house and neighborhood, family and relatives, habit and custom, our point of view, the subject who knows. The "cosmos," on the other hand, is the larger reality: the world, civilization, humanity, and what we contemplate and observe with the intention of understanding. Tuan encourages simultaneously embracing one's own culture and curiosity about the world, a way to be at home in the cosmos and conceiving the home as a place in the world. Topophilia, as a component of deep sustainability, contributes to resisting megaprojects—and, ultimately, to grassroots action—by strengthening the scale of experience and community, and contributes to the "landing on Earth," which, according to Bruno Latour, is what we need to "face Gaia" and begin to seriously tackle the perils of the new climate regime.[70]

Transdisciplinary Sustainability

Both socioeconomic and deep sustainability—that is, complex sustainability—require new analytical tools (or transforming the ones we have) and a more integrative thrust to capture and understand, without falling victim to reductionism, the full web of heterogeneous, dynamic, and changing assemblages that cause unpredictability due to uncertainty, indetermination, undecidability, and incompleteness of the Earth system in the Anthropocene.[71] With this understanding, viable strategies to contain systemic risks can be forged amid a "somber clarity of chaos," which does not invite us to expect a new order in the near future.[72] Both new materialism and transdisciplinary approaches point in this direction.[73] In both approaches, overcoming binary logic and the distancing of anthropocentric approaches are postulated, and the focus is on complexity, hybridization, nonlinearity, reflexivity, and heterogeneity. In both cases, it is assumed that sustainable development planning does not occur in the context of determinism, where control, causes, and effects can be used to establish predictions. Instead, we have a high probability of finding events yielding extremely negative results, which we observe in the case of the disruptive impacts of megaprojects. Sustainability entails the ability of systems to absorb disturbances and to evolve and co-evolve with other systems with which they interact. Therefore, it seems reasonable that sustainability-related policies (pursuing a normative transformation of social organization and economic activity) should be designed from a transdisciplinary perspective. This allows the formulation of questions relevant to addressing systemic problems in changing environments alongside analytical tools that contribute to fully understanding policies' complex nature in a pragmatic and integrative manner.[74]

Indeed, a transdisciplinary approach is advantageous for understanding the complexity inherent in sustainability science, as pursuing sustainability requires grasping and managing unprecedented and interconnected challenges. Increasingly, science and knowledge production is geared toward overcoming classic disciplinary questions and approaches, integrating the perspectives of different *stakeholders* (experts, citizens, academics, and professionals) and is particularly receptive to contextual differences and local knowledge.[75] In addition, transdisciplinary approaches to sustainability are typically focused on scientific research on new institutional and technical alternatives as well as cooperation and risk containment. Knowledge innovation initiatives are essential to better match resource use practices with diverse ecological and socioeconomic contexts and provide flexibility in the face of unforeseen changes.[76]

The practice of transdisciplinary research still needs to be significantly developed. There is no single research framework with a common glossary or shared communication platforms. A transdisciplinary attitude and practice seeking complexity and integration has not yet been capable of producing a shared instrumental canon. Nevertheless, it fulfills its function by raising awareness about the need to co-create knowledge in the interstices between disciplines. From these gaps, we can clearly observe the assemblages of material sustainability, which is not a problem to be solved but a complex normative strategy to be designed. We need to understand the mechanisms and purposes to manage them effectively and handle them appropriately in the context of socio-ecological concerns and alarms, structural uncertainty, and global risks.

Conclusion

This chapter has meaningfully framed responses (resilience and contestation) to megaproject disruptive complexity and has provided some foundations for sustainable policy frameworks within the notion of *complex sustainability*. We have built this notion as an assemblage of two main components, socioeconomic sustainability and deep sustainability, which are intended to address the urban question and the Earth system question in the face of relentless megaproject urbanization.

The question about *socioeconomic sustainability* has revolved around four axes: the need to go beyond green capitalism (or decoupling sustainability from profitability), sustainable innovation in the new AI urban regime (or framing innovation around sustainability goals), cohesion vs. competitiveness (or finding ways to design equity policies with a local focus), and public value (or developing bottom-up market-shaping and value-eco-creating strategies). These four axes represent fundamental vectors or directions for urban sustainability that dispute existing strategic public policies and priorities. They aimed at addressing priorities of urbanites away from the "megaproject solution" to urban development.

We have defined and articulated *deep sustainability* around the following discussions: urban ecology and nature (as self-organizing complex systems), low entropy (a desired sustainability state would be one of minimum entropy production), materialities (contributing to the nonanthropocentric thrust of complex sustainability), urban experience (on the role of agency in shaping human understanding and inquiry through civic frictions), and transdisciplinary sustainability (the integrative analytical approach for complex sustainability). The discussion on deep sustainability highlighted some key elements intended for a wider framework where megaprojects and capitalist urbanization align with the boundaries and limits of the Earth system.

Megaprojects are inherently unsustainable endeavors, and their construction compromises any attempt (even modest) to achieve urban sustainability. Complex sustainability (socioeconomic, deep, nonanthropocentric, resilient, and transdisciplinary) can work in combination with current global sustainability efforts, such as the UN SDGs.[77] As we know from climate science, several environmental risks have already materialized, others are rapidly materializing, and we are already in a world of impacts. In the current "green capitalism" regime characterized by the systematic use of environmental sustainability strategies for private profit, the official commitment by governments and megaconstruction companies to "sustainability" seems a feeble effort. Socioeconomic inequalities, labor market insecurity, and widespread inequity have increased after the 2008 financial crisis (the Great Recession), the covid-19 pandemic, and the consequences (steep inflationary pressures and supply chain blockages) of increased geopolitical confrontation, including war in Europe.

Our analysis of megaproject disruptive complexity (megaproject entropic impacts) contributes to further understanding the perils of continuing to build megastructures in cities. Building megaprojects contradicts the creation of effective solutions to the global impacts of both capitalism and climate change. When dealing with impacts, we must remember that the impact of human action far exceeds our current knowledge. Unintended consequences increase faster than we know in a scenario of complex assemblages and nonlinear science, where relationships, interdependencies, trajectories, and transitions matter more than structures and entities. Thus, our future knowledge can incorporate complex sustainability by implementing sustainable design proposals. While keeping in mind the planners' fallacy, this chapter has shown that "futuring" involves the use of biological cognitive capacity, the recognition that humans are inherently a part of the Earth system, that the design of the social ought to begin in the environmental frame, and that innovation in material culture needs to develop by need rather than supply. These conclusions—pointing toward planetary boundaries and the notion of limit more generally—unavoidably clash with megaprojects as a viable strategy for urban development.[78]

Part IV
Conclusions

Chapter 9
Contesting Megaprojects

In this book, we have argued that megaprojects are not merely transformative, but rather inherently disruptive. We have shown that they are incompatible with any approach to urban sustainability that prioritizes the well-being of urbanites and the boundaries of the Earth system. In fact, the megaproject solution to urban development undermines efforts to tackle the global impacts associated with both the destructive creation of capitalism and the new climate regime. The reason for this state of affairs is not cost overruns and poor megaproject management, but the disruptive impacts that megaprojects have on urban areas. Thus, we have emphasized the megaproject impacts imperative: the importance of identifying impacts as a prerequisite for critically assessing whether megaprojects should be built.

Within these pages, megaproject impacts have been examined empirically, conceptually framed, and positioned vis-à-vis urban sustainability. We have shown the similarities among the projects analyzed, despite very different political, social, geographic, and cultural contexts spanning the past 50 years. Megaproject development and impacts are linked to key causal forces, relevant governance systems, and related developmental discourses.

We have argued that these increasingly important and deeply disruptive projects need to be situated (embedded) within multiscalar political, economic, and sociocultural systems and contexts. We also stressed that large projects have highly complex and disruptive impacts that tend to damage or inhibit long-term urban sustainability. As demonstrated throughout these chapters, even when the best combination of planning, building, and management strategies has been implemented, megadevelopments will significantly disrupt urban structures and ways of life.

Megaprojects are insensitive to their environment. They are inherently disruptive. They exacerbate inequalities and destroy the connectivity that would otherwise encourage urban creativity and community. Megaprojects are complex adaptive systems. Thus, we contend that their embeddedness in societies and economies begets complexity—disruptive complexity—as the central organizing and explanatory

G. del Cerro Santamaría, *Contesting Megaprojects*,
https://doi.org/10.1007/978-3-031-98382-5_9

principle to describe the myriad, interrelated, cascading, and systemic impacts they have.

Through *disruptive complexity*, we showed that cost overruns and poor management—although both common—are not the only or worst effects of large projects; rather, their impacts are the primary reasons to criticize them. In turn, our focus on *disruptive complexity* has led us to assess a possible response scenario—*complex sustainability*—to the disruptions they cause, with two components: *socioeconomic sustainability* and *deep sustainability*. We have presented transdisciplinarity as the most robust and nonreductionist analytical strategy for understanding megaproject impacts, and the prospects for urban sustainability within the world economy.

While some projects are planned, designed, and built as megaprojects, we have departed from any mechanistic, quantitative, and universal definition based exclusively on cost, size, or scale. We believe this approach is a shortcut to bad theory. Instead, we emphasize embeddedness and impacts as constitutive elements in our approach and definition. From our relational and topological perspective, we have seen that projects become *mega*projects due to their regional economic and environmental impacts and the resulting complexity of its relational external stakeholder management challenges.

Just as humans have always strived to harness nature, governments and corporations continue to undertake significant construction projects that create and disrupt built environments, megacities, and megaregions; that transform the planet and the way humans move through it; and that deeply affect the processes of production, consumption, power, and meaning. At an earlier time, large infrastructure projects were constructed to meet the needs of a burgeoning industrial civilization. This impulse reached astounding levels in the 20th century and continues today with an even greater intensity, reach, and scope.

As we have shown, there are reasons to be highly concerned about these disruptive projects, particularly when their impacts and prospects for sustainability become the center of analysis. We have tried to show that the path of megaproject development is not compatible with any meaningful practice of urban sustainability and can in fact undermine it.

Repertoires of Disruption

Landscapes of Disruption

Chapter 3 presented and discussed urban megaprojects (UMPs) in three cities—Istanbul, Sydney, and Hong Kong. Evidence showed that, regardless of context or location, UMPs are disruptive and contentious enterprises that trigger important economic, sociopolitical, and environmental challenges that are often difficult to resolve for the benefit of urban communities. These cases are an illustration of the prevalence of various political and economic interests at the expense of most urbanites—through autocratic rule (Istanbul), erosion of intent (Sydney), and misrepresentation of targets (Hong Kong).

Megaproject impacts are particularly disruptive when the combined effects of simultaneous, interrelated, and interdependent impacts produce cumulative and cascading effects in the affected urban areas and populations. Would better management—based on avoiding cognitive bias among planners, developers, and managers—cancel the disruptive impacts that emerge after project implementation? The evidence presented in Chapter 3 shows that any application of optimal rationality in decision-making would not prevent a megaproject from negatively disrupting affected communities in significant ways. This is particularly so when communities and citizens are excluded from stakeholder roles and from management and governance roles after projects are built.

Moreover, we have shown that the impacts of UMPs on Istanbul, Sydney, and Hong Kong strongly suggest that the bioeconomy and biodiversity can be vulnerable to these developments. Several factors should be considered to enhance urban resilience and liveability: (1) mediating between private and public sector interests; (2) promoting greater involvement of local and regional governments for projects executed in their communities; (3) engaging citizens more proactively in consultation and direct involvement before, during, and after UMP development; (4) maintaining local heritage as a central component of any UMP agreement; and (5) preserving the environment across air, water, and land.

These are the main challenges for urban leaders and policymakers who share the goals of the economy, environment, and equity in urban governance. Under the rising *green capitalism* regime, recent megaprojects—such as the ones discussed in Chapter 3—are presented as sustainable. However, it remains unclear under what conditions these allegedly sustainable constructions can foster equity, liveability, and shared prosperity while mitigating ecological damage.

Design After Spectacle

Chapter 4 showed that UMPs function as emblems of global capitalism. By analyzing four cases of urban reimaging via iconic cultural projects, this chapter unveiled the controversies and disputes surrounding contested iconicity in Bilbao, Denver, Helsinki, and Abu Dhabi. We have seen that Bilbao Guggenheim's success is difficult to replicate and that the Bilbao Effect is fading due to diminishing returns of serialized megaconstructions and urban icons all over the world. The chapter also showed the main disruptive impacts of emblems (as a consequence of the need for visibility, the focus on spectacle and entertainment, and the traveling of icons via urban policy mobilities).

We then outlined the prospects for architecture after the neoliberal period, for which we suggest a focus on tectonics, matter, memory, and the practice of places, as well as a shift toward a fuller embrace of complexity and transdisciplinarity by researchers and practitioners. By embracing these tenets and their thrust of resistance, we believe that architecture and urbanism would be better prepared to fulfill their promise of improving human settlements. To be sure, the nature and shaping of projects is due to the developmental, institutional, and geographic contexts in which they emerge. However, UMPs constitute contentious enterprises that produce disruptive complexity, controversy, and criticism, which is clearly observable if we focus on impacts rather than risks. This is so regardless of the geographical location and socioeconomic context of UMPs.

Framing Innovation

In Chapter 5, we saw that smart UMPs that include digitalization and virtualization components have intensified since the advent of the coronavirus pandemic to form a perverse high-tech regime. Some refer to this as the rise of *disaster capitalism*, a *screen New Deal*, or *tech shock doctrine*. This shift has been justified by some as a resilient solution that bypass and minimizes the roles of humans seen as biological hazards.

High-tech innovation and AI continue to be presented as transformational tools for greater efficiency in megaproject management, economic growth, and urban sustainability. Researchers would benefit from developing analytical blueprints to account for the fact that smart construction produces negative externalities, including hugely negative societal impacts. Ultimately, human development and sustainability values should determine what technological innovations are required. In other words, technological innovation and AI alone do not necessarily foster urban sustainability, even if specific innovations (technological or otherwise) do.

Certainly, AI can help to improve megaproject management. However, if we privilege megaproject *embeddedness* in the wider economy and society, it is doubtful whether high-tech enhancements by themselves will contribute to greater urban sustainability. Urban sustainability *for all* must take precedence over high-tech innovation and AI schemes to preserve life on Earth. Thus, rather than standing by—either sleepwalking through the Anthropocene or accelerating technological innovation with no checks—we gain from working backward, identifying the kind of sustainability we want, and developing the types of innovations that can make it happen.

Intelligent disruption (destructive creation) produced by AI systems is no longer avoidable. But there should be discussions around the notion of smart and fair sustainability—a normative horizon from which we can decide what technological innovations and AI applications we need. As we know from Earth system science, we live on a planet that works as a complex, self-organizational, and self-regulating system. Sustainable life and development on Earth thus require a sustained and more robust global coordination effort for target setting.

Feeble Sustainability

Chapter 6 presented solid evidence of UMP impacts on the environment. It also analyzed some of the strategies used to develop allegedly sustainable UMPs to counter environmental damage. Owing to their considerable environmental effects, planners, developers, and other stakeholders are setting goals for acceptable levels of environmental sustainability as a top concern.

However, achieving environmental goals does not guarantee UMP sustainability. This objective must be approached holistically. UMPs epitomize a sustainability logic based on investing in the formation of urban regions as strategic scales for capital accumulation. However, this logic typically neglects socioeconomic sustainability issues and has been contributing to *feeble sustainability,* whereby several profit-making eco-friendly innovations and strategies, widely used in recent megaproject construction, coexist with the steady deterioration of our habitable climate while socioeconomic disparities and inequalities persist.

We proposed some basic elements to understand UMP sustainability as complex and multidimensional. It is key to pay attention to the complexity observed in the shifting interactor networks of decision-making—the development of preferences, perceptions, coalition-building activities, and choices by different stakeholder groups—and to the impact on the outcomes by structural and contextual contexts (regional and urban) in which they operate. Research on complex sustainability would aim to explore the actual significance and effects of decision-making when there are overlapping webs of interdependence between key groups of urban stakeholders—that is, political parties, urban officials, technical experts, advocates, and civil society groups seeking to transform local growth machines.

Disruptive Complexity

Chapter 7 defined impacts as *emergent properties* that cannot always be predicted in advance in any meaningful or comprehensive manner. We defined *disruptive complexity* as the complex web of interrelated and interdependent impacts that emerge after UMP construction and that affects the life of a large number of citizens, particularly those who are more exposed and vulnerable. The chapter offered a typology of UMP disruptive impacts: socioeconomic; environmental; spatial/territorial/visual; political (including geopolitical); financial; cultural; and systemic.

Megaprojects are complex systems (organizationally and scale-wise) embedded into urban, political, and socioeconomic systems. They are path-dependent and open systems. A characteristic of these complex systems is that they are driven by anti-entropic effort that facilitates their adaptation and survival. Such an anti-entropic effort can be viewed as resilience or systemic energy—the ability of any system to maintain continuity after shocks or catastrophes while contributing positively to adaptation and transformation.

Urban resilience is tied to sustainability as one of the forces working against the entropic impacts caused by UMPs. Similarly, contesting UMPs works toward a sustainability goal by integrating communities into the planning process. The three pillars of sustainable development—society/community, economy, and the environment—can be equally promoted through the concepts of intergenerational and intragenerational equity. Intergenerational equity maintains the quality of natural ecological systems and their services over time, while intragenerational equity promotes equal access to resources within the current generation and providing basic needs.

Green and high-tech capitalism claim to contribute to sustainability while neglecting its socioeconomic dimension. To plan and build sustainable projects, we need a multidimensional systems approach based on transdisciplinarity and a new knowledge generation agenda *vis-à-vis* the urgency to understand the challenges and opportunities in a rapidly urbanizing world. This effort needs to factor in the idea that both cities and nature belong to the realm of organized complexity and thus the notion of an urban ecology would deal with the emergence and self-organizing power of complex adaptive systems. This entails developing a nonanthropocentric notion of sustainability, as discussed in Chapter 8.

Contesting Megaprojects

Chapter 8 responded to megaproject disruptive complexity (resilience and contestation) and offered guidelines for sustainable policy frameworks. We crafted the notion of complex sustainability as a combination of two main components—socioeconomic sustainability and deep sustainability.

Socioeconomic sustainability revolved around four elements: (1) the need to go beyond green capitalism (or decoupling sustainability from profitability); (2) sustainable innovation in the new AI urban regime (or framing innovation around sustainability goals); (3) cohesion versus competitiveness (or finding ways to design equity policies with a local focus); and (4) public value (or developing bottom-up market-shaping and value-creating eco-strategies). These four elements represent fundamental vectors or directions for urban sustainability that dispute existing strategic public policies and private priorities.

Our discussion on deep sustainability highlighted some key elements intended for a wider framework where megaprojects and capitalist urbanization align with the boundaries and limits of the Earth system. We defined deep sustainability around (1) urban ecology and nature (as self-organizing complex systems); (2) low entropy (a desired sustainability state would be one of minimum entropy production); (3) materialities (contributing to the nonanthropocentric thrust of complex sustainability); (4) urban experience (on the role of agency in shaping human understanding and inquiry through civic frictions); and (5) transdisciplinary sustainability (the integrative analytical approach for complex sustainability).

Socioeconomic challenges and inequalities, labor market insecurity, and widespread inequity increased after the 2008 financial crisis (the Great Recession). Conditions worsened with the covid-19 pandemic, increased geopolitical confrontation, war in Europe, steep inflationary pressures and supply chain blockages. These conditions warrant our focus on socioeconomic sustainability. From a socioeconomic viewpoint, contesting megaprojects means contesting capitalism, given that UMPs constitute an expression of trickle-down urbanism that is closely tied to fundamental debates around the political economy of capitalist development—spectacularization, the right to the city, social justice and the crisis of democracy. Shifts and marked changes in urban, regional, and national political economies would be required to deflate the megaproject solution as a predominant developmental trajectory. Thus, our notion of complex sustainability goes beyond feeble sustainability (green capitalism's sustainability) and suggests growth-machine transformations and business model redesigns.

Maintaining megaproject development worldwide is incompatible with the limited resources of the Earth. The massive disruptions megaprojects cause make them a key constitutive element in the characterization of megaurbanization as a global risk—materializing into real impacts—of a magnitude similar to technological disruption, nuclear proliferation, or climate change. In this context, the official commitment by governments and megaconstruction companies to sustainability (environmental sustainability) clearly is a feeble effort. Instead, complex sustainability—socioeconomic, deep, nonanthropocentric, resilient, and transdisciplinary—aims at tackling capitalist urbanization's destructive impacts while protecting the Earth system, and thus enhancing, deepening, and renewing current global sustainability efforts such as the UN SDGs.

Looking to the Future

This book has shown that megaprojects are inherently unsustainable endeavors, and their construction compromises any attempt, however modest, to achieve urban sustainability. Our analysis of megaproject disruptive complexity can contribute to further understanding the perils of continuing to build megastructures in urban areas. This megaproject solution is contrary to any reasonable prospect of effectively addressing the global impacts associated with both the destructive creation of capitalism and the new climate regime. When dealing with impacts, we must consider that the impact of human action far exceeds our current knowledge. Unintended consequences increase faster than we know in a scenario of complexity and nonlinear science, where relationships, interdependencies, trajectories, and transitions matter more than structures and entities do. Thus, complex sustainability entails learning about the future by proposing a sustainable future design.

While keeping in mind the planners' fallacy, we tried to show that *futuring* involves the use of biological cognitive capacity and the recognition that humans are inherently a part of the Earth system. It also involves that the design of the social ought to begin with the environment, and that innovation in material culture needs to be developed by need rather than by supply. These conjectures—pointing toward planetary boundaries and the notion of limit more generally—unavoidably clash with megaproject endeavors. In addition to a problem of reach, we see one of scale. Megaprojects neglect and disrupt the lived experiences and dignity of urbanites in the name of megascale governance and profit. A period of social and socioeconomic experimentation like the current one may be propitious for trying out alternatives to megaproject development and to the underlying ideology that understands prosperity as development and development as growth.

It is paradoxical, but not surprising, that UMPs keep being built in the face of their massively disruptive impacts, their incompatibility with meaningful and deep sustainability, and the diminishing returns of icons. These disruptive artifacts represent a case of *costly persistence* in the face of a greater environmental disaster. Challenging megaprojects affects bottom-line costs. Owing to strict planning regulations, more available legal options, and a denser population, residents of areas where UMPs are being built are more likely to oppose them, making the expense of legal defense more expensive.

Similarly, stricter environmental restrictions frequently cause construction to take longer and incur more costs. However, the question about which projects should be built is not a cost overrun or a performance evaluation question, but rather a civil society and democracy question in which citizen criteria and resistance goals ought to be prioritized in the process of placemaking. In this desired scenario, civic participation, political conflict, and institutional creativity would coexist, and plans for UMPs would be subject to participatory urban democracy.

European and North American countries have passed laws that provide citizens with legal defense against megadevelopments. This translates into additional expenses, both directly and in terms of hiring legal counsel to guarantee compliance with law. This suggests that cost overruns are neither inherent nor the only or most fundamental problem regarding megaprojects. Attention ought to shift to the fact that global megaconstruction and other infrastructure seriously harm the environment and accentuate socioeconomic tensions in urban areas, thus undermining prospects for urban sustainability. At a minimum, people who are in charge of evaluating, supporting, and funding these programs must engage not only those who stand to gain from them, but also those who stand to lose. The essential elements include transparency, improved public participation, and a detailed examination of their direct and indirect impacts and repercussions. Such measures, aimed at improving democratic control, can help prevent significant political and economic interests from obscuring serious environmental and social issues.

As we move past neoliberalism and the covid-19 pandemic, science shows Earth to be advancing toward irreversible climatic tipping points while there seems to be no consistent, all-encompassing and effective enough platform to guide policy. However, some directions seem a priority—low-carbon, resilient, and sustainable development; measuring economic well-being away from GDP; reducing hyper-consumption and limiting consumerist behaviors; finding relevant ways of formulating environmental policy; and crafting effective pathways for collaboration among citizens and private and public actors. Many nations seem to be combining globally coordinated efforts—such as the UN SDGs—with the adoption of experimentalist, iterative, and evolutionary methods for policy change. This results in a landscape that may seem chaotic and incoherent, but that also makes it easier for nations to create policy solutions tailored to their own unique problems and settings. Contesting UMPs through the notion of complex sustainability dovetails this context of policy shifts and radical transformations.

It should be noted that current sustainability discourses rely heavily on sociopolitical norms, in the form of ethical, technological, and scientific standards. The sustainability community can thus better grasp how such proposals portray and shape the future of sustainability by analyzing the normative structure and dynamics of proposals, such as *complex sustainability*. By simulating normativity in sustainability discourse, we can better understand how our idea will ultimately impact the future trajectory of sustainable development. An important, new, and difficult area of research in sustainability science is the explicit representation of such normative functions in a model that is appropriate and considers the complexity of several diverse aspects and interconnections. To achieve qualitative change in the context of sustainability, normativity would need to be viewed as a complex framework and be integrated in a sophisticated and transparent way for knowledge and action toward *just sustainability*—encompassing both environmental and socioeconomic justice. Such efforts may contribute to developing planetary consciousness. Such consciousness would help us understand and take seriously the notion of limits and boundaries, and the endgame of building prosperity without unnecessary, disruptive, and unsustainable growth.

Notes

Chapter 1

1. G20 Research Group (2015) *Mobilizing Climate Finance*, Toronto, CA: University of Toronto.
2. Flyvbjerg, B. (2014) *What you should know about megaprojects and why*; See also Fay, M., H. I. Lee, M. Mastruzzi, S. Han and M. Cho (2019) *Hitting the Trillion Mark*.
3. Rockström, J. et al. (2023) *Safe and just Earth system boundaries*.
4. For a critical view of the Green New Deal, see Martin (2020) Abolish Oil: From Green New Deal to Green Reconstruction, *Places*, pp. 1–16.
5. See United Nations COP 27, and IPCC Synthesis Report (2022), https://www.ipcc.ch/ar6-syr/.
6. Beiser, V. (2019) *The World in a Grain*.
7. See Wendy Brown (2017) *Undoing the Demos. Neoliberalism's Stealth Revolution*, Cambridge, MA: Zone Books.
8. Keane, J. (2018) *Power and Humility*.
9. Buchoud, N. J. A., M. Keith, S. Parnell and M. van der Pütten (2022) *A G7/G20 Pact for Sustainable Urbanisation? Building on the Positive Legacy of Italy's G 20 Legacy*, Istituto Affari Internazionali, https://www.iai.it/en/pubblicazioni/g7g20-pact-sustainable-urbanization.
10. See, for example, Flyvbjerg, B. (2014), ibid.; Flyvbjerg, B. ed. (2017) *The Oxford Handbook of Megaproject Management*; Altshuler, A. and Luberoff, D. (2003) *Mega Projects: The Changing Politics of Urban Public Investment*.
11. del Cerro Santamaría, G. ed. (2013) *Urban Megaprojects. A Worldwide View*.
12. Hirschman 1967a: vii, xi, *Development Projects Observed*.
13. See, inter alia, Flyvbjerg, B. (2014) *What you should know about megaprojects and why*; Flyvbjerg, B. ed. (2017) *The Oxford Handbook of Megaproject Management*. Söderlund, J., S. Sankaran and C. Biesenthal (2017) *The Past and Present of Megaprojects*.

G. del Cerro Santamaría, *Contesting Megaprojects*,
https://doi.org/10.1007/978-3-031-98382-5

14. Some recent research on megaprojects has adopted a complexity perspective in dealing with sustainability. However, such research focuses on megaproject management issues. See, for example, Favari and Cantoni (eds.) (2023) *Complexity and Sustainability in Megaprojects: MeRIT Workshop 2022*.
15. Robinson, J. (2022) *Comparative Urbanism. Tactics for Global Urban Studies.*
16. Randolph, G. F., and M. Storper (2022) *The great theory debate on urbanization in the Global South.* See also Rokem, J., and Boano, C. (2020) *Urban Geopolitics. Rethinking Planning in Contested Cities*, a book with similar objectives of comparing across different cities and building global urban theories by focusing on common traits. Proposing or establishing a global theory of megaproject impacts does not exclude or contradict the possibility of refining the analysis by identifying regional differences within the global theory.

Chapter 2

1. Among the vast literature on urban projects, many works have had a direct or indirect influence on this book. A list of selected publications of particular note is as follows: Sorkin, M. (1992) *Variations on a Theme Park*; Hughes, T. (1998) *Rescuing Prometheus*; Fainstein, S. (2001) *The City Builders*; Olds, K. (2002) *Globalization and Urban Change*; Marshall (2002) *Emerging Urbanity*; Altshuler, A. and Luberoff, D. (2003), *Mega Projects: The Changing Politics of Urban Public Investment*; Moulaert, F., Rodriguez, A. and Swyngedouw, E. (2003), *The Globalized City*; Flyvbjerg, B., Bruzelius, N., Rothengatter, W. (2003) *Megaprojects and Risk*; Diaz Orueta, F. and S. Fainstein (2008) *The New Mega-Projects: Genesis and Impacts*; Brunn, S. D., ed. (2011) *Engineering Earth*; Galloway, P. D. et al. (2012) *Managing Gigaprojects*; OMEGA Center (2012) *Mega-Projects: Lessons for Decision-Makers*; del Cerro Santamaría, G. ed. (2013) *Urban Megaprojects. A Worldwide View*; Priemus, H., and van Wee, B. (2014) *International Handbook on Mega-Projects*; Flyvbjerg, B. ed. (2017) *The Oxford Handbook of Megaproject Management*; Hallock, B. E. and J. G. Zack Jr. (2018) *What Have We Learned From Megaprojects?*; Schindler, S., S. Fadaee, and D. Brockington (2021) *Contemporary Megaprojects*.
2. Harvey (1989) *From managerialism to Entrepreneurialism*
3. A concept discussed by Grubbauer (2013) *"Global" architecture as a contradictory signifier*.
4. Bunnell (2013) *Encountering Kuala Lumpur through the "travel" of UMPs*
5. Mazzucatto, M. (2015) *The Entrepreneurial State*; Castells, M, and Himanen, P. (2014) *Reconceptualizing Development in the Global Information Age*.
6. This Chapter section, as well as the next one (Growth Machines), draws on del Cerro Santamaría, G. (2019) Megaprojects, Development and Competitiveness. Building the Infrastructure for Globalization and Neoliberalism, *Athens Journal of Social Science, 6* (4), 263-291, October.

7. Harris, M. (2017) *Competitive Precinct Projects.*
8. Moulaert, F., Rodriguez, A. and Swyngedouw, E. (2003) *The Globalized City*, p. 551
9. Brenner, N., Theodore N. (2005) *Neoliberalism and the urban conditions.*
10. Narrated by Sklair in The Icon Project (2017).
11. del Cerro Santamaría, G., ed. (2013) *Urban Megaprojects. A Worldwide View.*
12. Harris, M., ibid. (2017), p. 65.
13. Siemiatycki, M. (2013). *Riding the wave.*
14. Zimmermann, J., & Eber, W. (2014) *Consideration of risk in PPP projects.*
15. Marcuse, P. (1997) *Glossy globalization.*
16. Logan J R and Molotch H (1987, second ed. 2007) *Urban Fortunes: The Political Economy of Place.*
17. Larson, S. (2013) *Building like Moses with Jacobs in Mind*; Berube, A., & Kneebone, E. (2013) *Confronting Suburban Poverty in America*; Zukin, S. (2008) *Consuming Authenticity*; Florida, R. (2007) *Megaregions*. Importance of Place.
18. Fields, D. (2015) *Contesting the commercialization of urban space*; Weber, R. (2015) *From Boom to Bubble*; Benner, C. & Pastor, M. (2016) *Equity, Growth, and Community*; Jonas, A. E. G. and D. Wilson (1999) *The Urban Growth Machine. Critical Perspectives, Two Decades Later.*
19. del Cerro Santamaría, G. (2013) *Urban Megaprojects. A Worldwide View.*
20. Hiller, H. H. (2000) *Mega-events, urban boosterism and growth strategies: An analysis of the objectives and legitimacy of the Cape Town 2004 Olympic Bid.*
21. Wood, A. (2004) *Domesticating Urban Theory? US Concepts, British Cities and the Limits of Cross-national Applications.*
22. While, A., Andrew E. G. Jonas and David Gibbs (2004) *The environment and the entrepreneurial city.*
23. del Cerro Santamaría (2007) *Bilbao. Basque Pathways to Globalization.*
24. Ibid.
25. Ibid.
26. Harris, ibid., 2017, p. 41.
27. See del Cerro Santamaría, ibid., 2013. Over the last three decades, many large-scale urban interventions carried out by government in collaboration with private interests have taken place outside the usual or traditional planning framework. As a response to this trend, strategic planning, which is increasingly common in cities worldwide, reflects the influence of globalization in achieving major changes in the institutional structure, processes, influence, and scope of urban planning, as observed by Friedmann. See Friedmann, J. (2005) *Globalization and the Emerging Culture of Planning.*
28. del Cerro Santamaría, ibid., 2013, p. 25.
29. Harris, ibid., 2017, p. 45.
30. A recent commentary on the importance of urban financialization in connection with tech megaprojects was provided by Patrick Le Galès (2022) *Newyorkology of Tech Capitalism.*
31. Fainstein, S. (2001) *The City Builders.*

32. Flyvbjerg, B. (2014) *What you should know about megaprojects and why*, p. 63.
33. Flyvbjerg, ibid., 2014, passim.
34. Hirschman (1967b) The principle of the hiding hand. The issue of whether megaprojects are on-time and under-budget (or the opposite) is much contested in the field of project management. Key aspects of this discussion are the prior perception of the scholar (for or against megaprojects), the sampling methodology and size of sample, the extent to which the sample is biased (systematically distorted by the research process), whether they are cherry-picking the cases being considered, and the definition of megaproject (which arguably should affect the sample). Another issue is whether it matters. An explanation for cost over-run relates to what is called "scope-creep" (the fact that a project's requirements increase over time). There are many other elements that contribute to cost and time overrun. Sometimes writers argue that the bigger issue is whether the megaproject is ultimately fit for purpose and inherently useful. This can modify prior assessments about time and cost overruns.
35. Ibid.
36. Flyvbjerg, ibid., 2014, p. 61.
37. Flyvbjerg, B., Bruzelius N., Rothengatter W. (2003) *Megaprojects and Risk.*
38. Flyvbjerg, ibid., 2014, passim.
39. OMEGA Center (2012) *Mega-Projects: Lessons for Decision-Makers.*
40. Ibid.
41. Martínez, M. A. & J. Gil (2022) *Grassroots struggles challenging housing financialization in Spain.*
42. Mega-Projects: The Changing Politics of Urban Public Investment by Alan Altshuler and David Luberoff.
43. Sagalyn, L. (2007) *Public/Private Development.*
44. Beauregard, R. and Pierre, (2000) *Disputing the Global*; Fainstein, S. (2001) *The City Builders.*
45. Harvey (1989) from managerialism to entrepreneurship; Sagalyn, L., and B. Frieden (1989) *Downtown Inc. How America Rebuilds Cities?*
46. Logan J R and Molotch H (1987, second ed. 2007) *Urban Fortunes: The Political Economy of Place.*
47. Harvey, ibid., 1989.
48. Fainstein, ibid., 2001, p. 218.
49. Ibid., p. 220.
50. Sagalyn, ibid., 2007, p. 377. Sagalyn, L. (2003). *Times Square Roulette.*
51. Moulaert, F., Rodriguez, A., and Swyngedouw, E. (2003) *The Globalized City*, p. 250.
52. Fainstein, S. (2008) *Mega-projects in New York, London, and Amsterdam*, p. 783.
53. In The Oxford Handbook of Megaproject Management (2017), Winch, AHLERS, Hodge and Greve.
54. One of the earliest analysis of urban projects as transnational or global projects is Olds, K. (2002). *Globalization and Urban Change*. Some chapters in

del Cerro Santamaría (2013) *Urban Megaprojects. A worldwide View* analyses urban projects as national projects. The concluding section (Conclusions and Theoretical Implications) interprets this multiscalar phenomenon.

55. See Rodrik, D., S. Naidu, and G. Zucman (2019) *Economics after Neoliberalism*, Boston Review Forum; See also Gertz, G., and H. Kharas (2019) *Beyond Neoliberalism. Insights from Emerging Markets*, Brookings Institution Report.
56. See the classic works by Sassen and Storper, Sassen, S. (2011) *Cities in a World Economy*; Storper, M. (1997) *The Regional World.*
57. Florida, R. (2007) *Megaregions. The Importance of Place.*
58. Harrison, J., Hoyler, M., eds. (2015) *Megaregions: Globalization´s New Urban Form?*
59. Khanna, P. (2016) *A New Map for America.*
60. Ibid.
61. Katz, B. (2018) *The New Localism.*
62. Khanna, ibid., 2016.
63. Ansar et al. (2016) *Does infrastructure investment lead to economic growth or economic fragility?* Wu F. (2017) *China´s Emergent City-Region Governance.*
64. Berry, C. and A. Giovannini (2018) *Developing England's North.*
65. Ross, C. and A. Amekudzi (2009) *Megaregions. Planning for Global Competitiveness.*
66. del Cerro Santamaría, G. (2019) *Megaprojects, Development and Competitiveness. Building the Infrastructure for Globalization and Neoliberalism.*
67. Devinney, T. M. and Hartwell, C. A. (2020) *Varieties of populism.*
68. Kobrin, S. J. (2017) *Bricks and mortality in a borderless world*; Acharya, A. (2017) *After Liberal Hegemony*; Grosse, R., Gamso, J., & Nelson, R. C. (2021). *China's Rise, World Order.*
69. Miller, A. (2013) *The Afghan Ring Road Megaproject.*
70. Hartwell, C. A. and Devinney, T. (2021). *Populism, political risk, and pandemics.*
71. See Castán Broto, V. & Bulkeley, H. (2013) *Maintaining Climate Change Experiments*, and Dodson, J. (2017) *The Global Infrastructure Turn and Urban Practice.*
72. Heydarian, R. (2021) *Global Infrastructure Development: China, Biden and the New Pivot of 21st Century Geopolitics*, Aspenia Online, October 19th, https://aspeniaonline.it/global-infrastructure-development-china-biden-and-the-new-pivot-of-21st-century-geopolitics
73. Chen, X. (2021) *Connectivity, Connectivity, Connectivity.*
74. Therefore, future initiatives in local sustainability should involve local actors to develop a shared sense of sustainability by acknowledging the features of regionalism, from which diverse local support for mega- infrastructure might develop. These results suggest that the New Silk Road is a subject of inquiry that requires interdisciplinary investigation at many spatial scales. See Thees H. (2020).
75. The issue of "debt trap diplomacy" is contentious. Some writers claim that there is no debt trap diplomacy. For an enlightening discussion see Ogwang, T. & Vanclay, F. (2021) Resource Financed Infrastructure: Thoughts on four

Chinese-financed projects in Uganda. *Sustainability 13* (6), 3259. https://doi.org/10.3390/su13063259

76. Jones ad Hameiri argue that China is not the one-dimensional, uniform actor that many think it to be. The fragmentation, decentralization, and internationalization of party-state apparatuses over the past 40 years have fundamentally altered the development and implementation of its foreign policy. Today's Chinese behavior overseas frequently is not the outcome of a well-coordinated grand strategy, but rather a sometimes-chaotic struggle for resources and influence between competing political interests under a remarkably lenient regulatory state in the Chinese manner. See Jones and Hameiri (2021) and Fractured China.
77. Khanna, P. (2021) *Asia's Megacities and the Future of Geopolitics*, The Chicago Council on Global Affairs, December 2nd, https://globalaffairs.org/commentary-and-analysis/blogs/asias-megacities-and-future-geopolitics
78. See Chen's analysis of SEZ's global influence in Chen, X. From Follower to Leader (2019). According to Khanna (2021), the role of SEZs in Asia is not dissimilar to the way cities were connected to global supply lines during the European colonial era and anchored a nation's expanding economic weight and gravity in the world. In other words, creating a supply chain empire that starts with creating cities is the first step in gaining geopolitical dominance.
79. Heydarian (2021), ibid.
80. See Godehardt, N. and K. Postel-Vinay (2020) *Connectivity and Geopolitics*.
81. It is known that geopolitics and geoeconomics are intertwined and can be interpreted as two sides of the process of development. In a study of the Panama Canal, Galiani et al. argue that "even if no effective competitor emerges, the appearance of a geopolitical challenger capable of credibly supporting the entrant [into the Canal] has a pro- competition economic effect which benefits consumers all over the world." See Galiani, S., J. M. Paz y Miño and G. Torrens (2021).
82. Moisio, S. (2018) *Geopolitics of the Knowledge-Based Economy*.
83. Ren, X. (2017) *Biggest Infrastructure Bubble Ever?*
84. Ansar, Flyvbjerg, Budzier, and Lunn, *Does infrastructure investment lead to economic growth or economic fragility?* Evidence from China.
85. Ansar et al, ibid.
86. Naamani, M. (2022) *Gulf Countries Race for Foreign Capital*, Orient XXI Magazine, https://orientxxi.info/magazine/gulf-countries-race-for-foreign-capital,5297
87. Ankura (2020) *Geopolitical Impacts on the Future of Construction*.
88. Arsel et al (2021). The authors contend that the policies of Trump (United States), Sisi (Egypt), Bolsonaro (Brazil), Orban (Hungary), and Duterte (Philippines) can be characterized as "authoritarian developmentalism." Although they come to power through democratic elections, these leaders have undermined some key tenets of a democratic society. While their policies are couched in nationalist terms and call for strengthening the state, they can also be seen as a continuation of the prevailing neoliberal global order. Other

world leaders could be added to the list, such as Turkey's Erdogan, the rulers of the United Arab Emirates and Saudi Arabia, and the leaders of the Chinese Communist Party, to mention the countries with megaprojects discussed in this book.

89. Morocco and Egypt can be included as examples of authoritarian developmentalism. For Morocco, see the 2018 book by Bogaert, *Globalized Authoritarianism.*
90. Morin, E. (1999) *Seven Complex Lessons in Education for the Future*, p. 15.
91. Edmonds B. (1996) *What is Complexity?* p. 45.
92. Batty, M. (2008) *Cities as Complex Systems*, p. 27.
93. Alexander, C. (1965) *A city is not a tree*; Barabasi, A-L. (2003) *Linked*; Bunge, M. (2014) *Emergence and Convergence*; Miller, J. (2016) *A Crude Look at the Whole.*
94. As is well known, the use of the idea of "embeddedness" in the social sciences dates back to Polanyi (1968). Later, Mark Granovetter (1985) contributed to its centrality in discussions about economic sociology. The use of the concept in this book does not convey Granovetter's understanding in full detail.
95. Stuart Kauffman has systematically analyzed the idea of "self-organization" in complex systems, particularly biological systems. See, inter alia, *At Home in the Universe*. The Conclusion to Chapter 3 in this book discusses Jane Jacobs's view of cities as "organized complexity."
96. We find them in world-system analyses, in the network society approach by Manuel Castells, in actor-network and assemblage theories (Latour; Deleuze & Guattari); cybernetics and systems theory (von Bertalanffy; Wiener; Luhmann); and in the new materialities approach to urban planning (Lieto and Beauregard). A recent monograph under the initiative of the Future Earth, Urban Planet. Knowledge Toward Sustainable Cities, advocates a systems approach to the field, as understood in the most contemporary notion of "systems." Here, complexity implies both a network approach to the urban as well as the building of new knowledge from the bottom up. In addition, the coproduction of knowledge among a variety of parties is key to the book's authors.
97. Wallerstein, I. (2013) The Itinerary of World-Systems Analysis; or, How to Resist Becoming a Theory. Chapter Four in Wallerstein, I., C. Lemert, and C. Aguirre Rojas, *Uncertain Worlds. World-Systems Analysis in Changing Times*, Boulder, CO: Paradigm Publishers.
98. See Wallerstein, I. (2013) *The Itinerary of World-Systems Analysis.*
99. Urry, J. (2002) *Global Complexity*; Taylor, P. J. (2010) *Unruly Complexity.*
100. Castells was invited to teach at U.C. Berkeley in the 1970s immediately after the instant success of his La question urbaine (The Urban Question), which meant the introduction to urban studies of a Marxist political economy analysis and a paradigmatic change in the field. Castells, Manuel (1972) *La question urbaine*, Paris: Maspero
101. Castells, Manuel (2000a) *The Information Age. The Rise of the Network Society*, London: Blackwell, 2nd edition.
102. Castells, Manuel (2004b) (ed.) *The Network Society. A Cross-Cultural Perspective*, Cheltenham, Glos & Northampton, MA: Edward Elgar, p. xvii.

103. Manuel Castells was one of my professors in the Doctoral Program at the School of Economics and Business of the Universidad Autónoma de Madrid (UAM). This was in the early 1990s, before I headed to New York for my PhD at the New School for Social Research. Castells combined his teaching and research activities at Berkeley with his responsibilities as Director of the Instituto de Sociología de Nuevas Tecnologías (Institute of Sociology of New Technologies) and Professor in the Doctoral Program at UAM. In his Madrid seminars, we discuss key Castellsian concepts and insights that are now familiar to his followers and the social-scientific community. These included his seminal distinction between "information society" and "network society," or the fascinating details and societal implications of the information technology revolution that had started in California in the 1980s. Castells then presented the core of his ideas by bridging his postulates in The Informational City (1989) with his more ambitious research, which was later published in his acclaimed trilogy on The Information Age.
104. Prigogine, I. (1997) *The End of Certainty.*
105. Flyvbjerg, B., Bruzelius, N. & Rothengatter, W. (2003). *Megaprojects and Risk. An Anatomy of Ambition*, New York: Cambridge University Press.
106. Kahneman, D. and Tversky, A. (1979). Prospect theory: An analysis of decisions under risk. *Econometrics 47*, 263–291. https://doi.org/10.2307/1914185
107. The statistics of complex systems is the statistics of power laws, where large and extreme events appear much more often than Gaussian statistics predicts.
108. Irimia-Diéguez, A. I. A. Sanchez-Cazorla, R. Alfalla-Luque (2014) Risk Management in Megaprojects, *Social and Behavioral Sciences 119*, 407–416, https://doi.org/10.1016/j.sbspro.2014.03.046; Li, Y., Y. Lu, J. E. Taylor, and Y. Han (2018) Bibliographic and Comparative Analyses to Explore Emerging Classic Texts in Megaproject Management, *International Journal of Project Management, 36* (2), February, pp. 241–396; Xie, L. T. Han and M. Skitmore (2019) Governance of Relationship Risks in Megaprojects: A Social Network Analysis, *Advances in Civil Engineering, 2019,* Article ID 1426139 https://doi.org/10.1155/2019/1426139
109. Beck, U. (1992) Risk Society. Towards a New Modernity, p. 21
110. Ibid., p. 183.
111. Ibid. p. 184.
112. Cartwright, N. (1999) *The Dappled World. A Study of the Boundaries of Science*, New York: Cambridge University Press.
113. Davis, D. (2015) *From Risk to Resilience and Back.*
114. Prigogine, I. (1997) *The End of Certainty*, New York: Free Press.
115. Davis, ibid., 2015.
116. Holling, C. S. (1973) Resilience and Stability of Ecological Systems, *Annual Review of Ecology and Systematics, 4*: 1–23, November, https://doi.org/10.1146/annurev.es.04.110173.000245.
117. Van Meerbeeck K., T. Jucker, J.-C. Svenning (2021) Unifying the concepts of stability and resilience in ecology, *Journal of Ecology 109* (9), 3114–3132.

118. Folke, C.; Carpenter, S.; Walker, B.; Scheffer, M.; Elmqvist, T.; Gunderson, L.; Holling, C.S. (2004). Regime Shifts, Resilience, and Biodiversity in Ecosystem Management, *Annual Review of Ecology, Evolution, and Systematics, 35*: 557–581, https://doi.org/10.1146/annurev.ecolsys.35.021103.105711; Scheffer, M. (2009). *Critical Transitions in Nature and Society*. Princeton NJ: Princeton University Press.
119. Berkes, F., J. Colding, and C. Folke (2003) *Navigating social-ecological systems: Building resilience for complexity and change*, Cambridge, UK: Cambridge University Press.
120. Gao J, Barzel B, & Barabási AL (2016). Universal resilience patterns in complex networks. *Nature 530*: 307.
121. Food Security Information Network (FSIN) (2014) *Resilience Measurement Principles. Toward an Agenda for Measurement Design*, Technical Series No.1, Word Food Program (WFP), and FAO http://www.fsincop.net/fileadmin/user_upload/fsin/docs/resources/FSIN_29jan_WEB_medium%20res.pdf
122. Hopkins, R. (2014) *The Transition Handbook. From Oil Dependency to Local Resilience*, Cambridge, UK: Green Books.
123. Jen, E. (2005) *Robust Design*, New York: Oxford University Press.
124. Taleb, N. N. (2014) *Antifragile: Things That Gain from Disorder*, New York: Random House.
125. Weichselgartner, J., I. Kelman (2015) Geographies of resilience: Challenges and opportunities of a descriptive concept, *Progress in Human Geography 39* (3), https://doi.org/10.1177/0309132513518834, p. 2-4.
126. Ibid. p. 9-12.
127. Davis, ibid., 2015.
128. Forman, R. T. T. (2019) *Towns, Ecology and the Land*, New York: Cambridge University Press.
129. Lee, Kai-Fu (2018) *AI Superpowers: China, Silicon Valley and the New World Order*.
130. Stiegler, B. (2019) *The Age of Disruption: Technology and Madness in Computational Capitalism*; see also the classic work by Kevin Kelly (1995) *Out of Control: The New Biology of Machines, Social Systems and the Economic World.*
131. Latour, B. (2016) *Facing Gaia. Eight Lectures on the New Climatic Regime*; Margulis, L. (1999) *Symbiotic Planet. A New Look at Evolution.*
132. Allen, C. R., D. G. Angeler, A. S. Garmestani, L. H. Gunderson, and C. S. Holling (2014) *Panarchy: Theory and Application.*
133. Pendall, R., Foster, K., & Cowell, M. (2007) *Resilience and Regions.*
134. Homer-Dixon, T. (2006) *The Upside of Down.*
135. Morin, E. (2002) *El Método II: La vida de la Vida*, Madrid: Cátedra.
136. Glaeser, E. (2021) *Urban Resilience.*
137. See Scott, W. R., R. E. Levitt, and R. J. Orr (2011) *Global Projects*. See also Davis, D. E. and O. Flores Dewey (2013) *How to Defeat an Urban Megaproject.*

138. The word "projectification" was initially used to characterize the spread of projects as a type of organizational structure for businesses and labor markets. Projects are now seen as a better approach to responding to unexpected and irregular circumstances, serving as an organization's emblem of flexibility and contingency. See, inter alia, Lundin, R.A. and Söderholm, A. (1995), *A theory of the temporary organization*, see also Sjöblom and Godenhjelm (2009), *Project proliferation and governance—implications for environmental management.*

Chapter 3

1. Yesilbag, M. (2022) Statecraft on cement: The politics of land-based accumulation in Erdoğan's Turkey, *Urban Studies 59* (13), October.
2. Marschall, M, Aydogan, A. (2015) *Mega Projects and the Limits of Urban Planning in Megacities: The Case of Istanbul in the AKP Era*, Middle East Institute, https://www.mei.edu/publications/mega-projects-and-limits-urban-planning-megacities-case-istanbul-akp-era, Accessed January 2022.
3. Sekhniashvili, A. (2017) *Authoritarian Infrastructure Complex: The Turkish Tale*, Center for Strategic and International Studies, December 19th, https://www.csis.org/npfp/authoritarian-infrastructure-complex-turkish-tale
4. Wilks, A. (2021) For Erdoğan's Istanbul Canal project, critics see few winners, Al Jazeera, April 16, https://www.aljazeera.com/economy/2021/4/16/for-erodgans-istanbul-canal-project-critics-see-few-winners
5. Kolisnichenko, V. (2023) *Türkiye has increased the fee for passage through the Bosphorus and Dardanelles by 8.3% since July.*
6. Çekmiş Görgülü, A., & Hacıhasanoğlu, I. (2012). Water-crossing utopias of Istanbul: Past and future. *ITU Journal of Faculty of Architecture, 9*(2), 67–88.
7. Eldem, T. (2021) *Canal Istanbul: Turkey's Controversial Megaproject*, German Institute for International and Security Affairs, https://www.swp-berlin.org/en/publication/canal-istanbul-turkeys-controversial-megaproject
8. Wilks, ibid., 2021.
9. Wilks, ibid., 2021. See also Baba et al. (2023) *Fear and Architecture*. The authors contend that Kanal Istanbul plays an important role in shaping urban fear in the city and examine the interrelationships between fear, uncertainty, and megaprojects.
10. Eldem, ibid., 2021.
11. Hammoura, J. (2021) *Istanbul Canal: A Megaproject With Strategic Implications*, Armenian Weekly, June 16, https://armenianweekly.com/2021/06/16/istanbul-canal-a-megaproject-with-strategic-implications/
12. Casareggio, A. (2021) *Turkey and the geopolitics of the Straits*, IARI, April 8th, https://iari.site/2021/04/08/turkey-and-the-geopolitics-of-the-straits/. As Hammoura argues, "Turkey's strategic value increases in the event of any serious conflict that poses a threat to the security of the region. In order to increase

Turkey's strategic value, the Istanbul Canal could be considered the cornerstone of a strategic project that Turkish military and strategist circles call "Mavi Vatan" (the blue homeland). This project aims to strengthen Turkey's maritime power, dominance, and sovereignty over its bordering seas" (Hammoura, ibid., 2021).

13. Wilks, ibid. According to Hammoura, "for Turkey, the international situation reflected according to the Montreux Convention. Russia can no longer solely dominate the Black Sea, the United States, and NATO are much more concerned about the Black Sea and Balkans affairs than before, and Turkish leadership no longer sees Turkey as a follower of the West, or as lost between East and West, but rather as a power that aspires to play a role greater than just as a marginal regional power. Turkey seeks to achieve this by increasing its strategic and geopolitical value to international powers through the use of geography, which includes the necessity of establishing the Istanbul Canal" (Hammoura, ibid., 2021).
14. Hammoura, ibid., 2021.
15. Eldem, ibid., 2021.
16. Wilks, ibid., 2021.
17. Dogan & Stupar, ibid., 2017, p. 286.
18. Ibid., p. 287. Displacement is closely associated with issues of land acquisition, expropriation or resettlement and the consequences of this (i.e., how the land is acquired and what happens to the relocated residents). See Vanclay, F. (2017) *Project induced displacement and resettlement: From impoverishment risks to an opportunity for development?*
19. Bar'el, Z. (2021) *Erdoğan's Environmental Mess Might Pale Compared to Washington's Ire*, Haaretz, June 24, https://www.haaretz.com/middle-eastnews/palestinians/premium-erdogan-s-environmental-mess-might-pale-compared-to-washington-s-ire-1.9934449
20. Bhattacharyya, S. (2021) *Turkey "crazy" and controversial Istanbul Canal project*, France June 29, https://www.france24.com/en/middle-east/20210629-turkey-s-crazy-and-controversial-istanbul-canal-project
21. ibid
22. Farooq, U. (2018) *Will Istanbul's Massive New Canal be an Environmental Disaster?* National Geographic, March 28th, https://www.nationalgeographic.com/science/article/istanbul-canal-project-Bosphorus-environmental-impacts
23. Ibid.
24. Ibid.
25. Mateescu, D. (2021) *Turkey's mega-projects and China's Growing Influence in the Bosphorus Area*, Veridica, May 31st, https://www.veridica.ro/en/analyses/turkeys-megaprojects-and-chinas-growing-influence-in-the-Bosphorus-area
26. Bhattacharyya, ibid., 2021.
27. Sonmez, M. (2020) *Turkey's megaprojects threaten 'mega' financial havoc*. Al- Monitor, June 1st, https://www.al-monitor.com/originals/2020/06/turkey-megaprojects-airport- bridge-threaten-financial-havoc.html

28. Ibid.
29. Wilks, ibid., 2021.
30. Mateescu, ibid., 2021.
31. Sonmez, ibid., 2020.
32. Mateescu, ibid., 2021.
33. Keskin, M., & M. Kurdak (2021) *Finally Enacted: Project Bonds, An Alternative Financing Method For Mega Project*, Essin Attorney Partnership, https://www.esin.av.tr/2021/07/26/finally-enacted-project-bonds-an-alternative-financing-method-for-mega-project/
34. Ibid.
35. Mateescu, ibid., 2021.
36. Dogan & Stupar, ibid., p. 287.
37. Owen, K. A. (2001) *The Local Impacts of the Sydney 2000 Olympic Games: Processes and Politics of Venue Preparation*. Sydney, Australia: Center for Olympic Studies.
38. Ibid.
39. Ibid.
40. Ibid. See also Harris (2019) *Australian cities pay the price for blocking council input to projects that shape them*, The Conversation, December 16th, https://theconversation.com/australian-cities-pay-the-price-for-blocking-council-input-to-projects-that-shape-them-127017
41. Tomlinson, R. (2016) *Metropolitan governance is the missing link in Australia's reform agenda*, The Conversation, April 4th, https://theconversation.com/metropolitan-governance-is-the-missing-link-in-australias-reform-agenda-55872
42. Westerman H. and S. Creagh (2012) *Barangaroo: Development interests counter the public interest*, The Conversation, November 22nd, https://theconversation.com/barangaroo-development-interests-counter-the-public-interest-10837
43. Lochhead, H. (2016) *Utzon Lecture: Re-imagining the Harbor City*, The Conversation, March 11th, https://theconversation.com/utzon-lecture-re-imagining-the-harbour-city-56127
44. Rolfe, M. (2012) *Barangaroo: politics, property and players– its business as usual*, The Conversation, November 20th, https://theconversation.com/barangaroo-politics- property-and-players-its-business-as-usual-10722
45. Pitsis, A. et al. (2017) Megaprojects redefined – complexity vs. cost and social imperatives, *International Journal of Managing Projects in Business, 11* (1), 7–34.
46. Hawken, S. and Mackenzie, B. (2016) *Barangaroo Reserve, Landscape Architecture Australia [1833-4814]*, 149, pp. 40–50.
47. Most contextual information about Barangaroo and other Sydney's megaprojects was provided by local informants at the University of Technology Sydney.
48. Harris, personal communication, Spring 2022.
49. For the notion of "erosion of intent" see Harris, M. (2018) Barangaroo: Machiavellian megaproject or erosion of intent? Ch 6, pp. 111–134 in *Urban*

Regeneration in Australia. Policies, Processes and Projects for Contemporary Urban Change, edited by Kristian Ruming, New York: Routledge.

50. Harris, ibid.
51. Johnston, J. and Clegg, S. (2012) Legitimate Sovereignty and Contested Authority in Public Management Organization and Disorganization: Barangaroo and the Grand Strategic Vision for Sydney as a Globalizing City, *Journal of Change Management, 12*(3), 279–299.
52. Harris, M. (2014) *Megaprojects: A Global Review and the Australian Context, Festival of Urbanism*, University of Sydney, http://static.squarespace.com/static/540680a6e4b0a7034afffb6f/t/543f572de4b0a72b21db9adc/1413437229478/Megaprojects+Review_Harris.pdf
53. Stickells, L. (2010) Instant urbanism– just add water. *Architecture Australia, 99* (3), 47 (2).
54. Stickells, ibid.
55. Reinmuth, G. (2012) *Barangaroo: The Loss of Trust?* The Conversation, November 20th, https://theconversation.com/barangaroo-the-loss-of-trust-10676
56. Ibid.
57. Ibid.
58. MacDonald, H. (2019) Planning for the Public Benefit in the Entrepreneurial City: Public Land Speculation and Financialized Regulation, *Journal of Planning Education and Research*, 1–13.
59. Pham, K. (2017) *Examining the development of public space in foreshore Sydney: Barangaroo case study*, Master of Research Thesis, Faculty of Design, Architecture and Building, UTS.
60. Legg, R. (2018) *Making developments green does not help with inequality*, The Conversation, November 1st, https://theconversation.com/making-developments- green- doesnt-help-with-inequality-104941
61. McGuirk, P. (2003) Producing the capacity to govern in global Sydney: A Multi-scaled Account. *Journal of Urban Affairs 25* (2), May, 201-223.
62. Barangaroo Remediation Fact Sheet (2019) Barangaroo: Hickson Road Remediation, https://www.seslhd.health.nsw.gov.au/sites/default/files/groups/Public_Health_Unit/Environmental%20Health/19p1669-factsheet-barangaroo-hickson-road-remediation.pdf
63. WKCD Authority (2017) *Annual Report 2016/17*, Hong Kong: West Kowloon Cultural District Authority.
64. This chapter section on the WKCD is an updated and expanded version of previously published material, del Cerro Santamaría, G. (2019) *Disruptive and Contentious Enterprises: Megaprojects in Bilbao, Istanbul, and Hong Kong*.
65. Lui, P. (2008) "Privatizing the Hong Kong Special Administrative Region Government: Revisiting the Case of The Link REIT from a Policy Network Perspective," presented at The International Conference on Promoting City Competitiveness, October 24, 2008, at Chung Hua University, Hsin Chu City, Taiwan.

66. Bradsher, K. (2006) *Hong Kong Halts Plans for Arts Center*. The New York Times, February 22. Retrieved May 27, 2015, from http://www.nytimes.com/2012/01/25/world/middleeast/abu-dhabi-reaffirms-its-grand-plan-for-museums.html
67. Chen, X. (2013) *Hong Kong Finds Its Footing in Art World*. The New York Times, May 27. Retrieved May 23, 2015, from http://www.nytimes.com/2013/05/23/arts/artsspecial/hong-kong-finds-its-footing-in-art-world.html
68. Chu, S. Y. (2010) Brand Hong Kong: Asia's World City as Method? *Visual Anthropology, 24*, 1–2.
69. Lui (2008), ibid.
70. Lau, J. (2011) *A Bid for Culture in a City of Commerce*. The New York Times. June 6. Retrieved May 23, 2015, http://www.nytimes.com/2011/06/07/world/asia/07iht-hong07.html
71. Wong, W. S. (2004) *Design Identity of Hong Kong: Colonization, de- colonization, and re-colonization*. European Academy of Design. University of the Arts, Bremen, Germany.
72. Chen (2013) ibid.
73. Lau, J. (2014) *Bringing Flagship of Contemporary Art to Hong Kong*. The New York Times. March 2. Retrieved May 23, 2015, http://www.nytimes.com/2014/03/21/arts/international/building-a-museum-in-Hong-Kong-from-scratch.html
74. Bradsher (2006), ibid.
75. Lau (2011), ibid.
76. Siu, H. and A. Ku (eds) (2008) *Hong Kong Mobile: Making a Global Population*. Hong Kong: Hong Kong University Press.
77. Siu and Ku (2008), ibid.
78. Raco, M. and Gilliam, K. (2012) Geographies of Abstraction, Urban Entrepreneurialism, and the Production of New Cultural Spaces: The West Kowloon Cultural District, Hong Kong, *Environment and Planning A, 44* no. 6, 1425–1442.
79. Chiu, A. (2014) *Legal Battles Mount for Former Chief Secretary Rafael Hui*, South China Morning Post, July 20, https://www.scmp.com/news/hong-kong/article/1298096/legal-battle-mount-former-chief-secretary-rafael-hui
80. Chow, V. (2013) *The Past and Present of West Kowloon Cultural District, Cultural Vision*, May, http://www.viviennechow.com/p/a-series-of- unfortunate-events-past- and.html
81. Bradsher (2006), ibid
82. Lui (2008), ibid.
83. Chu (2010), ibid.
84. Ibid.
85. Ibid.
86. Jian, M. (2018) *Social Exclusion: WKCD as a Study Case*, City University of Hong Kong, http://dspace.cityu.edu.hk/handle/2031/9105.

87. WKCD Authority (2013). *Environmental Impact Assessment*, Hong Kong: West Kowloon Cultural District Authority, https://ema.westkowloon.hk/eiare-port/Exec%20Summary%20RevC_html.htm.
88. Ibid.
89. Ibid.
90. Chan, Y. (2000) "Reclamation and Pollution in Hong Kong with Special Reference to Victoria Harbour." The HKU Scholars Hub. Stable URL: http://hdl.handle.net/10722/36319
91. WKCDA (2013) *Annual Report*, http://www.westkowloon.hk/en/the-authority/annual-report-20122013, Retrieved June 2015.
92. Chan (2000), ibid.
93. Ibid.
94. Clean Air Network (2018) *Clean Air Network Response to Policy Address*, http://www.hongkongcan.org/hk/press-release/
95. WKCDA (2013), ibid.
96. Chan (2000), ibid.
97. WKCDA (2013), ibid.
98. Dewolf, C. (2011). *The future of Hong Kong's harbor front*. CNN. October 24. Retrieved May 27, 2015, kong/life/future-hong-kong-129384
99. Yiu, M. H. L. (2018) *Everyday (Open) Cultural Space – An Approach to Public Space for West Kowloon Cultural District.*
100. Lorenz, E. & Shiqiao, L. (2015). *Kowloon Cultural District– an investigation into spatial capabilities in Hong Kong* in E. Lorenz, & L. Shiqiao, Eds. Hong Kong: MCCM Creations, Renfro, D. S.
101. Ibid.
102. Government of Hong Kong (2018) *Constitutional and Mainland Affairs Bureau*, https://www.cmab.gov.hk/en/home/index.htm.
103. Lam, Y. M. O. et al. (2021) *Strategic Solutions for Tourism Development Planning in Hong Kong – A Smart Plan for Resilient Tourism Development*, The University of Hong Kong, Faculty of Architecture, https://foa-media.arch.hku.hk/media/upload/2021/09/Final_Report_Planda_Tourism.pdf
104. Lam et al (2021), ibid.
105. See Jacobs (1961).
106. World Bank (2022) Urban Development Overview, https://www.worldbank.org/en/topic/urbandevelopment/overview
107. See Knudsen et al (2008).

Chapter 4

1. A previous version of this section was published by the Athens Journal of Architecture. See del Cerro Santamaría, G. (2020) The Fading Away of the Bilbao Effect: Bilbao, Denver, Helsinki, Abu Dhabi, *Athens Journal of Architecture* *6*(1), 25–52.

2. Healey, P., Davoudi, S., Tavsanoglu, S. (1992) *Rebuilding the City: Property-led Urban Regeneration*, London: Chapman & Hall; Fainstein, S. (2001) *The City Builders. Property Development in New York and London, 1980-2000*, Kansas: University Press of Kansas.
3. Harvey, D. (1989) From Managerialism to Entrepreneurialism: The Transformation in Urban Governance in Late Capitalism, *Geografiska Annaler 71*(1).
4. Plaza, B. and Haarich, N. (2009) Museums for urban regeneration? Exploring the conditions for their effectiveness. *Journal of Urban Regeneration and Renewal 2*(3): 259–271.88
5. Snyder, P. (2016) "Museums after Bilbao: Neoliberal Public Space at the Denver Art Museum," Wesleyan University; del Cerro Santamaría, G. ed. (2013) *Urban Megaprojects. A Worldwide View*. Bingley: Emerald.
6. McClellan, A. (2008) *The Museum from Boullée to Bilbao*, Berkeley: The University of California Press; Snyder, P. (2016) "Museums after Bilbao: Neoliberal Public Space at the Denver Art Museum," Wesleyan University.
7. Newhouse, V. (1998) *Toward a New Museum*, New York: Monacelli Press.
8. Pogrebin, R. (2007) *British Architect Wins 2007 Pritzker Prize*. The New York Times. March 28th. Retrieved May 2018.
9. Smyth, H. (2015) *Marketing the City. The role of flagship developments in urban regeneration*, London: Routledge.
10. del Cerro Santamaría, G. (2007) *Bilbao. Basque Pathways to Globalization*. London: Elsevier.
11. Alsdorf, B. (2002) *Collections Curatorial Assistant for the Guggenheim Foundation*, Guggenheim Foundation, http://www.guggenheim.org/exhibitions/past_exhibitions/global_gugg/global_gugg_bottom_index.html (consulted on 10/16/2017).
12. Klebnikov, A. (2001) *Museums Inc*. Forbes, January 8.
13. Wilson Lloyd, A. (2001) Architecture for Art's Sake, *Atlantic Monthly 287*: 85–8.
14. Ivy, R. A. (2001) cited in Wilson Lloyd, A. (2001) Architecture for art's sake, *Atlantic Monthly 287*: 85–8.
15. Woronkowicz, J. et al. (2015) *Building Better Arts Facilities: Lessons from a U.S. National Study*. (Routledge Research in Creative and Cultural Industries Management), London: Routledge, p. 16.
16. Blakemore, E. (2016) *Museum Building is Booming in the United States*, Smithsonian.com, April 11, https://www.smithsonianmag.com/smart-news/museum-building-booming-united-states-180958711/
17. Hune-Brown (2017) *Is This the End of Starchitecture?* Azure, August 18, https://www.azuremagazine.com/article/is-this-the-end-of-starchitecture/
18. Hune-Brown, Ibid.
19. Kamin, B. (2002) *How stellar are 'starchitects'?* Chicago Tribune, January 27, http://articles.chicagotribune.com/2002-01-27/news/0201270415_1_new-dorm-starchitects-new-york-city-designer
20. Rybczynski, W. (2002) *The Bilbao Effect*, Atlantic Monthly 290(2) 138–42, p. 35.

21. Rybczynski, W. (2008) *When Buildings Try Too Hard*, Wall Street Journal, November, http://www.wsj.com/articles/SB122731149503149341, p. 27.
22. Rybczynski, Ibid. p. 29.
23. Brown, J. (2006) "New Denver Art Museum Reflects Rocky Mountains," PBS Newshour, October 5, 2006. http://www.pbs.org/newshour/bb/entertainment-july-dec06-museum_10-05/
24. Webb, W. (2001) quoted in Wilson Lloyd, A. Architecture for art's sake, *Atlantic Monthly 287*: 85–8.
25. Snyder, P. (2016) "Museums after Bilbao: Neoliberal Public Space at the Denver Art Museum," Wesleyan University; Lindsay, G. (2013) "The Denver Art Museum and the Bilbao Effect," UC Berkeley.
26. Snyder, P. (2016) Ibid., p. 29.
27. McNeill, D. (2009), *The Global Architect: Firms, Fame and Urban Form.* New York: Routledge, p. 82.
28. Jencks, C. (2005) *The Iconic Building*, New York: Rizzoli, p. 54.
29. Lindsay, G. (2013) "The Denver Art Museum and the Bilbao Effect," UC Berkeley, p. 32.
30. Lindsay, Ibid., p. 36.
31. Lindsay, Ibid., p. 43.
32. Lindsay, Ibid., p. 56.
33. Snyder, P. (2016) "Museums after Bilbao: Neoliberal Public Space at the Denver Art Museum," Wesleyan University, p. 32.
34. Stephens, S. (2007) Studio Daniel Libeskind and the Davis Partnership shake downtown with a new addition to the Denver Art Museum, *Architectural Record, 195*, no. 1, p. 84.
35. Guash, A. M. and Joseba Zulaika (2005) *Learning from the Guggenheim Bilbao*. Reno University of Nevada Press, p. 16.
36. Newhouse, V. (1998) *Toward a New Museum.* New York: Monacelli Press, p. 194.
37. Snyder, P. (2016) "Museums after Bilbao: Neoliberal Public Space at the Denver Art Museum," Wesleyan University, p. 162.
38. Snyder, Ibid., p. 164.
39. Snyder, Ibid., p. 176.
40. Snyder, Ibid., p. 184.
41. Wainwright, O. (2014) *Helsinki v Guggenheim: The backlash against the global megabrand is on.* The Guardian, September 11. Retrieved May 29, 2018, from http://www.theguardian.com/artanddesign/architecture-design-blog/2014/sep/11/helsinki-guggenheim-outpost-battle
42. Taipale in Wainwright, Ibid., p. 26.
43. Ziabari, K., Urpilainen, J. (2014) *Finnish Economy is in the Middle of a "Lost Decade"*. Fair Observer, October 31. Retrieved May 29, 2018, from http://www.fairobserver.com/region/europe/finlands-economy-middle-lost-decade-01757/

44. YLE (2012) *Minister says there is no culture budget for Guggenheim*. YLE, December 1st. Retrieved May 28, 2018, https://yle.fi/uutiset/osasto/news/minister_says_no_culture_budget_for_guggenheim/5294388, p. 12.
45. YLE, Ibid., p. 16.
46. Volner, I. (2015) *Can Guggenheim Charm Finland?* The New Yorker, May 12. Retrieved May 28, 2018, http://www.newyorker.com/culture/culture-desk/can-the-guggenheim-charm-finland, p. 31.
47. Volner, Ibid., p. 29.
48. Volner, Ibid., p. 28.
49. The Guardian (2015) *Migrants building UAE cultural hub working in prison conditions*, April 4, Retrieved May 2018. https://www.theguardian.com/global-development/2015/apr/04/migrant-workers-uae-saadiyat-island-abu-dhabi-battery-hens
50. Wigley, cited in The Guardian, June 23, 2015.
51. Edelson, Z. (2015, January 27). *Michael Sorkin On The Guggenheim, Museum Culture, and "The Next Helsinki" Competition*. Arch Daily. Retrieved May 27, 2015, http://www.archdaily.com/591932/michael-sorkin-on-the- guggenheim-museum-culture-and-the-next-helsinki-competition/, p. 43.
52. TNH, *The Next Helsinki*, http://www.nexthelsinki.org.
53. TNH, Ibid.
54. Volner, I. (2015) *Can Guggenheim Charm Finland?* The New Yorker, May 12. Retrieved May 28, 2018, http://www.newyorker.com/culture/culture-desk/can-the-guggenheim-charm-finland, p. 43.
55. Guggenheim Helsinki (2014) *Guggenheim Helsinki Design Competition*. Retrieved May 28, 2015, from http://designguggenheimhelsinki.org/about/; http://yle.fi/uutiset/minister_says_no_culture_budget_for_guggenheim/5294388
56. Kalla, Z. (2014). *To Guggenheim or not to Guggenheim Helsinki ? This is the question that arises*. Arch2o. Retrieved May 29, 2018, http://www.arch2o.com/guggenheim-helsinki-3/, p. 51.
57. Kalla, Ibid., p. 53.
58. Dumortier, B. (2016) *The cultural imperative: Saadiyat cultural district in Abu Dhabi between public policy and architectural gesture. In Under construction: Logics of urbanism in the gulf region*, London: Routledge.
59. Vogel, C. (2014) *A New Art Capital, Finding Its Own Voice. Inside Frank Gehry's Guggenheim Abu Dhabi*, The New York Times, December 4, http://www.nytimes.com/2014/12/07/arts/design/inside-frank-gehrys-guggenheim-abu-dhabi.html?_r=0
60. Shadid, A. (2012) *An Ambitious Arab Capital Reaffirms Its Grand Cultural Vision*. New York Times, January 24. Retrieved May 27, 2015, from http://www.nytimes.com/2012/01/25/world/middleeast/abu-dhabi-reaffirms-its-grand-plan-for-museums.html
61. Batty, D. (2012) *Guggenheim delay raises a big question: Is Abu Dhabi ready for modern art?* Guardian, April 27. Retrieved May 23, 2018, http://www.

theguardian.com/world/2012/apr/17/abu-dhabi-guggenheim-delay-question, p. 71.
62. Cited in Mauk, B. (2014) *Abu Dhabi's High Cost of Culture*. The New Yorker, January 28. Retrieved May 27, 2015, from http://www.newyorker.com/business/currency/abu-dhabis-high-cost-of-culture
63. Cited in Mauk, B. (2014).
64. The Guardian (2015) *Migrants building UAE cultural hub working in prison conditions*, April 4, Retrieved May 2018. https://www.theguardian.com/global-development/2015/apr/04/migrant-workers-uae-saadiyat-island-abu-dhabi-battery-hens
65. Spiegel online International. (2008) *Krens' Museum for Global Contemporary Art: Guggenheim Abu Dhabi Will Be' Pharaonic*. Spiegel Online, March 27. Retrieved May 29, 2015, from http://www.spiegel.de/international/world/krens-museum-for-global-contemporary-art-guggenheim-abu-dhabi-will-be-pharaonic-a-543601.html
66. Skluzacek, C. (2010) *Universality and its Discontents: The Louvre and Guggenheim Abu Dhabi as a Case Study in the Future of Museums. Macalester College Art History Honor Projects*, 1-72. Retrieved May 23, 2015, http://digitalcommons.macalester.edu/art_honors/1/, p. 35.
67. Skluzacek, Ibid., p. 43.
68. Skluzacek, Ibid., p. 43.
69. Batty, D. (2015) *Migrants building UAE cultural hub 'working in prison conditions*, The Guardian, April 4, 2015, Retrieved, July 3rd, http://www.theguardian.com/global-development/2015/apr/04/migrant-workers-uae-saadiyat-island-abu-dhabi-battery-hens
70. Dhabi, A. (2011) *Why the Arab Spring Never Came to the U.A.E. Times*. July 18. Retrieved May 29, 2015, http://content.time.com/time/world/article/0,8599,2083768,00.html
71. Mcclellan, A. (2012) Museum Expansion in the Twenty-First Century: Abu Dhabi. *Journal of Curatorial Studies Journal of Curatorial Studies 1*(3), 271–293.
72. Wells cited in Batty, D. Ibid. (2015).
73. Taelman, E. (2009) Saadiyat Island Tourist Development Project: Dredging and Reclamation Works in an ecologically sensitive area. *Terra et Aqua, 116*, 3–11.
74. Taelman, Ibid., p. 31.
75. Taelman, Ibid., p. 36.
76. Sklair, L. (2017) *The Icon Project*.
77. del Cerro Santamaría, G. (2007) *Bilbao. Basque Pathways to Globalization*. London: Elsevier.
78. Kimmelman, M. (2012) *Why is This Museum Shaped Like a Tub?* The New York Times, Art and Design Section, December 23 http://www.nytimes.com/2012/12/24/arts/design/amsterdams-new-stedelijk-museum.html?_r=0

79. Campbell, R. (2007) *Does Gehry's Stata Center Really Work? Bloomberg Business*, June 19, http://www.bloomberg.com/bw/stories/2007-06-19/does-gehrys-stata-center-really-work-businessweek-business-news-stock-market-and-financial-advice
80. Koolhaas, 2013.
81. Woronkowicz, J. et al. (2015) *Building Better Arts Facilities: Lessons from a U.S. National Study*. (Routledge Research in Creative and Cultural Industries Management). London: Routledge.
82. Ibid., p. 56.
83. Ibid., p. 61.
84. Hune-Brown (2017) *Is This the End of Starchitecture?* Azure, August 18, https://www.azuremagazine.com/article/is-this-the-end-of-starchitecture/
85. Sklair, L. (2017) *The Icon Project.*
86. Broot, S. (2019) *Digital Monuments.*
87. See Chapter 7.
88. As an outstanding example, the recent book (2023) by Reinier de Graaf *Architect, Verb: The New Language of Building* is a systematic critique of the main tenets of architecture in the neoliberal period. De Graaf's convincing arguments demonstrate that architecture needs a new identity and a new language. In this respect, see also the last section of Broot's *Digital Monuments.*
89. See Flyvbjerg (2014), Harris (2017), del Cerro Santamaría (2013).
90. Frampton, K. (1995) *Studies in Tectonic Culture.*
91. Sennett, R. (1992) *The Conscience of the Eye*, p. 196.
92. Jacobs, J. (2000) *The Nature of Economies.*
93. Jullien, F. (2001). *Un sabio no tiene ideas*: O el otro de la filosofía. Siruela.
94. Lieto, L. (2017) *How material objects become urban things?*
95. de Certeau, M. (1984) *The Practice of Everyday Life*, pp. 59–67
96. de Certeau (1984) Ibid., p. 115–119.
97. de Certeau (1984) Ibid., p. 123–30.
98. Foucault, M. (1967) *Of Other Spaces. Utopias and Heterotopias.*
99. Sorkin, M. (2017) *Afterword: Architecture without Capitalism, in P. Deamer (ed.) Architecture without Capitalism*, London, Routledge, p. 217–220.
100. Jacobs (1961) *The Death and Life of Great American Cities.*
101. Jencks, C. (1997) *The Architecture of the Jumping Universe.*
102. See Bachman, L. (2008) *Architecture and the four counters with complexity, architecture, and design management*, 4(1), 15–30, https://doi.org/10.3763/aedm.2008.S407
103. See Hays, K. M. (2016) *Architecture's Appearance and the Practices of Imagination.* Log, vol. 37, 205–213, https://www.anycorp.com/store?category=Log
104. Hays (2016), ibid.
105. Berger and Luckman, 1967
106. Perry, B. (2002) *The Future of Urban Sociology*

107. See del Cerro Santamaría, G. (2018) Toward Transdisciplinary Urbanism: Megaprojects, Power and the Urban Imagination, *Transdisciplinary Journal of Engineering & Science, 9*, 23–36.
108. See, inter alia, the seminal work of Manfredo Tafuri, Architecture and Utopia
109. See Sorkin, M. (1992) *Variations on a Theme Park.*
110. See Stevens, G. 2002.
111. Latour, B. (1987) *Science in Action.*
112. Mostafavi and Doherty (2010) *Ecological Urbanism.* See del Cerro Santamaría, G. (2020a). *Building new knowledge about nature, ecology, and sustainability.*
113. See Bachman, L. (2008) *Architecture and the four counters with complexity.*
114. Mehaffy, M. W. and N. A. Salingaros (2015) *Design for a Living Planet: Settlement, Science and a Human Future.*

Chapter 5

1. Marcuse, P. (1989) *'Dual city': a muddy metaphor for a quartered city.*
2. del Cerro Santamaría, G. (ed.) (2013) *Urban Megaprojects. A Worldwide View.*
3. Geoffrey Hinton, who received the Nobel Prize in Physics (2024) for his work on machine learning, famously quit Google in 2023 to warn about the potential dangers of AI. Mustafa Suleyman, the CEO of Microsoft AI, proposes "containment" as an overall strategy to avoid the risks of AI in *The Coming Wave: Technology, Power, and the Twenty-First Century's Greatest Dilemma.*
4. Sassen, S. (2011) *Cities in a World Economy*, Thousand Oaks, CA: SAGE; Zukin, S. (2020b) Seeing like a city: How tech became urban. *Theory and Society, 49*(5-6), 941– 964.
5. Castells, M., and Hall, P. (1994) *Technopoles of the World. The Making of 21st Century Industrial Complexes*, London: Routledge; del Cerro Santamaria, G. (2020a) Innovation Districts, Complex Sustainability and Urban Redevelopment. Evaluating Multiple Success Factors, *IJEBMR 4*(6), 279–290.
6. ADB (2016) *Scaling New Heights. Vizag-Chennai Industrial Corridor. India's First Coastal Corridor*, Manila: Asian Development Bank; Li, Y., Wu, F., (2020) Territory, the state and geopolitics of mega city-region development in China, pp. 343–354, in *Handbook on the Changing Geographies of the State*, Edited by Sami Moisio, Natalie Koch, Andrew E.G. Jonas, Christopher Lizotte and Juho Luukkonen, London: Edward Elgar.
7. Zukin, S. (2020b) Seeing like a city: How tech became urban. *Theory and Society, 49* (5-6), 941–964.
8. Leigh, N., N. Hoelzel, B. Kraft, C. Dempwolf (2014) *Sustainable Urban Industrial Development.*
9. Jacobs, J. (1961) *The Death and Life of Great American Cities.*

10. Saxenian, A. L. (1994) *Regional Advantage*; Castells, M. (2009) *The Rise of the Network Society.*
11. Chib A. (2015) *Research on the Impact of the Information Society in the Global South.*
12. UNCTAD (2021) *Catching Technological Waves.*
13. WEF (2018) *Harnessing Artificial Intelligence for the Earth.*
14. Ibid.
15. IMF World Economic Outlook. (2018) *Cyclical Upswing, Structural Change.*
16. Ibid.
17. Ibid.
18. Korinek, A., M. Schindler, J. Stiglitz (2021) *Technological Progress, Artificial Intelligence and Inclusive growth.*
19. Moisio, S. (2018) *Geopolitics of the Knowledge-based Economy.*
20. Schwab, K. (2018) *Shaping the Future of the 4th Industrial Revolution.*
21. The Taiwan Semiconductor Manufacturing Company (TSMC) is now the world's largest chipmaker. Among its customers is Apple, which it supplies with the chips that power iPhones. A Chinese invasion of Taiwan would bring the global economy to a halt. Some tech firms are adding U.S. facilities: Intel, one of the world's largest semiconductor chip manufacturers, is building factories at a cost of more than $20 billion in Ohio; Micron, a producer of computer memory and computer data, will spend $100 billion on a chip factory in New York state; South Korea's Samsung Electronics, the world's largest information tech firm and chipmaker, is building a $25 billion plant in Texas; and TSMC itself is investing $40 billion for two plants in Arizona. See McGovern, G. & S. Branford (2024) *Critics fear catastrophic energy crisis as AI is outsourced to Latin America*, Mongabay Series: Planetary Boundaries https://news.mongabay.com/2024/03/critics-fear-catastrophic-energy-crisis-as-ai-is-outsourced-to-latinamerica/
22. Zukin, S. (2020b) *Seeing like a city: how tech became urban.*
23. Miao, J. T., P. Benneworth, and N. A. Phelps (2015), *Technopoles of the world: Changes, Dynamics and Challenges*, pp. 3-20 in Miao, J. T., P. Benneworth and N.A. Phelps (eds.) *Making 21st Century Knowledge Complexes. Technopoles of the World Revisited.* New York: Routledge.
24. Ibid.
25. Benneworth, P. J. T. Miao, and N. A. Phelps (2015), *Old and New Lessons for Technopoles*, pp. 275-295 in Miao, J. T., P. Benneworth, and N. A. Phelps (eds.) *Making 21st Century Knowledge Complexes. Technopoles of the World Revisited.* New York: Routledge.
26. Charles, D. (2015) *From technopoles to science cities.*
27. Safeguard Global (2024) *Top 10 manufacturing countries in 2023*, https://www.safeguardglobal.com/resources/top-10-manufacturing-countries-in-the-world-2023/
28. The Laboratory for Machine Tools and Production Engineering at RWTH Aachen University and BCG conducted a study in 2016 that focused on a global survey of more than 750 production managers from leading companies in three industrial sectors: automotive (which includes suppliers and original equip-

ment manufacturers, or OEMs), engineered products, and process industries. The goals were to define a vision for the factory in the future in 2030, assess the benefits, and create a roadmap for implementation. We partially relied on their findings. See Küpper et al. (2016) *The Factory of the Future*, https://www.bcg.com/publications/2016/leaning-manufacturing-operations-factory-of-future

29. Null, C. & Caulfield, B. (2003, June 1). Fade to black the 1980s vision of "lights-out" manufacturing, where robots do all the work, is no longer a dream. *CNN Money*.
30. Küpper et al, ibid.
31. Pinto, J. (2003, October 1). Fully automated factories approach reality. *AutomationWorld*.
32. Walker, C. R. (1982). Toward an automatic factory. *Electronic Servicing & Technology*.
33. Küpper et al, ibid. See also Govindarajan, V. & V. Venkatraman (2024) *Fusion Strategy. How Real-Time Data and AI Will Power the Industrial Future*, Cambridge, MA: HBR Press.
34. December 2013 article in *The Environmental Magazine*, Roddy Scheer and Doug Moss.
35. Masami, T. (2003). Realizing unattended hours of continuous operation of the machining center with the addition of an intelligent function. *Subaru Technical Review, 30*, 251–256.
36. Markoff, J. (2012, November 13). Technomy 2012: Where's my robot? *Techonomy*.
37. Walker, C. R. (1977). *Toward the automatic factory: A case study of men and machines*. Greenwood Press.
38. Morave, H. (2009, March). Rise of the robots—the future of artificial intelligence. *Scientific American*.
39. *CNN Money*. Fade to Black, the 1980s Vision of "Lights-Out' Manufacturing, where Robots Do All the Work, Is a Dream No More. http://money.cnn.com/magazines/business2/business2_archive/2003/06/01/343371/index.htm.
40. Montella, N. (2025) *ITIF Virtual Webinar, Scale Matters: Understanding the Economics of Global Semiconductor Innovation*, January 30th, https://itif.org/events/2025/01/30/understanding-economics-of-global-semiconductor-innovation/
41. Mickle, T. (2025) *TSMC, the Chip Giant, Is to Spend $100 billion in U.S. Over the Next 4 Years*, The New York Times, https://www.nytimes.com/2025/03/03/technology/tsmc-investment-trump.html
42. Atkinson, R. (2024) *China Is Rapidly Becoming a Leading Innovator in Advanced Industries, Information Technology and Innovation Foundation*, September 16, https://itif.org/publications/2024/09/16/china-is-rapidly-becoming-a-leading-innovator-in-advanced-industries/
43. Breckenfeld, E. (2025) *ITIF Virtual Webinar, Scale Matters: Understanding the Economics of Global Semiconductor Innovation*, January 30th, https://itif.org/events/2025/01/30/understanding-economics-of-global-semiconductor-innovation/

44. EDA (2023) *Biden-Harris Administration Designates 31 Tech Hubs Across America, U.S. Economic Development Administration*, October 23, https://www.eda.gov/news/press-release/2023/10/23/biden-harris-administration-designates-31-tech-hubs-across-america
45. This section of the chapter draws on material published by the author: del Cerro Santamaría, G. (2020) Innodistricts, Complex Sustainability and Urban Redevelopment: Evaluating Multiple Success Factors, *International Journal of Economics, Business and Management Research 4*(6) 279–290.
46. Zukin, S. (2021) *Planetary Silicon Valley.*
47. Katz, B. and J. Wagner (2019) *The Evolution of Innovation Districts.*
48. Hirschman, A. (1967) *Development Projects Observed,* p. 36. See also Helka Kalliomäki, Päivi Oinas & Teemu Salo (2023) Innovation districts as strategic urban projects: the emergence of strategic spatial planning for urban innovation, *European Planning Studies*, https://www.tandfonline.com/action/showCitFormats?doi=10.1080/09654313.2023.2216727
49. del Cerro Santamaría, G. (2020b) *Possibilities and Challenges for Intercultural Research in Global Urbanism.*
50. Katz, B. and J. Wagner (2014) *The Rise of Innovation Districts*, p. 1.
51. Katz, B., and Wagner, J. (2014) Ibid., p.7.
52. Katz, B. and Wagner, J. (2014) Ibid., p.11.
53. Katz, B., and J. Wagner (2014). Ibid., p. 14.
54. Ajuntament de Barcelona, 2013, 16.
55. Ajuntament de Barcelona, 2012, 19.
56. Carnes, S. (2016) *The Case for the Innovation District as a Sustainable Economic Development Tool in the Knowledge Economy*, p. 47. See also Gianoli, A., R. P. Henkes, R. (2020) *The Evolution and Adaptive Governance of the 22@ Innovation District in Barcelona*; and Battaglia, A. and D.-G. Tremblay (2011) *22@ and the Innovation District in Barcelona and Montreal: A Process of Clustering Development between Urban Regeneration and Economic Competitiveness.*
57. Boston Mayor´s Office, 2014, 33.
58. Ross, C. (2014) *Office rents soaring in city´s Innovation District*, p. 31.
59. Carnes, 2016, Ibid., p. 45.
60. Adelfio, M., I. Hamiduddin and E. Miederna (2021) *London's King's Cross redevelopment.*
61. For the replication of the King's Cross redevelopment concept and practice elsewhere, see Brill, F. and V. Conte (2019). Understanding project mobility: the movement of King's Cross to Brussels and Johannesburg, *Environment and Planning C: Politics and Space 38* (1), https://journals.sagepub.com/doi/10.1177/2399654419854458#con2. London as a whole is increasingly witnessing the creation and growth of several innovation districts, such as Tech City (studied by Max Nathan from UCL), Canary Wharf (centered around fintech innovations), and some sections in the Queen Elizabeth Olympic Park working around SHIFT and Here East.
62. del Cerro Santamaría, Gerardo (2019) *Review of Spaceship in the Desert. Energy, Climate Change and Urban Design in Abu Dhabi*, p. 221.

63. Günel, G. (2019) *Spaceship in the Desert.*
64. del Cerro Santamaría, ibid., 2019, p. 222.
65. del Cerro Santamaría, Ibid., 2019, p. 224.
66. del Cerro Santamaría, Ibid., 2019, p. 226.
67. Carnes, S. (2016) *The Case for the Innovation District as a Sustainable Economic Development Tool in the Knowledge Economy*, p. 61.
68. Katz, B. and J. Wagner (2019) *The Evolution of Innovation Districts.*
69. Gianoli, A., R. P. Henkes (2020) *The Evolution and Adaptive Governance of the @22 Innovation District in Barcelona.*
70. Marginson, S. (2016) cited in Cameron, H. (2016) *Innovation Districts – the way forward for sustainable growth?*, p. 21.
71. Zukin, S. (2020a) *The Innovation Complex. Cities, Tech and the New Economy.*
72. As described by Baily in Katz and Wagner, 2019. The Porter "Diamond model" uses four categories of private sector drivers, see the discussion in the next section.
73. Dall´Orso, M. (2019) *Promoting Sustainable Urban Development Through Impact Innovation.*
74. Van Dinteren, J. (2018) *The university as a catalyst in innovation district development*; Reichert, S. (2019) *The Role of Universities in Regional Innovation Ecosystems.*
75. Gao, T. and Lim, S. (2023) Socio-spatial integration in innovation districts: Singapore's mixed-use experiment, *CITIES* 140, September, https://doi.org/10.1016/j.cities.2023.104405
76. Wang, X., Sun, X., Liu, S., & Mu, C. (2021). *A Preliminary Exploration of Factors Affecting a University Entrepreneurial Ecosystem.*
77. Cameron, H. (2016) *Innovation Districts*, p. 53.
78. See Wagner, J. (2019) *Innovation Districts and Their Dilemmas With Place*, and da Cuhna, A. (2013), cited in *Ecodistricts: A sustainable utopia?*
79. Zukin, S. (2020b) Seeing like a city: How tech became urban. *Theory and Society*, 49 (5–6), 941–964.
80. Günel, 2019. See also Crot L. (2013) Planning for Sustainability in Non-Democratic Polities: The Case of Masdar City. *Urban Studies, 50*(13), 2809–2825. https://doi.org/10.1177/0042098012474697.
81. Keeton, R. (2021) *When Smart Cities Are Stupid.*
82. Worth, R. F. (2021) *The Dark Reality Behind Saudi Arabia's Utopian Dreams.*
83. Dall'Orso, M. (2017) *What characterises an ideal city, and how do we get there?*
84. Katz, B. and J. Wagner (2019) *The Evolution of Innovation Districts.*
85. Kalliomäki, H. et al. (2023) *Innovation districts as strategic urban projects: the emergence of strategic spatial planning for urban innovation.*
86. Harris, M. (2017) *Competitive Precinct Projects.*
87. Atkinson, R. (2012) *Innovation Economics.*
88. Paquot, T. (2013), cited in *Ecodistricts: A Sustainable Utopia?* Haeckel, E. (1866) *General Morphology of Organisms.*

89. Caragliu, A., C. Del Bo, P. Nijkamp (2009) *Smart Cities in Europe*; Mitchell, W. (2007) *Intelligent Cities*; Marrone, M., M. Hammerle (2018) *Smart cities: A review and analysis of stakeholders' literature.*
90. Schaffers, H., Komninos, N., Pallot, M.; Trousse, B., Nilsson M. (2011) *Smart Cities and the Future Internet*; Chan, K. (2017) *What Is A 'Smart City'?*; Komninos, N. (2013) *What makes cities intelligent?* Yovanof, G., G. N. Hazapis (2009) *An Architectural Framework and Enabling Wireless Technologies*; Nisenbaum, A. (2019) "What's Holding Smart Cities Back?".
91. Peris-Ortiz, M., D. R. Bennett, D. Pérez-Bustamante Yábar, eds., (2016) *Sustainable Smart Cities*; Zhao, K., L. Ge (2013) *A Survey on the Internet of Things Security*, Nam, T., T. A. Pardo (2018) *Conceptualizing Smart City with Dimensions of Technology, People, and Institutions.*
92. McLaren, D., J. Agyeman (2015) *Sharing Cities: A Case for Truly Smart and Sustainable Cities.*
93. Hollands, R. G (2008) *Will the real smart city please stand up?* Komninos, N. (2008) *Intelligent Cities and Globalisation of Innovation Networks.* Komninos, N. (2009) *Intelligent cities: toward interactive and global innovation environments.*
94. Cooke, P. (2012) *Complex Adaptive Innovation Systems*; Gu, Y., L. Hu, H. Zhang, C. Hou (2021) *Innovation Ecosystem Research.*
95. Augusto, J., Callaghan, V., Cook, D., Kameas, A., Satoh (2012) *Intelligent Environments*
96. Lehrmacher, W., M. Lind (2022) *Port cities' efforts in contributing to a sustainable port environment.*
97. De Oliveira, S. P. and E. G. Carayannis (2017) *The Quadruple Innovation Helix Nexus.*
98. Schiavone. F., F. P. Appio, L. Mora, and M. Risitano (2020) *The Strategic, Organizational and Entrepreneurial Evolution of Smart Cities.*
99. McGinty, S. (2016) *Top of the world: Why is Melbourne "the most liveable city"?*
100. Mortenson (2024) *Data Center Megaprojects Fuel Next-Level Construction Careers*, https://www.mortenson.com/newsroom/careers-data-center-construction
101. Bangalore, S. and B. Srivathsan (2023) *Investing in the Rising Data Center Economy*, McKinsey Report, https://www.mckinsey.com/industries/technology-media-and-telecommunications/our-insights/investing-in-the-rising-data-center-economy.
102. Zhang, M. (2024) *Cities and Regions with the Highest Concentration of Data Centers*, https://dgtlinfra.com/united-states-data-centers/
103. Waters, M. (2022) *Energy-hungry data centers are quietly moving into cities*, MIT Technology Review, https://www.technologyreview.com/2022/06/22/1053889/city-server-farms-energy/.
104. Hemment, D., and Townsend, A. (2013) *Smart Citizens*; Kitchin, R. and Dodge, M. (2014) *Code/Space*; Morozov E (2013) *To Save Everything.*

105. He and Wu (2009) *China's Emerging Neoliberal Urbanism: Perspectives from Urban Redevelopment.*
106. Ibid.
107. Zhang et al. (2022) *State-steered smart mentality in Chinese smart urbanism.*
108. He and Wu, ibid., 2009.
109. Zhang, J., J. Bates and P. Abbott (2022) *State-steered smartmentality in Chinese smart urbanism.*
110. Carr, C., and M. Hesse (2020) When Alphabet Inc. Plans Toronto's Waterfront: New Post-Political Mode of Urban Governance, *Urban Planning*, 5(1), 69–83; See also Walker, A. (2020) *Sidewalk Lab's 'Smart City' was destined to fail.*
111. See Martin, C. J., Evans, J., & Karvonen, A. (2018) *Smart and sustainable? Five tensions in the visions and practices of smart sustainable cities in Europe and North America.*
112. Jacobs, K. (2022) *Toronto Wants to Kill the Smart City Forever.* See also City of Toronto (2024) Quayside, https://www.toronto.ca/city-government/planning-development/waterfront/current-projects/quayside/
113. Jacobs, ibid., 2022.
114. Hu, Y., Le Y., Gao Y, Li Y., Liu M. (2018) *Grasping institutional complexity in infrastructure megaprojects through the multi-level governance system.*
115. Logan, J., H. Molotch (1992) *Urban Fortunes. The Political Economy of Place.*
116. WEF (2018) *Harnessing Artificial Intelligence for the Earth.*
117. Tomer, A. (2019) *Artificial intelligence in America's digital city.*
118. Berryhill, J., Heang, K., Clogher, R., & McBride (2019) *Hello, K. 2019.*
119. Guo, Z., F. Curugullo (2022) *Artificial Intelligence, Healthcare and Megaprojects in the Chinese Smart City.*
120. Open AI (2025) *Announcing the Stargate Project*, January 21st, https://openai.com/index/announcing-the-stargate-project/
121. Mazzucato, M. (2021) *Mission Economy. A Moonshot Guide to Changing Capitalism,* London: Penguin.
122. Vermesan, O., Bahr, R., Ottella, M., Serrano, M., Karlsen, T., Wahlstrøm, T., Sand, H., Ashwathnarayan, M. and Gamba (2020) *Internet of Robotics Things.*
123. Masami, T. (2003) *Realizing Unattended Hours of Continuous Operation of Machining Center with Addition of Intelligent Function*; Brumson, B. (2017) *Robotics Online.*
124. Zukin, S., Max Papadantonakis (2018) "Hackathons as Cooptation Ritual: Socializing Workers and Institutionalizing Innovation in the 'New' Economy." *Research in the Sociology of Work*, 31, 157–181. Zukin's work on "fictional expectatives of innovation" is portrayed in Griffith, E. (2018) *Sociologists Examine Hackathons and See Exploitation*, Wired (Business Section), March, https://www.wired.com/story/sociologists-examine-hackathons-and-see-exploitation/. Accessed December 2021.
125. Dimitriou, H., J. Ward, P. Wright (2013) *Mega transport projects*; Flyvbjerg, B. (2014) *What you should know about megaprojects and why.*
126. Sarkheyli, A., E. Sarkheyli (2022) *Dynamic Intelligent Risk Management in Megaprojects Case Study.*

127. Garza, A. (2018) *Sequencing Made Simpler.*
128. Chui, M., Manyika, J., Miremadi, M. (2018) *What AI can and cannot do (yet) for your business.*
129. Blanco et al (2018) *Artificial Intelligence.*
130. Skinner, 2020. See also Atif Ansar, *Why Megaprojects Need a Tech Revolution: Lessons from Carillion.*
131. Kahneman, D., Tversky, A., eds. (2000) *Choices, values, and frames.*
132. Linke, R. Jürgen K. Wilke, Özgür Öztürk, Ferdinand Schöpp & Eva Kassens-Noor (2022) *The future of the eHighway system.*
133. Decaminada, T. (2022) *Examining hyperloop hype on Twitter.*
134. Klein, N. (2020) *Screen New Deal.*
135. Tréguer, F. (2021) *The Virus of Surveillance.*
136. Gill, P. (2021) *Protecting the supply chain in mega- infrastructure projects.*
137. Ølgaard, D. (2020) *Reflections on Naomi Klein's Pandemic Shock Therapy.*
138. Greiman, V. (2020) *Artificial Intelligence in Megaprojects.*
139. del Cerro Santamaría, G. (ed.) (2013) *Urban Megaprojects. A Worldwide View.*
140. Sklair, L. (2017b) *Sleepwalking Through the Anthropocene.*
141. Jacobson, M. et al. (2000) *Earth System Science. From Biogeochemical Cycles to Global Changes.*
142. Lovelock, J. (2020) *Novacene. The Coming Age of Hyperintelligence.*

Chapter 6

1. Some of the material for this section draws from del Cerro Santamaría, G. (2021) Megaprojects, ecology and the environment: Dimensions of sustainability in global cities, in F. Trapani et al. (eds.) *Advanced studies in efficient environmental design and city planning*, Series on advances in science technology and innovation. Springer.
2. Béal, V. (2017) Greening the Neoliberal City? Urban sustainability and selective neo-liberalization in Manchester and Nantes, in Gilles Pinson, G. C. Morel Journel, Debating the Neoliberal City. Routledge. See Hodson M. and Marvin (2017):Intensifyingortransformingsustainablecities?andAnguelovski,I.(2022) *Rethinking urban environmental justice.*
3. Seto, K. C., and D. Satterthwaite (2010) Interactions between urbanization and global environmental change. See also Meyer, W. B. (2013). *Resource consumption.*
4. Seto, K. C., B. Güneralp, and L. R. Hutyra (2012) *Global forecasts of urban expansion to 2030 and direct impacts on biodiversity and carbon pools. The notion of "greenbelts" is relevant here.* See Amati, M. (2008). *Urban greenbelts in the 21st century.*

5. An analysis by the World Business Council for Sustainable Development (WBCSD) and professional services firm Arup, WBCSD (2021) *Net-zero buildings: Where do we stand?*
6. Sorensen, A. (2003) *Building World City Tokyo*. Embodied carbon is the carbon footprint of a building or infrastructure project before it becomes operational. It refers to emissions associated with materials and construction processes throughout the whole lifecycle of a building or infrastructure. It includes any CO_2 created during the manufacturing of building materials (material extraction, transport to manufacturer, manufacturing), the transport of those materials to the job site, and the construction practices used. It also refers to the CO_2 produced maintaining the building and eventually demolishing it, transporting the waste, and recycling it.
7. Sorensen, Ibid. 2003.
8. Lin, Z. (2007) *From Megastructure to Megalopolis*, p.34.
9. Dahlby, C. (2002) *Tokyo Bay*, p. 113.
10. Ibid., p. 114.
11. Kakuchi, S. (2006) *Environment: Japan swelters under urban heat island effect.*
12. Hooper, R. (2013) *Effects will become more obvious as Japan's climate changes.*
13. Hooper, ibid., 2013, p. 67.
14. Ibid.
15. Kakuchi, ibid., 2006, p. 82.
16. Kakuchi, ibid., 2006, p. 85.
17. Edahiro, J. (2008) *Efforts in Japan to mitigate the urban heat island effect.*
18. Hamilton, J. (2011) *In Japan, shaken soil turned soft after quake.*
19. Gaulter, S. (2011) *Why Japan is prone to earthquakes.*
20. Hamilton, J. (2011) *In Japan, shaken soil turned soft after quake.*
21. Japan Property Central (2015) *Reclaimed land in Japan.*
22. Bhattacharya, S. (2011) *Liquefaction of soil in the Tokyo Bay area.*
23. Hamilton, ibid., 2011, p. 64.
24. del Cerro Santamaria, G. (2019) London, in *Wiley Blackwell Encyclopedia of Urban and Regional Studies.*
25. Fainstein, S. (2008) *Megaprojects in New York, London and Amsterdam*, p. 121.
26. Port of London Authority (2013)
27. Duerksen, C., and Snyder, C. (2005) *Nature-Friendly Communities*, p. 26.
28. Centre for Cities (2015) *Cities Outlook*, https://www.centreforcities.org/publication/cities-outlook-2015.
29. Antebe, T. (2012) *The real environmental impacts of holding the Olympics.*
30. Kinver, M. (2005) *London's "One Planet Olympics" Vision.*
31. BBC News (2012) London 2012.
32. McNara, M. (2012) *The Great Olympic Greenwash.*
33. Ibid.
34. Antebe, ibid., 2012, p. 57.

35. Antebe, ibid., 2012, p. 54.
36. Antebe, ibid., 2012, p. 59.
37. Wells, M. (2010) *London 2012: A Rubbish Olympics.*
38. Griffiths, I. (2010) *Tons of radioactive waste casts doubt over London's Olympic stadium legacy.*
39. Wells, ibid., 2010, p. 142.
40. Wells, ibid., 2010.
41. Wells, ibid., 2010, p. 145.
42. Griffiths, ibid., 2010, 32.
43. Antebe, ibid., 2012, p. 65.
44. Antebe, ibid., 2012.
45. See "The controversy over Paris's Olympic village—and others" by Dominic Lutyens.
46. The Guardian (2022) "A massive betrayal": How London's Olympic legacy was sold out, June 30, https://www.theguardian.com/uk-news/2022/jun/30/a-massive-betrayal-how-londons-olympic-legacy-was-sold-out
47. Defacto (2015) *Guide to Architecture: Museum de la Défense.*
48. Defacto, ibid., 2015, p. 43.
49. Sklair, L. (2017) *The Icon Project.*
50. Sklair, Ibid.
51. Defacto, ibid., 2015, p. 58.
52. Langmead, D. and Christine G. (2001) *Encyclopedia of Architectural and Engineering Feats*, p. 31.
53. Sklair, The Icon Project, ibid.
54. Willsher, K. (2014) *France pledges €200 m to restore La Défense's crumbling cube.*
55. Defacto, ibid., 2015.
56. Rose, S. (2008) *Paris Triangle: beyond the shadows of doubt*, p. 38.
57. Rose, ibid., 2008, p. 62.
58. Rose, ibid., 2008, p. 61.
59. GBD Innovation Club (2021) *Paris La Défense: first post-carbon business district*, https://gbdinnovationclub.com/members-news/paris-la-defense-first-post-carbon-business-district/
60. Jacobs, K. (2016) *Calatrava Transportation Hub Finally Takes Flight, Architect Magazine*, https://www.architectmagazine.com/design/calatravas-transportation-hub-finally-takes-flight_o
61. World Trade Center (2024), *Building Facts*, https://www.wtc.com/about/buildings/1-world-trade-center
62. Icon Magazine (2020), *One World Trade Center: A fortress enveloped in glass*, https://www.iconeye.com/architecture/one-world-trade-center-fortress-glass-concrete-skidmore-owings-merrill
63. Schelmetic, T. (2012) *Green Facts About New York's New One World Trade Center.*
64. Smrckova, E. (2014) *Basic Characteristics of Green Cement of CEM V/A and V/B kind*
65. Physicians for Social Responsibility (2015) *Coal Ash: Hazardous to Human Health.*

66. Wells, M. (2010) *London 2012: A Rubbish Olympics*, p. 210.
67. West, J. (2014) *The Freedom Tower Was Supposed to Be the Greenest Building in America*, p. 211.
68. West, ibid., 2014, p. 215.
69. Ferro, S. (2013) *Why do Skyscrapers Melt Things?*
70. In Ferro, ibid., 2013, p. 93.
71. PRCMEE (2016) *China's National Plan on Implementation of the 2030 Agenda for Sustainable Development.*
72. Watkins, D., K. K. Lai, and K. Bradsher (2018), *The World, Built by China.*
73. World Bank (2018) *China - Systematic Country Diagnostic.*
74. Bachmann, J. and J. Burnett (2012) *Infrastructure and the Environment in Chinese Cities.*
75. Southerland, D. (2017) *China's Global Infrastructure Initiative Raises Environmental Concerns.*
76. Luo, Z. et al. (2017) *Pollution and Road Infrastructure in Cities of the People's Republic of China.*
77. Smith, B. (2018) *China: Environmental Issues.*
78. Schwärzel, K. (2017) *China's fight against desertification should not be done at the cost of water security.*
79. We partially follow den Hartog, H. (2021) *Shanghai's Regenerated Industrial Waterfronts: Urban Lab for Sustainability Transitions?*
80. den Hartog (2021), ibid.
81. Adam, H. (2020) *Awakening a Lifeline.*
82. den Hartog (2021), ibid.
83. Planete Energies (2016) *Shanghai, the Showcase for China's New Environmental Policy*, https://www.planete-energies.com/en/medias/close/shanghai-showcase-china-s- new-environmental-policy
84. den Hartog (2021), ibid.
85. Ibid.
86. Ibid.
87. Elsheshtawy, Y. (2008) *The Evolving Arab City.*
88. ICEX (2021) *Recopilación de los megaproyectos de la 'Visión 2030' de Arabia Saudí.*
89. Ibid.
90. Klingmann, A. (2021) *Rescripting Riyadh.*
91. ICEX, ibid., 2021.
92. Klingmann, ibid., 2021.
93. Ibid.
94. Elsheshtawy's book on Riyadh is titled Riyadh: Transforming a Desert City, published by Routledge in 2021.
95. Elsheshtawy (2019) *Transforming Riyadh: A New Urban Paradigm?* The Arab Gulf States Institute in Washington, https://agsiw.org/transforming-riyadh-a-new-urban-paradigm/, Accessed February 2022.
96. United Nations Environment Programme (2022), *Cities and Climate Change*, https://www.unep.org/explore-topics/resource-efficiency/what-we-do/cities/cities-and- climate-change

97. A panel on urban heat islands, co-convened by the International Science Reserve (ISR) and the University of California Disaster Resilience Network (UCDRN), took place during the Science Summit at the 79 UN General Assembly in September 2024, a testimony to the global interest and institutional support surrounding this widespread and increasing problem.
98. Güneralp, B. and K. C. Seto (2008) *Environmental impacts of urban growth from an integrated dynamic perspective.*
99. WBCSD (2021) *Net-zero buildings: Where do we stand?* See also Lamb, W. F., Creutzig, F., Callaghan, M. W., & Minx, J. C. (2019). *Learning about urban climate solutions from case studies.*
100. Meadows, D. H., J. Randers and D. L. Meadows (2004) *Limits to Growth.*
101. World Bank (2018) *China - Systematic Country Diagnostic*, p. 33.
102. Curran, W. and T. Hamilton (2017) *Just Green Enough.*
103. Curran, Ibid., 2017.
104. Grunert, K. G. and C. Ellegaard (1992) *The Concept of Key Success Factors.*
105. Ibid.
106. Harris, M. (2017) *Competitive Precinct Projects.*
107. World Bank, ibid., 2018.
108. Ness, D. (2018) *Sustainable urban infrastructure in China*
109. United Nations (2016) *Executive Summary of China's Actions on the Implementation of the 2030 Agenda for Sustainable Development.*
110. Ness, ibid., 2018.
111. XinhuaNet (2018) *China Approves Three Demonstration Zones on sustainable Development,* p. 22.
112. Carnes, S. (2016) *The Case for the Innovation District as a Sustainable Economic Development Tool in the Knowledge Economy.*
113. del Cerro Santamaría, G. (ed.) Urban Megaprojects (2013) *Urban Megaprojects. A Worldwide View.*
114. Flyvbjerg, B.(2014) *What you should know about megaprojects and why.*
115. del Cerro Santamaría (2020) *Complexity and Transdisciplinarity.*
116. We take inspiration here from the work on Transforming Urban Transport by Diane Davis and Alan Altshuler (2018).

Chapter 7

1. A growing body of literature on megaproject impacts deals with environmental, socioeconomic, and spatial impacts, with specific instances such as those listed in the body of the text. Our analysis in Chapter 7 takes into account all of these kinds of impacts but focuses and elaborates on the impacts discussed in Chapters 3 to 6. See, inter alia, Flyvbjerg, B.(2014) *What you should know about megaprojects and why*; Harris, M. (2017) *Competitive Precinct Projects*; del Cerro Santamaría, G., Ed. (2013) *Urban Megaprojects. A Worldwide View.*

2. We note the work of Robert Bullard, a leading campaigner against environmental racism, as well as the foremost scholar of the problem, and of the Environmental Justice Movement which sprung up in the United States in the 1980s. Bullard played a key role in the organizing of the First National People of Color Environmental Leadership Summit in 1991. See for example his 2005 book *The Quest for Environmental Justice: Human Rights and the Politics of Pollution*. San Francisco: Sierra Club Books.
3. Related to the issue of human rights in megaprojects are the requirements laid out by the 2011 *United Nations Guiding Principles on Business and Human Rights*. See also Vanclay, F. (2017) The potential contribution of social impact assessment to megaproject developments. In: Lehtonen, M., Joly, P. & Aparicio, L. (eds) *Socioeconomic Evaluation of Megaprojects: Dealing with Uncertainties*, Abingdon: Routledge, 181–198. https://doi.org/10.4324/9781315622125-18; van der Ploeg, L. & Vanclay, F. 2017 A human rights based approach to project induced displacement and resettlement. *Impact Assessment & Project Appraisal* 35(1), 34–52. https://doi.org/10.1080/14615517.2016.1271538
4. Sanyal, B. (Eds.). (2005) *Comparative Planning Cultures.*
5. Vygotsky, L. S. (1962) *Thought and Language;* Vygotsky, L. S. (1978) *Mind in Society;* Knorr Cetina, K. (2007) *Culture in global knowledge societies*; Guerin, C. (2013) *Rhizomatic research cultures.*
6. Guerin, C. (2013) *Rhizomatic research cultures.*
7. *Merton, R. K. (1936) The Unanticipated Consequences of Purposive Social Action, American Sociological Review, Vol. 1,* No. 6, December, pp. 894–904, https://doi.org/10.2307/2084615
8. It is relevant to mention the established field of social impact assessment. See, inter alia, Vanclay, F. (2002) Conceptualising social impacts. *Environmental Impact Assessment Review* 22(3), 183–211. https://doi.org/10.1016/S0195-9255(01)00105-6
9. Sassen, S. (2018) *Cities in a World Economy*; Storper, M. (1997) *The Regional World.*
10. Visual power through iconicity is related to the notion of "spatial power" discussed by Lefebvre (abstract and concrete space). See Lefebvre, H. (1991) *The Production of Space.*
11. Broot, S. (2019) *Digital Monuments.*
12. Virilio, P. (2002) *The Overexposed City.*
13. Vulnerability has recently been associated with discussions on resilience and risk, climate conditions, conflict, violence, and sustainability. See, inter alia, Cardona, O.D., M. K. van Aalst, J. Birkmann, M. Fordham, G. McGregor, R. Perez, R. S. Pulwarty, E. L. F. Schipper, and B. T. Sinh (2012) Determinants of risk: exposure and vulnerability. In: *Managing the Risks of Extreme Events and Disasters to Advance Climate Change Adaptation.*
14. Sassen, S. (2014) *Expulsions: Brutality and Complexity in the Global Economy.*
15. Sennett, R. (2018) *Building and Dwelling: Ethics for the City.*
16. Jacobs, J. (1961) *The Death And Life Of Great American Cities.*

17. Klein, N. (2020) *Screen New Deal.*
18. Zuboff, S. (2019) *The Age of Surveillance Capitalism.*
19. Carr, N. (2014) *The Glass Cage.*
20. Rudin, C., J. Radin (2019) Why Are We Using Black Box Models in AI When We Don't Need To?
21. Comiter, M. (2019) *Attacking Artificial Intelligence.*
22. Brundage, M. (2018) *The Malicious Use of Artificial Intelligence.*
23. Muro, M., R. Maxim, J. Whiton (2019) *Automation and Artificial Intelligence.*
24. See Suleyman's *The Coming Wave*. See also Gelles, D. (2025) What will power the A.I. revolution? The New York Times, January 5, https://www.nytimes.com/2025/01/07/climate/artificial-intelligence-power-emissions.html
25. I follow here the research by Steven Gonzalez Monserrate (2022) "The Cloud Is Material: On the Environmental Impacts of Computation and Data Storage." *MIT Case Studies in Social and Ethical Responsibilities of Computing*, no. Winter 2022 (January). https://doi.org/10.21428/2c646de5.031d4553
26. Valdivia, A. (2023) Rural Spain could end up hosting infrastructure hubs for AI—here's what the environmental cost could be, *The Conversation*, https://theconversation.com/rural-spain-could-end-up-hosting-infrastructure-hubs-for-ai-heres-what-the-environmental-cost-could-be-205504. See also Valdivia, A. (2024) The Supply Chain Capitalism of AI: A Call to (Re) think Algorithmic Harms and Resistance, Proceedings of the AAAI/ACM Conference on AI, Ethics, and Society.
27. Gonzalez Monserrate (2022) The Cloud Is Material: On the Environmental Impacts of Computation and Data Storage.
28. Ibid.
29. Ibid.
30. Lepawsky, J. (2024) Climate change induced water stress and future semiconductor supply chain risk. *iScience, 27*(2), 108,791. doi:https://doi.org/10.1016/j.isci.2024.108791
31. Crawford, K. (2024). Generative AI's environmental costs are soaring — and mostly secret. *Nature, 626*(8000), 693–693. doi:https://doi.org/10.1038/d41586-024-00478-x
32. McGovern, G. & S. Branford (2024) Critics fear catastrophic energy crisis as AI is outsourced to Latin America, *Mongabay Series: Planetary Boundaries* https://news.mongabay.com/2024/03/critics-fear-catastrophic-energy-crisis-as-ai-is-outsourced-to-latinamerica/
33. Ransbotham, S., Kiron, D., Gerbert, P., and M. (2017). *Reshaping Business with Artificial Intelligence.*
34. IISS (2022) *International Competition over Artificial Intelligence.*
35. Any algorithm becomes brittle and "fragile" when it is unable to generalize or adjust to situations that are not predicated on a small number of premises. Because machine learning algorithms do not genuinely learn to view the environment in a way that can generalize in the face of uncertainty, this perceptual brittleness is caused. For example, a machine learning algorithm needs to "see" tens of thousands of comparable images in order to recognize patterns of recur-

rence before it can "learn" to recognize a stop sign. The computer "learned" that a specific group of mathematical relationships should be used as a label for a specific object. See Cummings, M. L. (2020) *The Surprising Brittleness of AI.*

36. Mazzucato, M., R. Kattel and J. Entsminger (2021) *Let's imagine a different kind of platform economy.*
37. Mazzucato, M., J. Entsminger and R. Kattel (2020) *Public Value and Platform Governance.*
38. Polman, P. (2019) *A Business Model for Sustainability.*
39. Pyke, T. (2017) *The energy debate: renewable energy cannot replace fossil fuels.*
40. Healey, P. (2013) *Circuits of Knowledge and Techniques.*
41. Smith, M. P. (2005) *Transnational Urbanism Revisited.*
42. del Cerro Santamaría, G. (2017) *Urban Megaprojects in Global Context.*
43. Elsheshtawy, Y. (2014) *Dubaization.*
44. Elsheshtawy, Y. (2019) *Transforming Riyadh.*
45. "Verticality" has been used as a synonym of "Manhattanization." See *Urban Studies 61* (4) March 2024, Special Issue on Verticality.
46. Flyvbjerg et al. (2003) *Megaprojects and Risk.*
47. As shown in Alexandra Miller's work on the Afghan Ring Road (2013).
48. Davis, D. E. and O. Flores Dewey (2013) *How to Defeat an Urban Megaproject.*
49. Andrés Rodríguez-Pose, Federico Bartalucci, The green transition and its potential territorial discontents; See also Gould, K. and T. Lewis (2016) *Green Gentrification. Urban sustainability and the struggle for environmental justice*
50. Levin, S. et al. (2013) *Social-ecological systems as complex adaptive systems.*
51. de Boer, J., R. Muggah and R. Patel (2016) *Conceptualizing City Fragility and Resilience.*
52. Beall, J., T. Goodfellow and D. Rodgers (2011) Cities, conflict and state fragility.
53. Levin et al. (2013) *Social-ecological systems as complex adaptive systems.*
54. See, inter alia, Portugali, J. (1999) *Self-Organization and the City*; Florida, R. (2000) *The Rise of the Creative Class and How It's Transforming Work Leisure, Community and Everyday Life*; Shane, D.G. (2005) *Recombinant Urbanism—Conceptual Modeling in Architecture, Urban Design and City Theory.*
55. Rifkin, J. & T. Howard, *Entropy. A New World View.*
56. Xie, L. T. Han and M. Skitmore (2019) *Governance of Relationship Risks in Megaprojects.*
57. Kharrazi A, Rovenskaya E, et al. (2017). *Network structure impacts global commodity trade growth and resilience.*
58. Davis, D. (2015) *From Risk to Resilience and Back.*
59. Thurner S, Corominas-Murtra B, Hanel R (2017). *The three aspects of entropy for complex systems are information, thermodynamics, and the maximum entropy principle.*
60. The Second Law of Thermodynamics, which permits the determination of entropy variations, but not its absolute value, serves as the theoretical foundation for the idea of entropy. The field with the most structure for studying complicated systems is thermodynamics. According to the Second Law of Thermodynamics (Entropy Law), any transformation process results in the loss and irrecoverable

entropy of useful energy, which manifests itself as disorder and waste. The postulate of Clausius "Die Entropie der Weltstrebteinem Maximum zu" (the entropy of the world tends toward a maximum) is the basis of the Second Law of Thermodynamics. The scientific community has acknowledged the Second Law of Thermodynamics as a theoretical guide to the growth of urban systems. See inter alia, Bristow, & Kennedy, 2015; Prigogine, 1997; Rees, 2012; Rees & Wackernagel, 1996). As discussed in Chapter 2, Immanuel Wallerstein had a productive exchange with the Ilya Prigogine. Some key insights into world-systems analysis come from Prigogine's ideas on complexity. See Wallerstein, I. (2013). The Itinerary of World-Systems Analysis, in *Uncertain Worlds*.

61. Cook, R. (2000) *How Complex Systems Fail.*
62. Cook, R., Ibid.
63. Cook, R., Ibid.
64. Cook, R., Ibid. See also Cook (1999). *A Brief Look at the New Look in error, safety, and failure of complex systems.*
65. On resilience, see, inter alia, Davis, D. (2012) *Urban Resilience in Situations of Chronic Violence;* Davis, D. and M. Patvin (2017) *Resilience, Security and Spaces of Migrant Refuge*; Rus, K. K., K. Vojko, K. David (2018) *Resilience assessment of complex urban systems to natural disasters.*
66. Rees, W. and M. Wackernagel (1996) Urban ecological footprints; Lee, Y. (2014) Social vulnerability indicators as a sustainable planning tool.
67. Davis, D. (2015) *From Risk to Resilience and Back.*
68. Davis, Ibid.
69. Hou, J. & M. Rios (2003) *Community-Driven Place Making.*
70. Innes, J. E., Gruber, J. E., Neuman, M., and Thompson, R. (1994) *Coordinating Growth and Environmental Management Through Consensus Building.*
71. Sassen, S. (2014) *Expulsions.*
72. Morin, E. (2005). *On Complexity.*
73. A large portion of scientific study in the 20th century was motivated by reductionism. It teaches us that, in order to understand nature, we must first understand its parts. Once we comprehend the individual components, the total will become clear. As a result, we have been compelled to view the world through its parts for many years. To understand the universe, life, molecules, individual genes, complex human behavior, and origins of fads and religions, we studied atoms, superstrings, and individual genes. General System Theory, Cybernetics, Information Theory, and later Chaos and Complexity theories helped Western science in the 20th century to develop a new way of viewing the universe. These techniques share common thread in that they all sought to obtain past reductionism, analysis, and the paradigm of simplification.
74. Montuori, A. (2022) *Integrative Transdisciplinarity.* See also Del Cerro Santamaría, G. (2020b) *Complexity and Transdisciplinarity: The Case of Iconic Urban Megaprojects*, and del Cerro Santamaría, G. (2019a) *Rapid Urbanization, Ecology and Sustainability.*
75. Vygotsky, L. S. (1962) *Thought and Language.*

76. Vygotsky, L. S. (1978) *Mind in Society.*
77. Gruber, T. (2008) *Collective Knowledge Systems*, p. 5.
78. Kastenhofer, K. (2007) *Converging Epistemic Cultures?*
79. Knorr Cetina, K. (2007) *Culture in global knowledge societies*, p. 363.
80. Latour, B. (1987) *Science in Action.*
81. del Cerro Santamaría, G. (2018 a, b, c) *Building Transdisciplinary Urban Space.*
82. Homi Bhabha's notion hybridity/third space is akin to the notion of "assemblage" in Deleuze & Guattari, 1987.
83. Guerin, C. (2013) *Rhizomatic research cultures.*
84. Vojnovic (2014) *Urban Sustainability.*

Chapter 8

1. In a recent paper, Lauerman argues that the micro-scale of the individual and community, where they lived experience of dignity occurs, and the much larger scale of the megaproject, where most contemporary city making occurs, are inherently at odds with one another. The common good may take precedence over the dignity of particular urban residents in megascale governance, which places greater emphasis on big-picture thinking and economies of scale. In the "urban age" of rapid and widespread mega-urbanization, resolving these scalar conflicts is a fundamental issue. See Lauerman, J. (2022). *Dignity, megaprojects, and the problem of scale.*
2. Recent scholarship points out to the need to govern capabilities rather than places to implement sustainability goals in urban settings. See Janssen et al. (2023) *Governing capabilities, not places—how to understand social sustainability implementation in urban development.*
3. Peter Marcuse proposed many years ago that promoting "sustainability" may simply encourage the sustaining of the unjust status quo and how the attempt to suggest that everyone has common interests in "sustainable urban development" masks very real conflicts of interest. See Marcuse, P. (1998). Sustainability is not enough. *Environment and Urbanization. 10* (2), 103–112.
4. See Davidson, M. (2009). *Social sustainability: A potential for politics?*
5. See Glaeser, E. (2021) *Urban Resilience.*
6. Tollefson, J. & K. R. Weiss (2015) *Nations approve historic global climate accord.*
7. Jarvensivu, P. et al., (2018) *Global Sustainable Development Report.*
8. Järvensivu, P. et al., (2018), Ibid.
9. United Nations (1987) Bruntland Report, Chapter 2.
10. Yadong, Y (2013) *Eco-efficiency trends in China, 1978–2010.*
11. Barron, J. (2019) *How Big Business is Hedging Against the Apocalypse.*
12. Ferguson, P. (2018) *Post- Growth Politics.*

13. Klein, N. (2015) *This Changes Everything.*
14. Dario Gil, Chair of the National Science Board, https://issues.org/wp-content/uploads/2025/01/35-39-Interview-with-Dario-Gil-Winter-2025.pdf
15. As reported by MIT Professor and President of the Future of Life Institute Max Tegmark. See Tegmark in https://futureoflife.org/statement/agi-manhattan-project-max-tegmark/
16. Greenfield, A. (2013) *Against the Smart City.*
17. Zhang, X., & Li, H. (2018). *Urban resilience and sustainability.*
18. Kahneman, D. (2013) *Thinking Fast and Slow.*
19. Kahneman, D., Tversky, A., eds. (2000) *Choices, values, and frames.*
20. Harris, 2017, 34.
21. (1) Environmental sustainability (sustainable infrastructure and sustainable development zones), (2) sustainability in design and planning, (3) sustainability in management, (4) institutional sustainability, and (5) socioeconomic sustainability.
22. Dunn, P. (2022) *Whose interest is AI serving?.*
23. Sturup, S., N. Low (2019) *Sustainable development and mega infrastructure.* See also Köhler, J., Geels, F., Kern, F., Markard, J . (2019). *An agenda for sustainability transitions research.*
24. Ryan, M., & Gregory, A. (2019). *Ethics of Using Smart City AI and Big Data.*
25. Grubbauer, M., Čamprag, N. (2018) *Urban megaprojects, nation-state politics and regulatory capitalism in Central and Eastern Europe*; del Cerro Santamaría, G. (2019) *Rapid Urbanization, Ecology and Sustainability.*
26. Taleb, n. (2012) *Antifragile.*
27. del Cerro Santamaría, G. (2020c) *Building New Knowledge About Natures, Ecologies and Sustainability.*
28. Meadows, D. H., Randers J, Meadows DL (2004) *Limits to Growth.* See also the International Geosphere-Biosphere Program http://www.igbp.net.
29. Ness, D. (2018) *Sustainable urban infrastructure in China.*
30. XinhuaNet (2018) *China Approve s Three Demonstration Zones on Sustainable Development*, p. 22.
31. Carnes, S. (2016) *The Case for the Innovation District as a Sustainable Economic Development Tool in the Knowledge Economy.*
32. See Chapter 2 by Vranken in P. Ache et al. (eds.) *Cities between Competitiveness and Cohesion*, Springer Science+Business Media B.V. 2008.
33. Krueger, R. and L. Savage (2007) *City-Regions and Social Reproduction.*
34. See Jessop, B. (2002) *Liberalism, Neoliberalism, and Urban Governance*; While, A., A. Jonas EG, Gibbs D (2004) *The environment and the entrepreneurial city.*
35. Etzioni A (1998) 'Introduction', pp. ix–xxiv in Etzioni A (ed.) *The Essential Communitarian Reader,* p. xii.
36. Driver S, Martell L (1997) *'New Labour' s Communitarians'*, p. 29.
37. See Mazzucatto, M., and J. Ryan-Collins (2019) *Putting value creation back into public value*; See also Mazzucatto and Kattel (2019) *Getting Serious About Value.*

38. Mazzucatto (2019) *Putting value creation back into public value.*
39. Beauregard, *Planning Matter*, pp. 8–9, 36, 42–53, Chapter 5, Chapter 10
40. See Mazzucato, 2019, *Getting Serious About Value.* See also Janssen, C., Daamen, T.A. and Verheul, W. J. (2024) *Governing capabilities, not places.*
41. See Kumaraswamy, M., J. Chung, W. Zou and K. Wong (2015) Online paper, https://www.researchgate.net/publication/285512147
42. According to American poet and environmentalist Gary Snyder, we require an "urban practice of the wild" that combines, in transdisciplinary fashion, knowledge of Native American and Asian spiritual practices with those of Western philosophy, poetry, and the natural sciences. Snyder, G. (2008), cited in Dana Goodyear, Zen Master. *Gary Snyder and the Art of Live.*
43. Heynen, N. et al. (2006) *In the Nature of Cities. Urban Political Ecology and the Politics of Urban Metabolism.*
44. This occurs when eukaryotes emerge and fill the oxygen atmosphere, making it combustible and leading to the loss of millions of species due to atmospheric reduction of numerous dioxides. Millions of species were removed when the asteroids struck Earth. Morton, T. (2007) *Ecology Without Nature. Rethinking Environmental Aesthetics*, p. 46.
45. Morton, T. (2007) Ibid. p. 81.
46. Barriere, O. et al. (2019) *Coviability of Social and Ecological Systems: Reconnecting Mankind to the Biosphere in an Era of Global Change.*
47. Mayer, A. L., R. P. Donovan, Pawlowski CW. 2014. *Information and entropy theory for the sustainability of coupled human and natural systems.*
48. Addiscott, T. M. (2005) *Entropy and sustainability.*
49. Sengupta R. (2020) Entropy Law, Social Sustainability and Development Policy, ch. One in *Entropy Law, Sustainability and Third Industrial Revolution.*
50. Brundtland report, United Nations, 1987. Some conceptualizations of "environmental justice" follow this anthropocentric approach. See, inter alia, Anguelovski et al. (2020) *Expanding the Boundaries of Justice in Urban Greening Scholarship.*
51. According to Lovelock' s and Margulis's Gaia hypothesis, for much of the last 3.8 billion years, a holistic feedback system has developed in the biosphere, with life forms regulating the temperature and proportions of gases in the atmosphere for the benefit of the atmosphere. The authors also showed that Darwin was only half right. Life evolves in response to environmental changes; however, the environment also evolves in response to biological changes.
52. Harman, G. (2002). *Tool-Being: Heidegger and the Metaphysics of Objects.*
53. See Farías and Bender, 2010. ANT has been adopted and adapted in various fields, especially over the past decade. This includes planning, where the recent work of Robert Beauregard, Laura Lieto, and colleagues is at the forefront of attempts to reformulate planning theory around assemblage thinking and the new materialist, poststructuralist, and posthumanist thrusts.
54. Sennett, R. (1992). *The Conscience of the Eye,* p. 192; Jacobs, J. (2000) *The Nature of Economies.*
55. Lefevbre, *La production de l'espace.*

56. A key notion here is that of "affordances." It refers to the properties of matter, these properties being what influences how humans interact with things. Affordances, as applied to both things and places, are properties that allow a person to do some things and not others but are not fully constraining. See Lieto, 2017.
57. Beauregard, R., *Planning Matter*, 2015; Lieto, L., and R. A. Beauregard, eds. (2016) *Planning for a Material World.*
58. Whitehead, A. N. (1920) *The Concept of Nature.*
59. A similar focus on inquiry over conception, vision, plan, or critique is a major feature of assemblage thinking, of which actor-network theory (ANT) is perhaps the best- known variant. See Deleuze and Guattari (1987) *A Thousand Plateaus.*
60. See Harman, G., ibid., 2002; See also Bryant, L. (2011) *Onticology—A Manifesto for Object-Oriented Ontology*; Schrödinger, E. (1967) *What is Life?*
61. Jullien, F. (2001). *Un sabio no tiene ideas*: O el otro de la filosofía. Siruela.
62. Latour, B. (1992) *Where are the Missing Masses?*
63. Müller, M. (2015) *Assemblages and Actor-networks*, p. 36.
64. Mollenkopf, J. and Castells, M. (1992) *Dual City*; Marcuse, P. (1989) *Dual city: a muddy metaphor*; Mingione, E. (1991) *Fragmented Societies.*
65. Sennett, R. (1992) *The Conscience of the Eye.*
66. Sennett, R. (1992) *The Conscience of the Eye*, p. 196.
67. Tuan, Y. F. (1990) *Topophilia.*
68. Tuan, Y. F. (1979) *Landscapes of Fear.*
69. Tuan, Y. F. (1996) *Cosmos and Hearth: A Cosmopolite's Viewpoint.*
70. Latour, B. (2016) *Facing Gaia.* For a sharply critical view, see Davis (2010) *Who Will Build the Ark. For how ecological degradation and economic precarity proliferate in the 21st century*, see Tsing, A. L. (2015) *The Mushroom at the End of the World.* As explained in Chapter 1, this book does not focus on grassroots resistance to megaprojects for several reasons, even if we acknowledge its important role. For a discussion of forms of protest, see Hanna, P. et al. 2016 Conceptualizing social protest and the significance of protest action to large projects. *Extractive Industries & Society* 3(1), 217–239. https://doi.org/10.1016/j.exis.2015.10.006
71. For uncertainty, see Heisenberg, W. (2007) *Physics and Philosophy*; for indetermination, see Dennett, D. *Darwin' s Dangerous Idea*; for incompleteness, see Gödel, K. (1951) *Some basic theorems on the foundations of mathematics and their implications*, and Nicolescu, A.(2013) *From Modernity to Cosmodernity*: For undecidability, see Roy, A. (2015) *What is urban about critical urban theory.*
72. Castells, M. (2018) *Ruptura. La crisis de la democracia liberal.*
73. Gibbons, M. et al. (1994) *The New Production of Knowledge.*
74. For a pragmatist approach to transdisciplinary sustainability, see Popa et al. (2015). *For the strategy of knowledge integration within transdisciplinary sustainability*, see Cockburn (2022) *Knowledge integration in transdisciplinary sustainability science: Tools from applied critical realism.* "Convergence," rather than "integration," is a term used in earlier years. See the National

Research Council. (2014) *Convergence: Facilitating transdisciplinary integration of life sciences, physical sciences, engineering, and beyond*; National Academies of Sciences, Engineering, and Medicine (2019) *Fostering the culture of convergence in research: Proceedings of a workshop.*

75. Elmqvist, T. et al. (2018) *Urban Planet.* del Cerro Santamaría G. (2020a) *Complexity and Transdisciplinarity.*
76. Brandt, P. et al. (2013) *A review of transdisciplinary research in sustainability science.*
77. The SDG #11 specifically refers to the urban areas. Although the UN SDG represents an unprecedented global coordination effort for sustainability, it is not without criticism. See, inter alia, Stiftel, B. (2021) *Planners and the new urban agenda*, and Klopp, J. M., & Petretta, D. L. (2017) *Urban sustainable development goal.*
78. Meadows, D. H., Randers J, Meadows DL (2004) *Limits to Growth.* See also Escobar, A. (2015). *Degrowth, post-development, and transitions: a preliminary conversation.*

Bibliography

Acharya, A. (2017). After liberal hegemony: The advent of a multiplex world order. *Ethics & International Affairs, 31*(3), 271–285.

Ache, et al. (2008). *Cities between competitiveness and cohesion*. Springer.

Adam, H. (2020). Awakening a lifeline. *Topos Magazine*, October, https://toposmagazine.com/awakening-a-lifeline/

ADB. (2016). *Scaling new heights. Vizag-Chennai industrial corridor. India's first coastal corridor*. Asian Development Bank.

Addiscott, T. M. (2005). Entropy and sustainability. *European Journal of Soil Science, 46*(2), 161–168. https://doi.org/10.1111/j.1365-2389.1995.tb01823.x

Adelfio, M., Hamiduddin, I., & Miederna, E. (2021). London's king's cross redevelopment: A compact, resource efficient and 'liveable' global city model for an era of climate emergency? *Urban Research and Practice, 14*(2). https://doi.org/10.1080/17535069.2019.1710860

Ahlers, R. M. Z., & Bakker, K. (2017). Large dam development. From Trojan horse to Pandora's box. In B. Flyvbjerg (Ed.), *The Oxford handbook of megaproject management*. Oxford University Press.

Ajuntament de Barcelona. (2013). 22@Barcelona. *Background. Urban Planning Management.*. http://www.22barcelona.com/documentacio/Dossier22@/Dossier22@English_p.pdf, Retrieved July 15th, 2019

Alexander, C. (1965). A city is not a tree. *Archit Forum, 122*(1), 58–62.

Allen, C. R., D. G. Angeler, A. S. Garmestani, L. H. Gunderson, C. S. Holling (2014) Panarchy: Theory and application. Ecosystems. Springer, , 17(4): 578–589.

Alsdorf, B. (2002). Collections curatorial assistant for the Guggenheim Foundation, Guggenheim Foundation, http://www.guggenheim.org/exhibitions/past_exhibitions/global_gugg/global_gugg_bottom_index.html (consulted on 10/16/2017).

Altshuler, A., & Luberoff, D. (2003). *Mega projects: The changing politics of urban public investment*. Brookings Institution Press.

Amati, M. (2008). *Urban greenbelts in the 21st century*. Routledge.

Anguelovski, I. (2022). Rethinking urban environmental justice: Who benefits from green cities? *Harvard Design Magazine, 49*(2022), 117–121.

Anguelovski, I., et al. (2020). Expanding the boundaries of justice in urban greening scholarship: Toward an emancipatory, Antisubordination, intersectional, and relational approach. *Annals of the American Association of Geographers, 110*(6), 1743–1769. https://doi.org/10.1080/24694452.2020.1740579

Ankura (2020) Geopolitical Impacts on the Future of Construction, April 13th, https://angle.ankura.com/post/102hbvm/geopolitical-impacts-on-the-future-of-construction.

G. del Cerro Santamaría, *Contesting Megaprojects*,
https://doi.org/10.1007/978-3-031-98382-5

Ansar, A. (2018) Why megaprojects need a tech revolution: Lessons from Carillion, future of construction (mar 8) https://www.futureofconstruction.org/blog/why-megaprojects need-a-tech-revolution-lessons-from-carillion/.

Ansar, A., Flyvbjerg, B., Budzier, A., & Lunn, D. (2016). Does infrastructure investment lead to economic growth or economic fragility? *Evidence from China, Oxford Review of Economic Policy, 32*(3), 360–390.

Antebe, T. (2012) The real environmental impacts of holding the Olympics in East London. *Corporate Watch*, July 27. Retrieved June 15, 2015, from http://www.corporatewatch.org/news/2012/jul/26/real-environmental-impacts-holding- olympics-East-London.

Arsel, M. F. A., & Saad-Filho, A. (2021). Authoritarian developmentalism: The latest stage of neoliberalism? *Geoforum, 124*, 261–266.

Atkinson, R. (2012). *Innovation economics: The race for global advantage*. Yale University Press.

Augusto, J., Callaghan, V., Cook, D., Kameas, A., & Satoh, I. (2012). Intelligent environments. A manifesto. *Human-centric Computing and Information Sciences, 3*, 12.

Baba, E. C., et al. (2023). Fear and architecture: Learning from mega-projects and canal Istanbul as a case. *Journal of Contemporary Urban Affairs, 7*(1), 19–37.

Bachman, L. (2008). Architecture and the four encounters with complexity. *Architecture and Design Management, 4*(1), 15–30. https://doi.org/10.3763/aedm.2008.S407

Bachmann, J., & Burnett, J. (2012). Infrastructure and the environment in Chinese cities: Prospects for improvement. *China Business Review*. July 2012, https://www.chinabusinessreview.com/infrastructure-and-the-environment-in-chinese-

Baily, M. N., N. Montalbano, (2018) Clusters and innovation districts: Lessons from the U.S. Experience, Economic Studies at Brookings.. https://www.brookings.edu/wp-content/uploads/2018/05/es_20180508_bailyclustersandinnovation.pdf

Bangalore, S., & Srivathsan, B. (2023). Investing in the rising data center economy. *McKinsey Report*. https://www.mckinsey.com/industries/technology-media-and-telecommunications/our-insights/investing-in-the-rising-data-center-economy

Bar'el, Z. (2021) Erdoğan's environmental mess might pale compared to Washington's ire, Haaretz, June 24, https://www.haaretz.com/middle-east- news/palestinians/.Premium-erdogan-s-environmental-mess-might-pale-compared- to- Washington-s-ire-1.9934449.

Barabasi, A.-L. (2003). *Linked. How everything is connected to everything else and what it means for business, science and everyday life*. Penguin.

Barangaroo Remediation Fact Sheet. (2019). *Barangaroo: Hickson Road Remediation*. https://www.seslhd.health.nsw.gov.au/sites/default/files/groups/Public_Health_Unit/Environmental%20 Health/19p1669-factsheet-barangaroo-hickson-road-remediation.pdf.

Barriere, O., et al. (2019). *Coviability of social and ecological systems: Reconnecting mankind to the biosphere in an era of global change*. Springer.

Barron, J. (2019) How big business is hedging against the apocalypse, the New York times, April 11, https://www.nytimes.com/interactive/2019/04/11/magazine/climate- change-exxon-renewable-energy.Html.

Battaglia, A., & Tremblay, D. G. (2011). 22@ and the Innovation District in Barcelona and Montreal: A process of clustering development between urban regeneration and economic competitiveness. *Urban Studies Research, 2011*, Article ID 568159. https://doi.org/10.1155/2011/568159

Batty, D. (2012). Guggenheim delay raises big question: Is Abu Dhabi ready for modern art? *The Guardian*. April 27. Retrieved May 23, 2018, from http://www.theguardian.com/world/2012/apr/17/abu-dhabi-guggenheim-delay-question

Batty, D. (2013) In Abu Dhabi, they call it Happiness Island. But for the migrant workers, it is a place of misery. *The Guardian*, December 21. Retrieved May 28, 2018, from http://www.theguardian.com/world/2013/dec/22/abu-dhabi-happiness-island-misery

Batty, D. (2015) Migrants building UAE cultural hub 'working in prison conditions'. *The Guardian*, April 4, 2015, Retrieved, July 3rd, http://www.theguardian.com/global-development/2015/apr/04/migrant-workers-uae-saadiyat-island-abu-dhabi-battery-hens

Batty, M., & Longley, P. A. (1994). *Fractal cities: A geometry of form and function*. Academic Press.

Batty, M. (2008) Cities as complex systems: Scaling, interaction, networks, dynamics and urban morphologies, UCL Working Papers Series, Paper 131,. February.

Batty, M. (2016). How disruptive is the smart cities movement? *Environment and Planning B Planning and Design, 43*(3), 441–443. https://doi.org/10.1177/0265813516645965

BBC News (2012). London 2012: Olympic Games 'greenest ever'. *BBC News*, August 11th. Retrieved June 15, 2015, from http://www.bbc.com/news/uk-19220847

Béal, V. (2017). Greening the Neoliberal City? Urban sustainability and selective neo-liberalization in Manchester and Nantes. In G. Gilles Pinson & C. M. Journel (Eds.), *Debating the Neoliberal City*. Routledge.

Beall, J., T. Goodfellow and D. Rodgers (2011) Cities, conflict and state fragility, Working Paper 85, London School of Economics and Political Science, available at http://eprints.lse.ac.uk/39766/1/Cities,_conflict_and_state_fragility_wp852(author).pdf.

Beauregard, R., & Pierre, J. (2000). Disputing the global: A Sceptical view of locality- based international initiatives. *Policy and Politics, 28*(4), 465–478.

Beauregard, R. A. (2015). *Planning matter. Acting with things*. University of Chicago Press.

Beck, U. (1992 [1986]) *Risk society, towards a new modernity*. Trans. From the German by Mark Ritter, and with an Introduction by Scott Lash and Brian Wynne. Sage Publications.

Beiser, V. (2019). *The world in a grain: The story of sand and how it transformed civilization*. Riverhead Books.

Benner, C., & Pastor, M. (2016). *Equity, growth, and community: What the nation can learn from America's metro areas*. University of California Press.

Berger, P. L., & Luckmann, T. (1967). *The social construction of reality. A treatise in the sociology of knowledge*. Anchor Books.

Berkes, F., Colding, J., & Folke, C. (2003). *Navigating social-ecological systems: Building resilience for complexity and change*. Cambridge University Press.

Berry, C., & Giovannini, A. (2018). *Developing England's north. The political economy of the Northern Powerhouse*. Palgrave Macmillan.

Berryhill, J., Heang, K., Clogher, R., & McBride, K. (2019). Hello, world: AI and its use in the public sector, OECD working papers on public governance, no 36. https://doi.org/10.1787/726fd39d-en

Berube, A., & Kneebone, E. (2013). *Confronting suburban poverty in America*. Brookings Institution.

Bhabha, H. (1994). *The location of culture*. Routledge.

Bhattacharya, S. (2011). Liquefaction of soil in the Tokyo Bay area from the 2011 Tohoku (Japan) earthquake. *Soil Dynamics and Earthquake Engineering., 31*(11), 1618–1628.

Bhattacharyya, S. (2021). Turkey "crazy" and controversial Istanbul Canal project, France 24, June 29, https://www.france24.com/en/middle-east/20210629-turkey-s-crazy-and-controversial-istanbul-canal-project.

Bird, E., Fox-Skelly, J., Jenner, N., Larbey, R., Weitkamp, E., & Winfield, A. (2020). The ethics of artificial intelligence: Issues and initiatives, panel for the future of science and technology (STOA). *European Parliament*. https://www.europarl.europa.eu/RegData/etudes/STUD/2020/634452/EPRS_STU(2020)634452_EN.pdf. Accessed October, 2021

Blakemore, E. (2016) Museum building is booming in the United States, Smithsonian.com, April 11, https://www.smithsonianmag.com/smart-news/museum-building-booming-United-States-180958711/.

Blanco, J. L., et al. (2018). Artificial Intelligence – construction technology's next frontier, McKinsey & Co. (2018) https://www.mckinsey.com/business-functions/operations/our-insights/artificial-intelligence-construction-technologys-next-frontier. Accessed November 2021.

Bogaert, K. (2018). *Globalized authoritarianism. Megaprojects, slums and class relations in urban Morocco*. University of Minnesota Press.

Bonshoms, E. (2007). Personal interview with Eudald Carbonell. *WinMagazine*, 53–61, April, https://web.archive.org/web/20070929025915 / http://www1.winterthur.es/puertaabierta/lists/lists_abril/images/interview_abril.PDF.

Boston Mayor's Office. (2014). Boston's Innovation District, Boston Mayor's Office, August 1st, http://www.innovationdistrict.org/2010/10/15/.

Bradsher, K. (2006). Hong Kong halts plans for arts center. *The New York times*, February 22. Retrieved May 27, 2015, from http://www.nytimes.com/2012/01/25/world/middleeast/abu-dhabi-reaffirms-its- grand-plan-for-museums.Html.

Brandt, P., et al. (2013). A review of transdisciplinary research in sustainability science. *Ecological Economics, 92*, 1–15. August.

Brenner, N., & Schmid, C. (2011). Planetary urbanization. In M. Gandy (Ed.), *Urban constellations*. Jovis.

Brenner, N., & Theodore, N. (2005). Neoliberalism and the urban condition. *City, 9*(1), 101–107. https://doi.org/10.1080/13604810500092106

Brill, F., & Conte, V. (2019). Understanding project mobility: The movement of king's cross to Brussels and Johannesburg. *Environment and Planning C: Politics and Space, 38*(1). https://doi.org/10.1177/2399654419854458#con2

Bristow, D., & Kennedy, C. (2015). Why do cities grow? Insights from nonequilibrium thermodynamics at the urban and global scales. *Journal of Industrial Ecology, 19*(2), 211–221. https://doi.org/10.1111/jiec.12239

Broot, S. (2019). *Digital monuments. The uses and abuses of iconic architecture*. Routledge.

Brown, J. (2006) "New Denver art Museum reflects Rocky Mountains," *PBS Newshour*, October, 5, 2006. http://www.pbs.org/newshour/bb/entertainment- july-dec06-museum_10- 05/.

Brown, W. (2017). *Undoing the demos. Neoliberalism's Stealth Revolution* (Vol. 14). Zone Books.

Brumson, B. (2017). *Robotics Online*. Robotic Industries Association, https://roboticsonline.wordpress.com/tag/robotic-industries-association/. Accessed November, 2019.

Brundage, M. (2018) The Malicious Use of Artificial Intelligence: Forecasting, prevention and Mitigation, Future of Humanity Institute paper, February, https://arxiv.org/pdf/1802.07228.pdf. Accessed November, 2021.

Brundtland, G. H. (1987). United Nations, Bruntland Report, Chapter Two, United Nations World Commission on Environment and Development. https://en.wikisource.org/wiki/Brundtland_Report/Chapter_2._Towards_SustainabLe_Development.

Brunn, S. D. (Ed.). (2011). *Engineering earth. The impacts of megaengineering projects*. Springer.

Bryant, L. (2011). Onticology—A manifesto for object-oriented ontology, part 1. *Larval Subjects*. Retrieved August 3rd, 2019, https://larvalsubjects.wordpress.com/2010/01/12/object-oriented-ontology-a-manifesto-part-i/

Buchoud, N. J. A., M. Keith, S. Parnell and M. van der Pütten (2022) A G7/G20 Pact for Sustainable Urbanisation? Building on the Positive Legacy of Italy's G 20 Legacy, Istituto Affari Internazionali, https://www.iai.it/en/pubblicazioni/g7g20-pact-sustainable- urbanisation.

Bullard, R. (2005). *The quest for environmental justice: Human rights and the politics of pollution*. Sierra Club Books.

Bunge, M. (2014). *Emergence and convergence. Qualitative novelty and the unity of knowledge*. University of Toronto Press.

Bunnell, T. (2013). Encountering Kuala Lumpur through the "travel"of UMPs. In G. del Cerro Santamaría (Ed.), *Urban megaprojects: A worldwide view* (pp. 61–82). Emerald Group Publishing Limited.

Cameron, H. (2016) Innovation districts—The way forward for sustainable growth? The knowledge exchange blog, January 25th, https://theknowledgeexchangeblog.com/2016/01/25/innovation-districts-the-way-forward-for-sustainable-growth/.

Campbell, R. (2007) Does Gehry's Stata center really work? Bloomberg business, June 19, http://www.bloomberg.com/bw/stories/2007-06-19/does-gehrys-stata-center-eally-work-businessweek-business-news-stock-market-and-financial-advice.

Caragliu, A., Del Bo, C., & Nijkamp, P. (2009). Smart cities in Europe. In *Serie research memoranda 0048*. VU University Amsterdam, Faculty of Economics, Business Administration and Econometrics.

Cardona, O. D., van Aalst, M. K., Birkmann, J., Fordham, M., McGregor, G., Perez, R., Pulwarty, R. S., Schipper, E. L. F., & Sinh, B. T. (2012). Determinants of risk: Exposure and vulnerability. In C. B. Field, V. Barros, T. F. Stocker, D. Qin, D. J. Dokken, K. L. Ebi, M. D. Mastrandrea, K. J. Mach, G.-K. Plattner, S. K. Allen, M. Tignor, & P. M. Midgley (Eds.), *Managing the risks of extreme events and disasters to advance climate change adaptation* (A special report of Working Groups I and II of the Intergovernmental Panel on Climate Change (IPCC)) (pp. 65–108). Cambridge University Press.

Carnes, S. (2016) The case for the Innovation District as a sustainable economic development tool in the knowledge economy, Georgia Tech Center for Urban Innovation, February 1st, https://gtcui.wordpress.com/2016/02/01/the-case-for-the-innovation-district-as-a-sustainable-economic-development-tool-in-the-knowledge-economy/.

Carr, C., & Hesse, M. (2020). When alphabet Inc. plans Toronto's waterfront: New post political modes of urban governance. *Urban Planning, 5*(1), 69–83.

Carr, N. (2014). *The glass cage. Where automation is taking us*. W. W. Norton.

Cartwright, N. (1999). *The dappled world. A study of the boundaries of science*. Cambridge University Press.

Casareggio, A. (2021) Turkey and the geopolitics of the straits,. *IARI*, April 8th, https://iari.site/2021/04/08/turkey-and-the-geopolitics-of-the-straits/.

Castán Broto, V., & Bulkeley, H. (2013). Maintaining climate change experiments: Urban political ecology and the everyday reconfiguration of urban infrastructure. *International Journal of Urban and Regional Research, 37*(6), 1934–1948. https://doi.org/10.1111/1468-2427.12050

Castells, M. (1972). *La question urbaine*. Maspero.

Castells, M., & Hall, P. (1994). *Technopoles of the world. The making of 21st century industrial complexes*. Routledge.

Castells, M. (2000). *The information age. The rise of the network society* (2nd ed.). Blackwell.

Castells, M. (2004). *The network society. A cross-cultural perspective* (p. xvii). Edward Elgar.

Castells, M. (2009). *The rise of the network society* (2nd ed.). Wiley- Blackwell.

Castells, M., & Himanen, P. (2014). *Reconceptualizing development in the global information age*. Oxford University Press.

Castells, M. (2018). *Ruptura. La crisis de la Democracia liberal*. Alianza.

Catanzaro, M. (2017) Interview with Isabelle Anguelovski,. *Barcelona Metropolis, 102*, January.

Çekmiş Görgülü, A., & Hacıhasanoğlu, I. (2012). Water crossing utopias of Istanbul: Past and future. *ITU Journal of Faculty of Architecture, 9*(2), 67–88.

Chan, K. (2017). What is a 'Smart City'?. *Expatriate Lifestyle*, January 24. Accessed November, 2021.

Chan, Y. (2000) "Reclamation and pollution in Hong Kong with special reference to Victoria harbour." The HKU Scholars Hub. Stable URL: http://hdl.handle.net/10722/36319.

Charles, D. (2015) From technopoles to science cities: Characteristics of a new phase of science cities, https://www.researchgate.net/publication/323308836_From_technopoles_to_science_cities_characteristics_of_a_new_phase_of_science_cities.

Chen, X. (2013) Hong Kong finds its footing in art world. *The New York Times,* May 27. Retrieved May 23, 2015, from http://www.nytimes.com/2013/05/23/arts/artsspecial/hong- kong-finds-its-footing- in-art-world.html.

Chen, X. (2019) From follower to leader: China's development of special economic zones and its global impact and lessons, *The European Financial Review*, December, https://www.europeanfinancialreview.com/from-follower-to-leader-chinas- development-of-special-economic-zones-and-its-global-impact-and-lessons/.

Chen, X. (2021) Connectivity, connectivity, connectivity: Has the China-Europe freight train become a winning run? *The European Financial Review*, August-September, https://digitalrepository.trincoll.edu/cgi/viewcontent.cgi?article=1320&context=facpub.

Chib, A. (2015). Research on the impact of the information Society in the Global South: An introduction to SIRCA. In A. Chib, J. May, & R. Barrantes (Eds.), *Impact of information society research in the global south*. Springer. https://doi.org/10.1007/978-981-287-381-1_1

Chiu, A. (2014) Legal battles mount for former chief secretary Rafael hui, *South China Morning Post*, July 20, https://www.scmp.com/news/hong-kong/article/1298096/legal-battle-mount-former-chief-secretary-rafael-hui.

Chow, V. (2013) The past and present of West Kowloon cultural center, *Cultural Vision*, May, http://www.viviennechow.com/p/a-series-of-unfortunate-events-past-and.html.

Chu, S. Y. (2010). Brand Hong Kong: Asia's World City as method? *Visual Anthropology, 24*(1–2), 46.

Chui, M., Manyika, J., & Miremadi, M. (2018). What AI can and can't do (yet) for your business. *McKinsey Quarterly*. January. cities-prospects-for-improvement/.

City of Toronto. (2024). Quayside. https://www.toronto.ca/city-government/planning-development/waterfront/current-projects/quayside/.

Clean Air Network (2018) Clean Air Network Response to Policy Address, http://www.hongkong-can.org/hk/press-release/.

CNN Money.. Fade to Black the 1980s Vision of "Lights-Out" Manufacturing, WhereRobots Do All the Work, Is a Dream No More. http://money.cnn.com/magazines/business2/business2_archive/2003/06/01/343371/index.htm.

Cockburn, J. (2022). Knowledge integration in transdisciplinary sustainability science: Tools from applied critical realism. *Sustainable Development, 30*, 358–374.

Columbia University (2015) Istanbul: Planning for Transparency and Inclusion. Retrieved 26 April 2018. http://www.arch.columbia.edu/files/gsapp/imceshared/Istanbul_Final_report_Spread.pdf.

Comiter, M. (2019). *Attacking artificial intelligence: AI's security vulnerability and what policy makers can do about IT*. Belfer Center for Science and International Affairs, Harvard Kennedy School. https://www.belfercenter.org/publication/AttackingAIcompetition-over-artificial-intelligence

Cook, R. (1999). *A brief look at the new look in error, safety, and failure of complex systems*. Ct Lab. https://www.adaptivecapacitylabs.com/BriefLookAtTheNewLook.pdf

Cook, R. (2000) How complex systems fail, https://how.complexsystems.fail.

Cooke, P. (2012). *Complex adaptive innovation systems*. Routledge.

Crawford, K. (2024). Generative AI's environmental costs are soaring — And mostly secret. *Nature, 626*(8000), 693–693. https://doi.org/10.1038/d41586-024-00478-x

Crot, L. (2013). Planning for sustainability in non-democratic polities: The case of Masdar City. *Urban Studies, 50*(13), 2809–2825. https://doi.org/10.1177/0042098012474697

Cummings, M. L. (2020) The surprising brittleness of AI, *Women Corporate Directors*. https://www.womencorporatedirectors.org/WCD/News/JAN-Feb2020/Reality%20Light.pdf.

Curran, W., & Hamilton, T. (2017). *Just green enough. Urban development and environmental Grentrification*. Routledge.

da Cuhna, A. (2013), Cited in Ecodistricts: A sustainable utopia? Paris innovation review, April, http://parisinnovationreview.com/articles-en/ecodistrict-a- sustainable-utopia.Edn. Accessed, March, 2020.

Dahlby, C. (2002) Tokyo Bay. *National Geographic*, October. Retrieved July 3, 2015, from http://ngm.nationalgeographic.com/print/features/world/asia/japan/tokyo-bay-text.

Dall'Orso, M. (2017) What characterises an ideal city, and how do we get there?,. *Urbanet*, June 22nd, https://www.urbanet.info/ideal-city/.

Dall'Orso, M. (2019) Promoting sustainable urban development through impact innovation, *Urbanet*, January 10th, https://www.urbanet.info/sustainable-urban-development-through-impact-innovation/.

Davidson, M. (2009). Social sustainability: A potential for politics? *Local Environment, 14*(7), 607–619. https://doi.org/10.1080/13549830903089291

Davis, D. (2012). *Urban resilience in situations of chronic violence*. Center for International Studies, Massachusetts Institute of Technology. http://web.mit.edu/cis/urban_resilience.html

Davis, D. (2015). *From risk to resilience and back. New design assemblages for confronting unknown futures* (p. 90). European Landscape Magazine.

Davis, D. E., & Dewey, O. F. (2013). How to defeat an urban megaproject. Lessons from the Mexico City Airport controversy. In G. Del Cerro Santamaría (Ed.), *Urban megaprojects. A worldwide view*. Emerald.

Davis, D., & Patvin, M. (2017). Resilience, security and spaces of migrant refuge. In P. Bourbeau (Ed.), *Chapter Ten in handbook on migration and security* (pp. 191–209). Edward Elgar Publishing.

Davis, D., & Altshuler, A. (2018). *Transforming urban transport*. Oxford University Press.

Davis, M. (2010). Who will build the ark? *New Left Review, 61*, 29–46.

de Boer, J., Muggah, R., & Patel, R. (2016). *Conceptualizing city fragility and resilience* (p. 2). United Nations University, Centre for Policy Research.

De Certeau, M. (2011). *The practice of everyday life*. University of California Press.

De Oliveira, S. P., & Carayannis, E. G. (2017). *The quadruple innovation helix nexus. A smart growth model, quantitative empirical validation and operationalization for OECD countries*. Palgrave Macmillan.

De Roo, G. Hillier, J. van Wezemael, J., Eds. (2012) Complexity and planning: Systems, assemblages and simulations. New directions in planning theory, : Ashgate.

Decaminada, T. (2022). Examining hyperloop hype on twitter. *Journal of MegaInfrastructure & Sustainable Development*. https://doi.org/10.1080/24724718.2022.2131095

Defacto. (2015). Guide to Architecture: Museum de la Défense. *Defacto*. Retrieved July 5, 2015, from http://www.ladefense.fr/sites/default/files/kiosque/pdf/guide_architecture_uk.pdf

del Cerro Santamaría, G. (2007). *Bilbao. Basque pathways to globalization*. Elsevier.

del Cerro Santamaría, G. (Ed.). (2013). *Urban megaprojects. A worldwide view*. Emerald Publishing.

del Cerro Santamaría, G. (2017). Iconic urban megaprojects in a global context. Revisiting Bilbao. In B. Flyvbjerg (Ed.), *The Oxford handbook of megaproject management*. Oxford University Press.

del Cerro Santamaría, G. (2018). Toward transdisciplinary urbanism: Megaprojects, power and the urban imagination. *Transdisciplinary Journal of Engineering & Science, 9*, 23–36.

del Cerro Santamaría, G. (2018a). Building transdisciplinary urban space: Theoretical transurbanism. *Transdisciplinary Journal of Engineering and Science, 9*, 57–74.

del Cerro Santamaría, G. (2018b). Building transdisciplinary urban space: Phenomenological Transurbanism. *Transdisciplinary Journal of Engineering and Science, 9*, 75–87.

del Cerro Santamaría, G. (2018c). Building transdisciplinary urban space: Experimental Transurbanism. *Transdisciplinary Journal of Engineering and Science, 9*, 88–106.

del Cerro Santamaria, G. (2019). London. *Wiley Blackwell Encyclopedia of Urban and Regional Studies*. https://doi.org/10.1002/9781118568446.eurs0188

del Cerro Santamaría, G. (2019a). Review of *Spaceship in the Desert. Energy, Climate Change and Urban Design in Abu Dhabi*, by Gökçe Günel. *International Journal of Urban and Regional Research, 43*(5), September.

del Cerro Santamaría, G. (2019b). Megaprojects, development and competitiveness. Building the infrastructure for globalization and neoliberalism. *Athens Journal of Social Science, 6*(4), 263–291. October.

del Cerro Santamaría, G. (2019c). Rapid urbanization, ecology and sustainability: The need for a broad strategy, holism and Transdisciplinarity. *Transdisciplinary Journal of Engineering and Science, 10*, 181–198. http://atlas-tjes.org/index.php/tjes/article/view/142/134

del Cerro Santamaría, G. (2019d). Disruptive and contentious enterprises: Megaprojects in Bilbao, Istanbul and Hong Kong. In *XIII CTV 2019 proceedings: XIII international conference on Virtual City and territory: Challenges and paradigms of the contemporary city: UPC*, Barcelona, October 2–4, 2019. p. 8503. E- ISSN 2604-6512. https://doi.org/10.5821/ctv.8503

del Cerro Santamaría, G. (2020a). The fading away of the Bilbao effect: Bilbao, Denver, Helsinki, Abu Dhabi, Athens. *Journal of Architecture, 6*(1), 25–52.

del Cerro Santamaría, G. (2020c). Building new knowledge about natures. *Ecologies and Sustainability, Urban Studies, August*. https://doi.org/10.1177/0042098020945954

del Cerro Santamaria, G. (2020). Innovation districts, complex sustainability and urban redevelopment. Evaluating multiple success factors. *IJEBMR, 4*(6), 279–290.

del Cerro Santamaría, G. (2020b). Complexity and Transdisciplinarity: The case of iconic urban megaprojects. *Transdisciplinary Journal of Engineering and Science, 11*(January), 17–31. http://atlas-tjes.org/index.php/tjes/article/view/139/136

del Cerro Santamaría, G. (2020d). Possibilities and challenges for intercultural research in global urbanism. In R. Brunet Thornton (Ed.), *Examining cultural perspectives in a globalized world*. IGI-Global.

del Cerro Santamaría, G. (2022). Urbanism, megaprojects and the world economy. In R. Baikady et al. (Eds.), *The Palgrave handbook of global social change*. Springer Nature.

Deleuze, G., & Guattari, F. (1987). *A thousand plateaus. Capitalism and schizophrenia*. University of Minnesota Press.

Deloitte. (2017). *Project Bonds. An alternative source of financing infrastructure projects*. https://www2.deloitte.com/za/en/pages/finance/articles/project-bonds-an-alternative-to- financing-infrastructure-projects.html

den Hartog, H. (2021). Shanghai's regenerated industrial waterfronts: Urban lab for sustainability transitions? *Urban Planning, 6*(3), 181–196. https://doi.org/10.17645/up.v6i3.4194

Dennett, D. (2014). *Darwin' s dangerous idea. Evolution and the meaning of life*. Simon and Schuster.

Devereux, G. (1969). *From anxiety to method in the behavioral sciences*. Mouton. 1968; A. Maslow, *The psychology of science: A reconnaissance*. Regnery.

Devinney, T. M., & Hartwell, C. A. (2020). Varieties of populism. *Global Strategy Journal, 10*(1), 32–66.

Dewolf, C. (2011). The future of Hong Kong's harbor front. *CNN*. October 24. Retrieved May 27, 2015, from http://travel.cnn.com/hong-kong/life/future-hong-kong-129384

Dhabi, A. (2011). Why the Arab spring never came to the U.A.E. Times, July 18. Retrieved May 29, 2015, from http://content.time.com/time/world/article/0,8599,2083768,00.html.

Diaz Orueta, F., & Fainstein, S. (2008). The new mega-projects: Genesis and impacts. *International Journal of Urban and Regional Research, 32*(4), 759–767.

Dimitriou, H., Ward, J., & Wright, P. (2013). Mega transport projects—Beyond the iron triangle: Findings from the OMEGA research programme. *Progress in Planning, 86*(November), 1–43.

Dodson, J. (2017). The global infrastructure turn and urban practice. *Urban Policy and Research, 35*(1), 87–92. https://doi.org/10.1080/08111146.2017.1284036

Dogan, E., & Stupar, A. (2017). The limits of growth: A case study of three mega- projects in Istanbul. *Cities, 60 Part A*, 281–288.

Douglass, M. (2010). Globalization, mega-projects and the environment: Urban form and water in Jakarta. *Environment and Urbanization ASIA, 1*(1). https://doi.org/10.1177/097542530900100105

Driver, S., & Martell, L. (1997). New labour' s communitarians. *Critical Social Policy, 17*(52), 27–46.

Duerksen, C., & Snyder, C. (2005). *Nature-friendly communities: Habitat protection and land use*. Island Press.

Dumortier, B. (2016). The cultural imperative: Saadiyat cultural district in Abu Dhabi between public policy and architectural gesture. In *Under construction: Logics of urbanism in the gulf region*. Routledge.

Dunn, P. (2022). Whose interest is AI serving? *Sustainability vs competitive advantage, Journal of Mega Infrastructure and Sustainable Development, Special Issue on Artificially Intelligent Megaprojects, Benefits, Risks and Sustainable Development, forthcoming in Journal of Mega Infrastructure and Sustainable Development.*

Edahiro, J. (2008). *Efforts in Japan to mitigate the urban heat Island effect*. Japan for Sustainability. September. Retrieved July 3, 2015, from http://www.japanfs.org/en/news/archives/news_id027856.html

Edelson, Z. (2015). Michael Sorkin on the Guggenheim, museum culture, and "the next Helsinki" competition. *Arch Daily*. Retrieved May 27, 2015, from http://www.archdaily.com/591932/michael-sorkin-on-the-guggenheim-museum-culture-and-the-next-helsinki-competition/

Edmonds, B. (1996). What is complexity? In F. Heylighen & D. Aerts (Eds.), *The evolution of complexity*. Kluwer.

Eldem, T. (2021). *Canal Istanbul: Turkey's controversial megaproject*. German Institute for International and security Affairs. https://www.swp-berlin.org/en/publication/canal-istanbul-turkeys-controversial-megaproject/

Elmqvist, T., et al. (Eds.). (2018). *Urban planet. Knowledge toward sustainable cities*. Cambridge University Press.

Elsheshtawy, Y. (2008). *The evolving Arab City. Tradition, modernity and urban development*. Routledge.

Elsheshtawy, Y. (2014). *Dubaization*, online blog, https://dubaization.com/dubaization.

Elsheshtawy, Y. (2019). Transforming Riyadh: A new urban paradigm? *The Arab Gulf States Institute in Washington*. https://agsiw.org/transforming-riyadh-a-new-urban-paradigm/. Accessed February 2022

Elsheshtawy, Y. (2021). *Riyadh: Transforming a Desert City*. Routledge. Emerald Publishing.

Escobar, A. (2015). Degrowth, post-development, and transitions: A preliminary conversation. *Sustainability Science, 10*(2015), 451–462.

Etzioni, A. (1998). Introduction. In A. Etzioni (Ed.), *The essential communitarian reader* (pp. ix–xxiv). Rowman & Littlefield.

Fainstein, S. (2001). *The City builders. Property development in New York and London, 1980–2000* (2nd ed.). University Press of Kansas.

Fainstein, S. (2008). Megaprojects in New York, London, and Amsterdam. *International Journal of Urban and Regional Research, 32*(4), 768–784.

Farías, I., & Bender, T. (Eds.). (2010). *Urban assemblages. How actor-network theory changes urban studies*. Routledge.

Farooq, U. (2018) Will Istanbul's massive New Canal be an environmental disaster? National Geographic, March 28th, https://www.nationalgeographic.com/science/article/istanbul-canal-project-Bosphorus-environmental-impacts.

Favari, E., & Cantoni, F. (Eds.). (2023). *Complexity and sustainability in megaprojects: MeRIT workshop 2022*. Springer Nature Switzerland.

Fay, M., Lee, H. I., Mastruzzi, M., Han, S., & Cho, M. (2019). *Hitting the trillion mark. A look at how much countries are spending on infrastructure* (Policy research working paper 8730). World Bank Group.

Ferguson, P. (2018). *Post-growth politics. A critical theoretical and policy framework for Decarbonization* (p. 53). Springer.

Ferro, S. (2013) Why do skyscrapers melt things? Popular Science, September 4. Retrieved July 6, 2015, from http://www.popsci.com/science/article/2013-09/why-do-skyscrapers-melt-things.

Fields, D. (2015). Contesting the Financialization of urban space: Community organizations and the struggle to preserve affordable rental housing in new York City. *Journal of Urban Affairs, 37*(2), 144–165.

Florida, R. (2000). *The rise of the creative class and how it's transforming work leisure, community and everyday life*. Basic Books.

Florida, R. (2007) Megaregions. The importance of place, Harvard business review, march 2008, citing research by Tim Goulden.

Flyvbjerg, B. (Ed.). (2017). *The Oxford handbook of megaproject management*. Oxford University Press.

Flyvbjerg, B., Bruzelius, N., & Rothengatter, W. (2003). *Megaprojects and risk: An anatomy of ambition*. Cambridge University Press.

Flyvbjerg, B. (2014). What you should know about megaprojects and why. An overview. *Project Management Journal, 45*(2), 6–19. https://doi.org/10.1002/pmj

Folke, C., Carpenter, S., Walker, B., Scheffer, M., Elmqvist, T., Gunderson, L., & Holling, C. S. (2004). Regime shifts, resilience, and biodiversity in ecosystem management. *Annual Review of Ecology, Evolution, and Systematics, 35*, 557–581. https://doi.org/10.1146/annurev.ecolsys.35.021103.105711

Forman, R. T. T. (2019). *Towns, ecology and the land*. Cambridge University Press.

Foucault, M. (1967). Of other spaces. Utopias and heterotopias, ("Des Espace Autres," March 1967, Translated from the French by Jay Miskowiec), Architecture/Mouvement/ Continuité, October, 1984. https://web.mit.edu/allanmc/www/foucault1.pdf

Frampton, K. (1995). *Studies in tectonic culture: The poetics of construction in nineteenth and twentieth century architecture*. The MIT Press.

Friedmann, J. (2005a) Globalization and the emerging culture of planning. Revised version of a paper originally prepared for UN Habitat (Nairobi), in March 2004, found at http://www.scarp.ubc.ca/faculty%20profiles/friedmann-paper2.htm

Friedmann, J. (2005b). Planning cultures in transition. In B. Sanyal (Ed.), *(2005) comparative planning cultures*. Routledge.

FSIN (Food Security Information Network). (2014). Resilience measurement principles. Toward an agenda for measurement design. Technical series no.1, Word Food Programme (WFP) and FAO. http://www.fsincop.net/fileadmin/user_upload/fsin/docs/resources/FSIN_29jan_WEB_medium%20res.pdf.

G20 Research Group. (2015). *Mobilizing climate finance*. University of Toronto.

Galiani, S., J. M. Paz y Miño and G. Torrens (2021) Geopolitics and international trade infrastructure. NBER Working Paper No. 29026. https://www.nber.org/papers/w29026.

Galvin, M. (2025) Interview with Dario Gil, Issues in Science and Technology, Winter. https://issues.org/wp-content/uploads/2025/01/35-39-Interview-with-Dario-Gil-Winter-2025.pdf.

Galloway, P. D., et al. (2013). *Managing Gigaprojects; advice from those who have been there, done that*. American Society of Civil Engineers.

Gao, J., Barzel, B., & Barabási, A. L. (2016). Universal resilience patterns in complex networks. *Nature, 530*, 307.

Gao, T., & Lim, S. (2023). Socio-spatial integration in innovation districts: Singapore's mixed-use experiment. *Cities, 140*(September). https://doi.org/10.1016/j.cities.2023.104405

Garza, A. (2018). Sequencing made simpler: Bechtel is turning to artificial intelligence for mega-projects. *PM Network, 32*(11), 14.

Gaulter, S. (2011). Why Japan is prone to earthquakes. *Al Jazeera*. March 11. Retrieved July 3, 2015, from http://www.aljazeera.com/news/asiapacific/2011/03/201131105046675615.html

GBD Innovation Club. (2021). *Paris La Défense: first post-carbon business district*. https://gbdinnovationclub.com/members-news/paris-la-defense-first-post-carbon-business-district/

Gelles, D. (2025) What will power the a.I. Revolution?. *The New York Times*, January 5, https://www.nytimes.com/2025/01/07/climate/artificial-intelligence-power-emissions.html.

Gertz, G. and H. Kharas (2019) *Beyond Neoliberalism. Insights from emerging markets*. Brookings Institution Report. https://www.brookings.edu/wp-content/uploads/2019/05/beyond-neoliberalism-final-05.01.pdf.

Gianoli, A., & Henkes, R. P. (2020). The evolution and adaptive governance of the @22 Innovation District in Barcelona. *Urban Science, 4*(16). https://doi.org/10.3390/urbansci4020016

Gibbons, M., et al. (1994). *The new production of knowledge: The dynamics of science and research in contemporary societies*. Sage.

Gill, P. (2021). Protecting the supply chain in mega infrastructure projects. https://roadsonline.com.au/protecting-the-supply-chain-in-mega-infrastructure-projects/. Accessed November 2021.

Glaeser, E. (2021). *Urban resilience*. NBER working paper no. 29261 September.

Godehardt, N. and K. Postel-Vinay (2020) Connectivity and geopolitics: Beware the "new wine in old bottles" approach, Stiftung Wissenschaft und Politik Publications, Comment 2020/C 35, https://www.swp-berlin.org/en/publication/connectivity-and-geopolitics-beware-the-new-wine-in-old-bottles-approach.

Gödel, K. (1951) Some basic theorems on the foundations of mathematics and their implications, in Solomon Feferman, ed., 1995. *Kurt Gödel Collected works, Vol. III*, Oxford University Press, pp. 304–323.

Gold, J. (2007). *The practice of modernism: Modern architects and urban transformation, 1954–1972*. Taylor & Francis.

Gould, K., & Lewis, T. (2016). *Green gentrification. Urban sustainability and the struggle for environmental justice*. Routledge.

Government of Hong Kong. (2018). *Constitutional and Mainland Affairs Bureau*. https://www.cmab.gov.hk/en/home/index.htm.

Granovetter, M. (1985). Economic action and social structure: The problem of embeddedness. *The American Journal of Sociology., 91*(3), 487.

Greenfield, A. (2013). *Against the Smart City*. Verso.

Greiman, V. (2020). Artificial intelligence in megaprojects. *The Next Frontier*, European Conference on Cyber Warfare and Security. https://doi.org/10.34190/EWS.20.123

Griffiths, I. (2010). Tons of radioactive waste casts doubt over London's Olympic stadium legacy. *The Guardian*. June 20. Retrieved June 16, 2015, from http://www.theguardian.com/business/2010/jun/20/radioactive-waste-olympic-site

Grosse, R., Gamso, J., & Nelson, R. C. (2021). China's rise, world order, and the implications for international business. *Management International Review, 61*(1), 1–26.

Grubbauer, M. (2013). "Global" architecture as a contradictory signifier: Lessons from Hamburg's and Vienna's urban megaprojects. In G. Del Cerro (Ed.), *Urban megaprojects: A worldwide view* (pp. 185–210). Emerald Group Publishing Limited.

Grubbauer, M., & Čamprag, N. (2018). Urban megaprojects, nation-state politics and regulatory capitalism in central and Eastern Europe: The Belgrade waterfront project. *Urban Studies, 56*(4), 649.

Gruber, T. (2008). Collective knowledge systems: Where the social web meets the semantic web. *Web Semantics: Science, Services and Agents on the World Wide Web, 6*(1) February, 4.

Grunert, K. G. and C. Ellegaard (1992) The concept of key success factors. Theory and method. MAPP working paper no. 4, October, ISSN 09072101, https://pure.au.dk/portal/files/32299581/wp04.pdf.

Gu, Y., Hu, L., Zhang, H., & Hou, C. (2021). Innovation ecosystem research: Emerging trends and future research. *Sustainability, 13*(20), 11458. https://doi.org/10.3390/su132011458

Guash, A. M., & Zulaika, J. (2005). *Learning from the Guggenheim Bilbao*. University of Nevada Press.

Guerin, C. (2013). Rhizomatic research cultures, writing groups and academic researcher identities. *International Journal of Doctoral Studies, 8*, 137–150. https://doi.org/10.28945/1897

Guggenheim Helsinki. (2014). About. *Guggenheim Helsinki Design Competition*. Retrieved May 28, 2015, from http://designguggenheimhelsinki.org/about/. http://yle.fi/uutiset/minister_says_no_culture_budget_for_guggenheim/5294388.

Guillén, M. (2006). *The Taylorized beauty of the mechanical. Scientific management and the rise of modernist architecture*. Princeton University Press.

Gunderson, L., & Holling, C. S. (Eds.). (2002). *Panarchy: Understanding transformations in human and natural systems*. Island Press.

Günel, G. (2019). *Spaceship in the desert. Energy, climate change and Urban Design in Abu Dhabi*. Duke University Press.

Güneralp, B., & Seto, K. C. (2008). Environmental impacts of urban growth from an integrated dynamic perspective: A case study of Shenzhen, South China. *Global Environmental Change, 18*(4), 720–735.

Guo, Z., & Curugullo, F. (2022). Artificial intelligence, healthcare and megaprojects in the Chinese Smart City: The case of Guangzhou. *Journal of Mega Infrastructure and Sustainable Development*. Special Issue on *Artificially Intelligent Megaprojects, Benefits, Risks and Sustainable Development*, in Journal of Mega Infrastructure and Sustainable Development, forthcoming.

Haeckel, E. (1866). *General morphology of organisms* (original in German, *Generelle Morphologie des Organismen*). G. Reimer.

Hallock, B. E. and J. G. Zack Jr. (2018). What have we learned from megaprojects? 8th International Society of Construction Law, Chicago.

Hamilton, J. (2011) In Japan, shaken soil turned soft after quake. *National Public Radio*, April 7. Retrieved July 3, 2015, from http://www.npr.org/2011/04/07/135181474/in-japan-shaken-soil-turned-soft-after-quake.

Hammoura, J. (2021) Istanbul canal: A megaproject with strategic implications. *Armenian weekly*, June 16, https://armenianweekly.com/2021/06/16/istanbul-canal-a-megaproject-with-strategic-implications/.

Hanna, P., et al. (2016). Conceptualizing social protest and the significance of protest action to large projects. *Extractive Industries & Society, 3*(1), 217–239. https://doi.org/10.1016/j.exis.2015.10.006

Hannan, S., & Sutherland, C. (2015). Mega-projects and sustainability in Durban, South Africa: Convergent or divergent agendas? *Habitat International, 45*(P3), 205–212.

Harman, G. (2002). *Tool-being: Heidegger and the metaphysics of objects*. Open Court.

Harris, M. (2014). *Megaprojects: A global review and the Australian context*. Festival of Urbanism, University of Sydney. http://static.squarespace.com/static/540680a6e4b0a7034afffb6f/t/543f572de4b0a72b21db9adc/1413437229478/Megaprojects+Review_Harris.pdf

Harris, M. (2017). Competitive precinct projects. The five consistent criticisms of "global" mixed-use megaprojects. *Project Management Journal, 48*(6), 76–92.

Harris, M. (2018) Ch 6 Barangaroo: Machiavellian megaproject or erosion of intent? , pp. 111–134, in Urban Regeneration in Australia. Policies, Processes and Projects for Contemporary Urban Change, edited by Kristian Ruming, New York: Routledge.

Harris, M. (2019). Australian cities pay the price for blocking council input to projects that shape them. *The Conversation*, December 16th. https://theconversation.com/australian-cities-pay-the-price-for-blocking-council-input-to-projects-that-shape-them-127017

Harrison, J., & Hoyler, M. (Eds.). (2015). *Megaregions: Globalization's new urban form?* Edward Elgar.

Hartwell, C. A., & Devinney, T. (2021). Populism, political risk, and pandemics: The challenges of political leadership for business in a post-COVID world. *Journal of World Business, 56*(4), 101225.

Harvey, D. (1989a). From managerialism to entrepreneurialism: The transformation in urban governance in late capitalism. *Geografiska Annaler: Series B, Human Geography, 71*(1), 3–17.

Harvey, D. (1989b). From managerialism to entrepreneurialism: The transformation in urban governance in late capitalism. *Geografiska Annaler, 71*(1), 3.

Haskins, C. (1989). Kant and the autonomy of art. *The Journal of Aesthetics and Art Criticism, 47*(1 (Winter), 43–54), 43.

Hawken, S. & Mackenzie, B. (2016). Barangaroo reserve, landscape architecture Australia [1833-4814], issue:149 (pp. 40–50).

Hays, K. M. «Architecture's appearance and the practices of imagination.» The Log, vol. 37, 2016, 205–213, https://www.anycorp.com/store?category=Log.

He, S., & Wu, F. (2009). China's emerging neoliberal urbanism: Perspectives from urban redevelopment. *Antipode, 41*(2), 282.

Healey, P. (2007). *Urban complexity and spatial strategies. Towards a relational planning for our times*. Routledge.

Healey, P. (2013). Circuits of knowledge and techniques: The transnational flow of planning ideas and practices. *International Journal of Urban and Regional Research, 37*(5), 1510–1526. https://doi.org/10.1111/1468-2427.12044

Healey, P., Davoudi, S., & Tavsanoglu, S. (1992). *Rebuilding the City: Property-led urban regeneration*. Chapman & Hall.

Heisenberg, W. (2007). *Physics and philosophy. The revolution in modern science*. Harper Perennial Modern Classics.

Hemment, D. & Townsend, A. (2013). *Smart citizens*. Available at: https://discovery.dundee.ac.uk/ws/portalfiles/portal/20381872/smartcitizens1.pdf. Accessed 2 August 2022.

Heydarian, R. (2021). Global Infrastructure Development: China, Biden and the New Pivot of 21st Century Geopolitics. *Aspenia Online*, October 19th, https://aspeniaonline.it/global-infrastructure-development-china-biden-and-the-new- pivot-of-21st-century-geopolitics/.

Heynen, N., et al. (2006). *In the nature of cities. Urban political ecology and the politics of urban metabolism*. Routledge.

Hiller, H. H. (2000). Mega-events, urban Boosterism and growth strategies: An analysis of the objectives and legitimations of the Cape Town 2004 Olympic bid. *International Journal of Urban and Regional Research, 24*(2), 449–458.

Hirschman, A. (1967). *Development projects observed.* Brookings Institution.

Hirschman, A. O (1967b). The principle of the hiding hand. *The public interest,* Winter (pp. 10–23).

Hodge, G & C. Greve (2017), Private finance: What problems does it solve, and how well?, in Flyvbjerg, B. (ed.) The Oxford handbook of megaproject management, London: Oxford University Press,. Chapter 16.

Hodson, M., & Marvin, S. (2017). Intensifying or transforming sustainable cities? Fragmented logics of urban environmentalism. *Local Environment, 22*(sup1), 8–22. https://doi.org/10.1080/13549839.2017.1306498

Hollands, R. G. (2008). Will the real smart city please stand up? *City, 12*(3), 303–320. https://doi.org/10.1080/13604810802479126. S2CID 143073956.

Holling, C. S. (1973). Resilience and stability of ecological systems. *Annual Review of Ecology and Systematics, 4*, 1–23. https://doi.org/10.1146/annurev.es.04.110173.000245

Homer-Dixon, T. (2006). *The upside of down: Catastrophe, creativity, and the renewal of civilization.* Island Press.

Hooper, R. (2013) Effects will become more obvious as Japan's climate changes. The Japan Times, July 13. Retrieved July 3, 2015 from http://www.japantimes.co.jp/news/2013/07/13/national/science-health/effects-will-become-more-obvious-as-japans-climate-changes/#.VaFvQflVhHw.

Hopkins, R. (2014). *The transition handbook. From oil dependency to local resilience.* Green Books.

Hou, J., & Rios, M. (2003). Community-driven place making: The social practice of participatory Design in the Making of union Point Park. *Journal of Architectural Education, September*, 19–27. https://doi.org/10.1162/104648803322336557

Hu, Y., Le, Y., Gao, X., Li, Y., & Liu, M. (2018). Grasping institutional complexity in infrastructure mega-projects through the multi-level governance system: A case study of the Hong Kong-Zhuhai-Macao Bridge construction. *Frontiers in Engineering Management, 5*(1), 52–63. https://doi.org/10.15302/J-FEM-2018074

Hughes, T. P. (1998). *Rescuing Prometheus: Four monumental projects that changed the modern world.* Vintage.

Hune-Brown (2017) Is this the end of Starchitecture? Azure,. August 18.

ICEX. (2021). Recopilación de los megaproyectos de la 'Visión 2030' de Arabia Saudí, https://www.icex.es/icex/es/navegacion-principal/todos-nuestros-servicios/informacion-de-mercados/paises/navegacion-principal/noticias/gigaproyectos-arabia-saudi-new2021890486.html?idPais=SA

Icon Magazine. (2020). One World Trade Center: A fortress enveloped in glass, https://www.icon-eye.com/architecture/one-world-trade-center-fortress-glass-concrete-skidmore-owings-merrill

IISS. (2022). International competition over artificial intelligence, Strategic Comments 28 (11), https://www.iiss.org/publications/strategic-comments/2022/international-.

IMF World Economic Outlook. (2018) Cyclical upswing, structural change, the IMF, , www.imf.org/en/Publications/WEO/Issues/2018/03/20/ world-economic-outlook-april-2018.

Innes, J. E., Gruber, J., Neuman, M., & Thompson, R. (1994). CPS report: A policy research program report. In *Coordinating growth and environmental management through consensus building*. University of California at Berkeley, California Policy Seminar.

IPCC (2022) Synthesis report of the sixth assessment report, https://www.ipcc.ch/ar6- syr/.

Irimia-Diéguez, A. I. A., & Sanchez-Cazorla, R. A.-L. (2014). Risk Management in Megaprojects. *Social and Behavioral Sciences, 119*, 407–416. https://doi.org/10.1016/j.sbspro.2014.03.046

Ivy, R. (2001). Quoted in Wilson Lloyd, A. (2001) architecture for art's sake. *Atlantic Monthly, 287*, 85–88. https://www.azuremagazine.com/article/is-this-the-end-of-starchitecture/

Jacobs, J. (1961). *The death and life of great American cities.* Random House.

Jacobs, J. (2000). *The nature of economies.* Modern Library.

Jacobs, K. (2016) Calatrava transportation hub finally takes flight, architect magazine, https://www.architectmagazine.com/design/calatravas-transportation-hub-finally-takes-flight_o.

Jacobs, K. (2022) Toronto wants to kill the Smart City forever, MIT technology review, June 29, https://www.technologyreview.com/2022/06/29/1054005/toronto- kill-the-smart-city/.

Jacobson, M., et al. (2000). *Earth system science. From biogeochemical cycles to global changes* (2nd ed.). Elsevier Academic Press.

Janssen, C., et al. (2023). Governing capabilities, not places—How to understand social sustainability implementation in urban development. *Urban Studies, 61*(2), 331–349. https://doi.org/10.1177/00420980231 1795

Japan Property Central (2015) Reclaimed Land in Japan. *Japan Property Central.* Retrieved July 3, 2015, from http://japanpropertycentral.com/real-estate-faq/reclaimed-land-in- japan/.

Jarvensivu, P. et al, (2018) Global sustainable development report 2019 drafted by the group of independent scientists, BIOS Research Unit, , https://bios.fi/bios- governance_of_economic_transition.pdf.

Jen, E. (2005). *Robust design*. Oxford University Press.

Jencks, C. (1997). *The architecture of the jumping universe: A polemic: How complexity science is changing architecture and culture*. Academy Press.

Jencks, C. (2005). *The iconic building*. Rizzoli.

Jessop, B. (2002). Liberalism, neoliberalism, and urban governance: A state–theoretical perspective. *Antipode, 34*(3), 452.

Jian, M. (2018) Social exclusion: WKCD as a study case, City University of Hong Kong, http://dspace.cityu.edu.hk/handle/2031/9105.

Johnston, J., & Clegg, S. (2012). Legitimate sovereignty and contested Authority in Public Management Organization and Disorganization: Barangaroo and the grand strategic vision for Sydney as a Globalizing City. *Journal of Change Management, 12*(3), 279–299.

Jonas, A. E. G., & Wilson, D. (1999). *The urban growth machine. Critical perspectives, two decades later*. SUNY Press.

Jones, L., & Hameiri, S. (2021). *Fractured China: How state transformation is shaping China's rise*. Cambridge University Press.

Jullien, F. (2001). *Un sabio no tiene ideas*: O el otro de la filosofía. Siruela.

Kahneman, D., & Tversky, A. (1979). Prospect theory: An analysis of decision under risk. *Econometrics, 47*, 263–291. https://doi.org/10.2307/1914185

Kahneman, D., & Tversky, A. (Eds.). (2000). *Choices, values, and frames*. Cambridge University Press.

Kahneman, D. (2013). *Thinking fast and slow*. Farrar, Straus and Giroux.

Kakuchi, S. (2006) Environment: Japan swelters under urban Heat Island effect. *Inter press service news agency*, august 31. Retrieved July 3, 2015, from http://www.ipsnews.net/2006/08/environment-Japan-swelters-under-urban-heat-Island- effect/.

Kalla, Z. (2014). To Guggenheim or not to Guggenheim Helsinki ?… That is the question. Arch2o. Retrieved May 29, 2018, from http://www.arch2o.com/guggenheim-helsinki-3/

Kalliomäki, H., et al. (2023). Innovation districts as strategic urban projects: The emergence of strategic spatial planning for urban innovation. *European Planning Studies*. https://doi.org/10.1080/09654313.2023.2216727

Kamin, B. (2002) How stellar are 'starchitects'? *Chicago tribune*, January 27, http://articles.chicagotribune.com/2002-01-27/news/0201270415_1_new-dorm- starchitects-new-york- city-designer.

Kang, W. (2019). Perceived barriers to implementing education for sustainable development among Korean teachers. *Sustainability, 11*(9), 26. https://doi.org/10.3390/su11092532

Kastenhofer, K. (2007). Converging epistemic cultures? *Innovation, 20*(4), 359–373. https://doi.org/10.1080/13511610701767908

Katz, B. (2018). *The new localism. How cities can thrive in an age of populism*. Brookings Press.

Katz, B. and J. Wagner (2014) The rise of innovation districts, Brookings institution, metropolitan policy program, May, https://c24215cec6c97b637db6-9c0895f07c3474f6636f95b6bf3db172.

ssl.cf1.rackcdn.com/content/metro-innovation-districts/~/media/programs/metro/images/innovation/innovationdistricts1.pdf.

Katz, B. and J. Wagner (2019) The evolution of innovation districts. *The New Geography of Global Innovation, Brookings Institution.* https://www.giid.org/the-evolution-of-innovation-districts-download/. Accessed March, 2020.

Kauffman, S. (1996). *At home in the universe. The search for the laws of self-organization and complexity.* Oxford University Press.

Keane, J. (2018). *Power and humility. The future of monitory democracy.* Cambridge University Press.

Keeton, R. (2021) When Smart Cities Are Stupid, International New Town Institute, http://www.newtowninstitute.org/spip.php?article1078, Accessed November, 2021.

Kelly, K. (1995). *Out of control: The new biology of machines, social systems and the economic world.* Basic Books.

Keskin, M., & Kurdak, M. (2021). Finally enacted: Project bonds, an alternative financing method for mega project. *Essin Attorney Partnership.* https://www.esin.av.tr/2021/07/26/finally-enacted-project-bonds-an-alternative-financing-method-for-mega-project/

Khanna, P. (2016) A new map for America, *New York times*, April 15. https://www.nytimes.com/2016/04/17/opinion/sunday/a-new-map-for-america.html.

Khanna, P. (2021) Asia's megacities and the future of geopolitics, the Chicago council on global affairs, December 2nd, https://globalaffairs.org/commentary-and- analysis/blogs/asias-megacities-and-future-geopolitics.

Kharrazi, A., Rovenskaya, E., & Fath, B. D. (2017). Network structure impacts global commodity trade growth and resilience. *PLoS One, 12*(2), e0171184.

Kimmelman, M. (2012). Why is this museum shaped like a tub? *The New York Times*, Art and Design Section, December 23, http://www.nytimes.com/2012/12/24/arts/design/amsterdams-new-stedelijk-museum.html?_r=0.

Kinver, M. (2005) London's "one planet Olympics" vision. BBC News, October 24. Retrieved June 15, 2015, from http://news.bbc.co.uk/2/hi/science/nature/4299732.stm.

Kitchin, R., & Dodge, M. (2014). *Code/space: Software and everyday life.* MIT Press.

Klebnikov, A. (2001). Museums Inc. *Forbes*, (January 8).

Klein, N. (2015). *This changes everything. Capitalism vs. the climate.* Simon and Schuster.

Klein, N. (2020). Screen new deal. *The Intercept.* https://theintercept.com/2020/05/08/andrew-cuomo-eric-schmidt-coronavirus-tech-shock-doctrine/

Klingmann, A. (2021). Rescripting Riyadh: How the capital of Saudi Arabia employs urban mega-projects as catalysts to enhance the quality of life within the city's neighborhoods. *Journal of Place Management and Development.* https://doi.org/10.1108/JPMD-06-2021-0062

Klopp, J. M., & Petretta, D. L. (2017). The urban sustainable development goal: Indicators, complexity and the politics of measuring cities. *Cities, 63*, 92–97. https://doi.org/10.1016/j.cities.2016.12.019

Knorr Cetina, K. (1999). *Epistemic cultures. How the sciences make knowledge.* Harvard University Press.

Knorr Cetina, K. (2007). Culture in global knowledge societies: Knowledge cultures and epistemic cultures. *Interdisciplinary Science Reviews, 32*(4), 361–375. https://doi.org/10.1179/030801807X163571

Knudsen, B., Florida, R., Stolarick, K., & Gates, G. (2008). Density and creativity in U.S. regions. *Annals of the Association of American Geographers, 98*(2), 461–478.

Kobrin, S. J. (2017). Bricks and mortar in a borderless world: Globalization, the backlash, and the multinational Enterprise. *Global Strategy Journal, 7*(2), 159–171.

Köhler, J., Geels, F., Kern, F., & Markard, J. (2019). An agenda for sustainability transitions research: State of the art and future directions. *Environmental Innovation and Societal Transitions*, https://www.researchgate.net/publication/330785717_An_agenda_for_sustainability_transitions_research_State_of_the_art_and_future_directions. Accessed October, 2021, *31*, 1.

Kolisnichenko, V. (2023) Türkiye has increased the fee for passage through the Bosphorus and Dardanelles by 8.3% since July, GMK Center, June 6[th]https://gmk.center/en/news/turkiye-has-increased-the-fee-for-passage-through-the- bosphorus-and-dardanelles-by-8-3-since-july/.

Komninos, N. (2008). *Intelligent cities and globalisation of innovation networks*. Routledge.

Komninos, N. (2009). Intelligent cities: Toward interactive and global innovation environments. *International Journal of Innovation and Regional Development., 1*(4), 337. https://doi.org/10.1504/ijird.2009.022726

Komninos, N. (2013). What makes cities intelligent? In M. Deakin (Ed.), *Smart cities: Governing, modelling and analysing the transition* (pp. 77–91). Taylor and Francis.

Koolhaas, R. (2013). Quoted in Dunham-Jones, E. The irrational exuberance of Rem Koolhaas. *Places Journal*. April https://placesjournal.org/article/the-irrational- exuberance-of-rem-koolhaas/

Korinek, A., Schindler, M., & Stiglitz, J. (2021). Technological progress, artificial intelligence and inclusive growth. *IMF eLibrary, Issue 166*. https://www.elibrary.imf.org/view/journals/001/2021/166/article-A001-en.xml. Accessed November, 2021.

Krueger, R., & Savage, L. (2007). City-regions and social reproduction: A ' place' for sustainable development? *International Journal of Urban and Regional Research, 31*(1), 215.

Kumaraswamy, M., J. Chung, W. Zou and K. Wong (2015) , Online paper, https://www.researchgate.net/publication/285512147.

Lam, Y. M. O., et al. (2021). *Strategic solutions for tourism development planning in Hong Kong—A smart plan for resilient tourism development*. The University of Hong Kong, Faculty of Architecture. https://foa-media.arch.hku.hk/media/upload/2021/09/Final_Report_Planda_Tourism.pdf

Lamb, W. F., Creutzig, F., Callaghan, M. W., & Minx, J. C. (2019). Learning about urban climate solutions from case studies. *Nature Climate Change, 9*(4), 279–287. https://doi.org/10.1038/s41558-019-0440-x

Langmead, D. and Christine G. (2001) Encyclopedia of architectural and engineering feats. Retrieved July 5, 2015, from https://books.google.com/books?id=T5J6GKvGbmMC&dq=la+defense+forest+of+towers &source=gbs_navlinks_s.

Larson, S. (2013). *Building like Moses with Jacobs in mind: Contemporary planning in new York City*. Temple University Press.

Latour, B. (1987). *Science in action. How to follow scientists and engineers through society*. Harvard University Press.

Latour, B. (1992). Where are the missing masses? The sociology of a few mundane artifacts in Wiebe Bijker and John law, eds. In M. A. Cambridge (Ed.), *Shaping technology / building society: Studies in sociotechnical change* (pp. 225–258). MIT Press.

Latour, B. (2016). *Facing Gaia. Eight lectures on the new climatic regime*. Polity.

Lau, J. (2011). A bid for culture in a City of Commerce. *The New York Times*. June 6. Retrieved May 23, 2015, from http://www.nytimes.com/2011/06/07/world/asia/07iht- hong07.html

Lau, J. (2014). Bringing flagship of contemporary art to Hong Kong. *The New York Times*. March 2. Retrieved May 23, 2015, from http://www.nytimes.com/2014/03/21/arts/international/building-a-museum-in- Hong-Kong-from-scratch.html, *03*, 23.

Lauerman, J. (2022). Dignity, mega-projects, and the problem of scale. *Dialogues in Human Geography*, 1–5. https://doi.org/10.1177/20438206221121650

Le Galès, P. (2022). Newyorkology of tech capitalism. *Sociologica, 16*(22). https://doi.org/10.6092/issn.1971-8853/15458

Lee, K.-F. (2018). Silicon Valley and the New World order. In *AI superpowers: China*. Harper Business.

Lee, Y. (2014). Social vulnerability indicators as a sustainable planning tool. *Environmental Impact Assessment Review, 44*, 31–42. https://doi.org/10.1016/j.eiar.2013.08.002

Lefebvre, H. (1991). *The production of space. Translated by Donald Nicholson- Smith*. Blackwell.

Legg, R. (2018) Making developments green does not help with inequality, the conversation, November 1st, https://theconversation.com/making-developments-green-doesnt-help-with-inequality-104941.

Lehrmacher, W., & Lind, M. (2022). Port cities' efforts in contributing to a sustainable port environment. *Journal of Mega Infrastructure and Sustainable Development.* Special Issue on *Artificially Intelligent Megaprojects, Benefits, Risks and Sustainable Development*, forthcoming in Journal of Mega Infrastructure and Sustainable Development.

Leigh, N., Hoelzel, N., Kraft, B., & Dempwolf, C. (2014). *Sustainable urban industrial development, PAS Report 577.* American Planning Association.

Lepawsky, J. (2024). Climate change induced water stress and future semiconductor supply chain risk. *iScience, 27*(2), 108791. https://doi.org/10.1016/j.isci.2024.108791

Levin, S., Xepapadeas, T., Crépin, A.-S., Norberg, J., de Zeeuw, A., Folke, C., hughes, T., Arrow, K., Barrett, S., Daily, G., Ehrlich, P., Kautsky, N., Mäler, K.-G., Polasky, S., Troell, M., Li, Y., Xiang, P., You, K., Guo, J., Lui, Z., & Ren, H. (2021). Identifying the key risk factors of mega infrastructure projects from an extended sustainable development perspective. *International Journal of Environmental Research and Public Policy, 18*, 7515. https://doi.org/10.3390/ijerph18147515

Li, Y., & Wu, F. (2020). Territory, the state and geopolitics of mega city-region development in China. In S. Moisio, N. Koch, A. E. G. Jonas, C. Lizotte, & J. Luukkonen (Eds.), *Handbook on the changing geographies of the state* (pp. 343–354). Edward Elgar.

Li, Y., Lu, Y., Taylor, J. E., & Han, Y. (2018). Bibliographic and comparative analyses to explore emerging classic texts in megaproject management. *International Journal of Project Management, 36*(2), 241–396.

Lieto, L. (2017). How material objects become urban things? *City, 21*(5), 568–579. https://doi.org/10.1080/13604813.2017.1374782

Lieto, L., & Beauregard, R. A. (Eds.). (2016). *Planning for a material world.* Routledge.

Lin, Z. (2007). From megastructure to megalopolis: Formation and transformation of megaprojects in Tokyo Bay. *Journal of Urban Design., 12*(1), 73–92.

Lindsay, G. (2013). *The Denver art Museum and the Bilbao effect.* UC Berkeley.

Linke, R. J., Wilke, K., Öztürk, Ö., Schöpp, F., & Noor, E. K. (2022). The future of the eHighway system: A vision of a sustainable, climate- resilient, and artificially intelligent mega-project. *Journal of Mega Infrastructure & Sustainable Development.* https://doi.org/10.1080/24724718.2022.2131087

Lochhead, H. (2016). Utzon lecture: Re-imagining the Harbor City. *The Conversation*, March 11th, https://theconversation.com/utzon-lecture-re- imagining-the-harbour-city-56127.

Logan, J., & Molotch, H. (1992). *Urban fortunes. The political economy of place.* University of California Press.

Lorenz, E., & Shiqiao, L. (2015) Kowloon Cultural District - an investigation into spatial capabilities in Hong Kong in E. Lorenz, & L. Shiqiao, Eds. *Hong Kong: MCCM Creations*, Renfro, D. S.

Lovelock, J. (2020). *Novacene. The coming age of hyperintelligence.* The MIT Press.

Lovelock, J. E., & Margulis, L. (1974). Atmospheric homeostasis by and for the biosphere: The Gaia hypothesis. *Tellus. Series A. Stockholm: International Meteorological Institute., 26*(1–2), 2–10.

Lui, P. (2008) "Privatizing the Hong Kong Special Administrative Region Government: Revisiting the Case of The Link REIT from a Policy Network Perspective," presented at The International Conference on Promoting City Competitiveness, October 24, 2008 at Chung Hua University, Hsin Chu City, .

Lundin, R. A., & Söderholm, A. (1995). A theory of the temporary organization. *Scandinavian Journal of Management, 11*(4), 437–455.

Luo, Z., et al. (2017). Pollution and road infrastructure in cities of the People's Republic of China. *Asian Development Bank Institute Working Paper no. 717*, 41. https://www.adb.org/sites/default/files/publication/239581/adbi-wp717.pdf

MacDonald, H. (2019). Planning for the public benefit in the Entrepreneurial City: Public land speculation and financialized regulation. *Journal of Planning Education and Research, 43*, 1–13.

Marcuse, P. (1989). Dual city: A muddy metaphor for a quartered city. *International Journal of Urban and Regional Research, 13*(4), 697.

Marcuse, P. (1997). Glossy globalization. In P. Droege (Ed.), *Intelligent environments: Spatial aspects of the information revolution* (pp. 29–47).

Marginson, S. (2016) cited in Cameron, H. (2016) Innovation Districts—the way forward for sustainable growth? The Knowledge Exchange Blog,. January 25th, https://theknowledgeexchangeblog.com/2016/01/25/innovation-districts-the-way-forward-for-sustainable-growth/.

Margulis, L. (1999). *Symbiotic planet. A new look at evolution*. Basic Books.

Marrone, M., & Hammerle, M. (2018). Smart cities: A review and analysis of stakeholders' literature. *Business & Information Systems Engineering, 60*(3), 197–213. https://doi.org/10.1007/s12599-018-0535-3. S2CID 207433624.

Marschall, M & Aydogan, A. (2015) Mega projects and the limits of urban planning in megacities: The case of Istanbul in the AKP era, Middle East institute, December 1st, https://www.mei.edu/publications/mega-projects-and-limits-urban-planning-megacities-case-istanbul-akp-era.

Marshall. (2002). *Emerging urbanity. Global urban projects in the Asia Pacific rim*. Routledge.

Martin, C. J., Evans, J., & Karvonen, A. (2018). Smart and sustainable? Five tensions in the visions and practices of the smart-sustainable city in Europe and North America. *Technological Forecasting and Social Change, 133*, 269–278. https://doi.org/10.1016/j.techfore.2018.01.005

Martin, R. (2020). Abolish oil: From green new Deal to green reconstruction. *Places*, 1–16.

Martínez, M. A., & Gil, J. (2022). Grassroots struggles challenging housing financialization in Spain. *Housing Studies*. https://doi.org/10.1080/02673037.2022.2036328

Masami, T. (2003). Realizing unattended hours of continuous operation of machining center with addition of intelligent function. *Subaru Technical Review, 30*, 251–256. (in Japanese).

Mateescu, D. (2021) Turkey's Megaprojects and China's Growing Influence in the Bosphorus Area, Veridica, May 31st, https://www.veridica.ro/en/analyses/turkeys-megaprojects-and-chinas-growing-influence-in-the-Bosphorus-area.

Mauk, B. (2014) Abu Dhabi's high cost of culture. The New Yorker, January 28. Retrieved May 27, 2015, from http://www.newyorker.com/business/currency/abu- dhabis-high-cost-of-culture.

Mayer, A. L., Donovan, R. P., & Pawlowski, C. W. (2014). Information and entropy theory for the sustainability of coupled human and natural systems. *Ecology and Society, 19*(3), 11. https://doi.org/10.5751/ES-06626-190311

Mazzucatto, M. (2015). *The entrepreneurial state. Debunking public vs private sector myths*. Public Affairs.

Mazzucato, M., & Ryan-Collins, J. (2019). Putting value creation back into public value: From market fixing to market shaping, UCL Institute for Innovation and Public Purpose. *Working Paper no. 5*.

Mazzucato, M., & Kattel, R. (2019). Getting serious about value, UCL Institute for Innovation and Public Purpose. *Working Paper no. 7*.

Mazzucato, M., Entsminger, J., & Kattel, R. (2020). Public value and platform governance. *Mapping Value Creation and Extraction in the Platform Economy, UCL Institute for Innovation and Public Purpose, WP9*.

Mazzucato, M., R. Kattel and J. Entsminger (2021) Let's imagine a different kind of platform economy, https://sifted.eu/articles/platform-economy-mariana-mazzucato/.

Mazzucato, M. (2021). *Mission economy. A moonshot guide to changing capitalism*. Penguin.

McCarthy, D. E. (1996). *Knowledge as culture. The new sociology of knowledge*. Routledge.

McClellan, A. (2008). *The museum from Boullée to Bilbao*. The University of California Press.

Mcclellan, A. (2012). Museum expansion in the twenty-first century: Abu Dhabi. *Journal of Curatorial Studies Journal of Curatorial Studies, 1*(3), 271–293.

McGinty, S. (2016). Top of the world: Why is Melbourne "the most liveable city"? *The Knowledge Exchange*. November, https://theknowledgeexchangeblog.com/2016/11/21/top-of-the-world-why-is-melbourne-the-most-liveable-city/. Accessed November, 2021

McGovern, G., & Branford, S. (2024). Critics fear catastrophic energy crisis as AI is outsourced to Latin America. *Mongabay Series: Planetary Boundaries*. https://news.mongabay.com/2024/03/critics-fear-catastrophic-energy-crisis-as-ai-is-outsourced-to-latinamerica/

McGuirk, P. (2003). Producing the capacity to govern in global Sydney: A multi-scaled account. *Journal of Urban Affairs, 25*(2), 201–223.

McLaren, D., & Agyeman, J. (2015). *Sharing cities: A case for truly smart and sustainable cities*. MIT Press.

McNara, M. (2012). The great Olympic greenwash. *Aljazeera*. June 20. Retrieved June 15, 2015, from http://www.aljazeera.com/programmes/peopleandpower/2012/06/201262075757520106.h

McNeill, D. (2009). *The global architect: Firms, fame and urban form*. Routledge.

Meadows, D. H., Randers, J., & Meadows, D. L. (2004a). *Limits to growth. The 30-year update*. Chelsea Green Publishing Company.

Meadows, D. H., Randers, J., & Meadows, D. L. (2004b). *Limits to growth: The 30-year update*. Chelsea Green Publishing.

Mehaffy, M. W., & Salingaros, N. A. (2015). *Design for a living planet: Settlement, science and a human future*. Sustasis Foundation.

Merton, R. K. (1936) The unanticipated consequences of purposive social action, American sociological review, Vol. 1, No. 6, December, pp. 894–904, https://doi.org/10.2307/2084615

Meyer, W. B. (2013) Resource consumption in *the environmental advantages of cities: Countering commonsense Antiurbanism*, pp. 37–56.

Mickle, T. (2025). TSMC, the Chip Giant, is to spend $100 billion in U.S. over the next 4 years. *The New York Times*. https://www.nytimes.com/2025/03/03/technology/tsmc-investment-trump.html

Miller, A. (2013). The afghan ring road megaproject. In G. Del Cerro Santamaría (Ed.), *Urban megapeojects. A worldwide view*. Emerald Group Publishing Limited.

Miller, J. (2016). *A crude look at the whole. The science of complex Systems in Business, life and society*. Basic Books.

Mingione, E. (1991). *Fragmented societies. A sociology of economic life beyond the market paradigm*. Blackwell.

Mitchell, W. (2007). Intelligent cities. *e-Journal on the Knowledge Society*. Inaugural Lecture of the UOC's 2007-2008 Academic Year. https://www.uoc.edu/uocpapers/5/dt/eng/mitchell.html. Accessed November, 2021

Moisio, S. (2018). *Geopolitics of the knowledge-based economy*. Routledge.

Mollenkopf, J., & Castells, M. (1991). *Dual City. Restructuring New York*. Russell Sage.

Monserrate, Steven Gonzalez (2022) "The Cloud Is Material: On the Environmental Impacts of Computation and Data Storage." *MIT Case Studies in Social and Ethical Responsibilities of Computing*, no. Winter 2022 (January). https://doi.org/https://doi.org/10.21428/2c646de5.031d4553.

Montella, N. (2025) ITIF virtual webinar, scale matters: Understanding the economics of global semiconductor innovation, January 30th https://itif.org/events/2025/01/30/understanding-economics-of-global-semiconductor-innovation/.

Montuori, A. (2022). Integrative Transdisciplinarity: Explorations and experiments in creative scholarship. *Transdisciplinary Journal of Engineering and Science, 13*. Special Issue on Complex Resilience and Sustainability, Edited by Gerardo del Cerro Santamaría. https://doi.org/10.22545/2022/00209

Moravec, H. (2009). Rise of the robots—The future of artificial intelligence. *Scientific American*. https://www.scientificamerican.com/article/rise-of-the-robots/

Morin, E. (1984). *Ciencia con conciencia*. Anthropos.

Morin, E. (1999). *Seven complex lessons in education for the future*. UNESCO.

Morin, E. (2002). *El Método II: La Vida de la Vida*. Cátedra.

Morin, E. (2005). *On complexity*. Hampton Press.

Morozov, E. (2013). *To save everything, click here:Technology, solutionism, and the urge to fix problems that Don't exist*. Allen Lane.

Mortenson. (2024). Data Center Megaprojects Fuel Next-Level Construction Careers, https://www.mortenson.com/newsroom/careers-data-center-construction.

Morton, T. (2007). *Ecology without nature. Rethinking environmental aesthetics* (p. 46). Harvard University Press.

Mostafavi, M., & Doherty, G. (Eds.). (2010). *Ecological urbanism*. Lars.

Moulaert, F., Rodriguez, A., & Swyngedouw, E. (2003). *The Globalized City: Economic restructuring and social polarization in European cities*. Oxford University Press.

Müller, M. (2015). Assemblages and actor-networks: Rethinking socio-material power. *Politics and Space, Geography Compass, 9*(1), 27–41.

Muro, M., R. Maxim, J. Whiton (2019) Automation and artificial intelligence. How machines are affecting people and places, brooking institution report, January, https://www.brookings.edu/wp- content/uploads/2019/01/2019.01_BrookingsMetro_Automation-AI_Report_Muro-Maxim-Whiton-FINAL-version.pdf. Accessed November, 2021.

Naamani, M. (2022), Gulf countries race for foreign capital, in orient XXI magazine, https://orientxxi.info/magazine/gulf-countries-race-for-foreign-capital,5297.

Nam, T., & Pardo, T. A. (2018). Conceptualizing Smart City with Dimensions of Technology, People, and Institutions. Center for Technology in Government University at Albany, State University of New York, U.S. *The Proceedings of the 12th Annual International Conference on Digital Government Research*, https://inta-aivn.org/images/cc/Urbanism/background%20documents/dgo_2011_smartcity.pdf. Accessed November, 2020.

National Academies of Sciences, Engineering, and Medicine. (2019). *Fostering the culture of convergence in research: Proceedings of a workshop*. National Academies Press. https://doi.org/10.17226/25271

National Research Council. (2014). *Convergence: Facilitating transdisciplinary integration of life sciences, physical sciences, engineering, and beyond*. National Academies Press. https://doi.org/10.17226/18722

Ness, D. (2018a). Sustainable urban infrastructure in China: Towards a factor 10 improvement in resource productivity through integrated infrastructure systems. *The International Journal of Sustainable Development and World Ecology, 15*(4).

Ness, D. (2018b). Sustainable urban infrastructure in China: Towards a factor 10 improvement in resource productivity through integrated infrastructure systems. *The International Journal of Sustainable Development and World Ecology, 15*(4), 288–301.

Newhouse, V. (1998). *Toward a new museum*. Monacelli Press.

Nicolescu, B. (2013). *From modernity to cosmodernity. Science, culture and spirituality*. SUNY Press.

Nisbett, R. E., Peng, K., Choi, I., & Norenzayan, A. (2001). Culture and systems of thought: Holistic versus analytic cognition. *Psychological Review, 108*(2), 291.

Nisenbaum, A. (2019) "What's Holding Smart Cities Back?", *Scientific American Blog Network*. Archived from the original on 29 June 2019. Accessed June, 2020.

Ogwang, T., & Vanclay, F. (2021). Resource financed infrastructure: Thoughts on four Chinese-financed projects in Uganda. *Sustainability, 13*(6), 3259. https://doi.org/10.3390/su13063259

Olds, K. (2002). *Globalization and urban change: Capital, culture and Pacific Rim megaprojects*. Oxford University Press.

Ølgaard, D. (2020). Reflections on Naomi Klein's pandemic shock therapy. *E-International Relations*. https://www.e-ir.info/2020/09/15/reflections-on-naomi-kleins-pandemic-shock-doctrine/. Accessed November 2021

Olsen, D. J. (1986). *The city as a work of art. London, Paris, Vienna* (p. 21). Yale University Press.

OMEGA Center. (2012). *Mega-projects: Lessons for decision-makers. An analysis of selected international large-scale transport infrastructure projects*. Bartlett School of Planning, University College London.

Open AI. (2025). *Announcing the stargate project*, January 21st, https://openai.com/index/announcing-the-stargate-project/.

Owen, K. A. (2001). *The local impacts of the Sydney 2000 Olympic games: Processes and politics of venue preparation*. Centre for Olympic Studies.

Paquot, T. (2013), cited in Ecodistricts: a sustainable utopia? Paris Innovation Review, April, http://parisinnovationreview.com/articles-en/ecodistrict-a-sustainable-utopia.edn. Accessed March, 2020.

Pendall, R., Foster, K., & Cowell, M. (2007). *Resilience and regions: Building understanding of the metaphor*. Working paper of The Institute of Urban and Regional Development.

Peris-Ortiz, M., Bennett, D. R., & Pérez-Bustamante Yábar, D. (Eds.). (2016). *Sustainable smart cities: Creating spaces for technological, social and business development*. Springer.

Perry, B. (2002). The future of urban sociology: Report of joint sessions of the British and American sociological associations. *International Journal of Urban and Regional Research, 26*(4), 844–853.

Pham, K. (2017). *Examining the development of public space in foreshore Sydney: Barangaroo case study*. Master of Research Thesis, Faculty of Design, Architecture and Building, UTS.

Physicians for Social Responsibility (2015) Coal ash: Hazardous to human health. *Physicians for social responsibility*. Retrieved July 6, 2015, from http://www.psr.org/assets/pdfs/coal-ash-hazardous-to-human-health.pdf.

Pitsis, A., et al. (2017). Megaprojects redefined—Complexity vs. cost and social imperatives. *International Journal of Managing Projects in Business, 11*(1), 7–34.

Planete Energies (2016) Shanghai, the Showcase for China's New Environmental Policy, https://www.planete-energies.com/en/medias/close/shanghai-showcase-china-s-new-environmental-policy.

Plaza, B., & Haarich, S. N. (2009). Museums for urban regeneration? Exploring conditions for their effectiveness. *Journal of Urban Regeneration and Renewal, 2*(3), 259–271.

Pogrebin, R. (2007). British architect wins 2007 Pritzker prize. *The New York Times*. March 28th. Retrieved May 2018.

Polanyi, K. (1944 [2001]) The great transformation. The political and economic origins of our time. Beacon Press.

Polanyi, K. (1968). The economy as instituted process. In H. Schneider (Ed.), *Economic anthropology E LeClair*. Holt, Rinehart and Winston.

Polman, P. (2019) A business model for sustainability, world economic forum, https://www.weforum.org/agenda/2019/01/a-business-model-for-sustainability/.

Popa, F., et al. (2015). A pragmatist approach to transdisciplinarity in sustainability research. From complex systems theory to reflexive science. *Futures, 65*, 45–56.

Porter, M. (1998). *The competitive advantage of nations*. Palgrave.

Portugali, J. (1999). *Self-organization and the City*. Springer.

PRCMEE. (2016). China's National Plan on Implementation of the 2030 Agenda for Sustainable Development, Ministry of Ecology and Environment of the People's Republic of China, http://english.mee.gov.cn/Resources/Plans/Plans/201803/P020180316511712813464.pdf.

Priemus, H., & van Wee, B. (2014). *International handbook on mega-projects*. Edward Elgar.

Prigogine, I. (1997). *The end of certainty: Time, chaos and the new laws of nature*. The Free Press.

Prigogine, I. and E. Stengers (2018 [1978]) *Order Out of Chaos: Man's New Dialogue With Nature*, : Verso.

Prior, J., & Blessi, G. T. (2012). Social capital, local communities and culture- led urban regeneration processes: The Sydney Olympic Park experience. *Cosmopolitan Civil Societies: An Interdisciplinary Journal, 3*(4), 78–96. Retrieved on 26 April 2015.

Pyke, T. (2017) The energy debate: Renewable energy cannot replace fossil fuels, development education, https://developmenteducation.ie/feature/the-energy-debate-renewable-energy-cannot-replace-fossil-fuels/.

Raco, M. and K. Gilliam, (2012) Geographies of Abstraction, Urban Entrepreneurialism, and the Production of New Cultural Spaces: The West Kowloon Cultural District, Hong Kong, Environment and Planning A June 2012 vol. 44 no. 6, 1425–1442.

Randolph, G. F. and M. Storper (2022) The great theory debate on urbanization in the global south—And why it matters, Urban Studies DOI: https://doi.org/10.1177/00420980211067926 https://www.urbanstudiesonline.com/resources/resource/the-great-theory-debate-on-urbanisation-in-the-global-southand-why-it-matters/.

Ransbotham, S., Kiron, D., Gerbert, P., & Reeves, M. (2017). Reshaping business with artificial intelligence: Closing the gap between ambition and action. *MIT Sloan Management Review.*

Rees, W. (2012). Cities as dissipative structures: Global change and the vulnerability of urban civilization. In M. P. Weinstein & R. E. Turner (Eds.), *Sustainability science: The emerging paradigm and the urban environment.* Springer.

Rees, W., & Wackernagel, M. (1996). Urban ecological footprints: Why cities cannot be sustainable - and why they are a key to sustainability. *Environmental Impact Assessment Review, 16*(4–6), 223–248. https://doi.org/10.1016/S0195-9255(96)00022-4

Reichert, S. (2019) The role of universities in regional innovation Ecosystems, European University Association paper, https://www.eua.eu/downloads/publications/eua%20innovation%20ecosystem%20re port_final_digital.pdf. Accessed November, 2021.

Reinmuth, G. (2012) Barangaroo: the loss of trust?, The Conversation, November 20th, https://theconversation.com/barangaroo-the-loss-of-trust- 10676.

Ren, X. (2017). Biggest infrastructure bubble ever? City and nation building with debt-financed megaprojects in China. In B. Flyvbjerg (Ed.), *The Oxford handbook of megaproject management.* Oxford University Press.

Rifkin, J., & Howard, T. (1980). *Entropy. A New World view.* Viking Press.

Rizzo, A. (2016). Sustainable urban development and green megaprojects in the Arab states of the Gulf region: Limitations, covert aims, and unintended outcomes in Doha, Qatar. *International Planning Studies.* https://doi.org/10.1080/13563475.2016.1182896

Robinson, J. (2022). *Comparative urbanism. Tactics for global urban studies.* Wiley.

Rockström, J., et al. (2023). Safe and just earth system boundaries, nature, May 31st. https://doi.org/10.1038/s41586-023-06083-8

Rodríguez-Pose, A., & Bartalucci, F. (2023). The green transition and its potential territorial discontents. *Cambridge Journal of Regions, Economy and Society, rsad039.* https://doi.org/10.1093/cjres/rsad039

Rodrik, D., Naidu, S., & Zucman, G. (2019). Economics after neoliberalism. *Boston Review Forum.* https://www.bostonreview.net/forum/suresh-naidu-dani-rodrik-gabriel-zucman-economics-after-neoliberalism/

Rokem, J., & Boano, C. (2020). *Urban geopolitics. Rethinking planning in contested cities.* Routledge.

Rolfe, M. (2012) Barangaroo: politics, property and players—It's business as usual. *The Conversation*, November 20th, https://theconversation.com/barangaroo-politics-property-and-players-its-business-as-usual-10722.

Rose, S. (2008) Paris triangle: Beyond the shadows of doubt. The Guardian, October. Retrieved July 5, 2015, from http://www.theguardian.com/artanddesign/2008/oct/07/triangle.herzog.demeuron.paris.

Ross, C. (2014). Office rents soaring in city's Innovation District. *The Boston Globe*, January 10th, https://www.bostonglobe.com/business/2014/01/10/rents-soaring-city-innovation-district/nqeKNcRiLJiyjKEEGog8GP/story.html.

Ross, C., & Amekudzi, A. (2009). *Megaregions. Planning for global competitiveness.* Island Press.

Roy, A. (2015). What is urban about critical urban theory? *Urban Geography.* https://doi.org/10.1080/02723638.2015.1105485

Rudin, C., J. Radin (2019a) Why are we using black box models in AI when we Don't need to? A lesson from an explainable AI competition, HDSR 1.2, Fall, https://hdsr.mitpress.mit.edu/pub/f9kuryi8/release/6. Accessed November, 2021.

Rudin, C., J. Radin (2019b) Why are we using black box models in AI when we Don't need to? A lesson from an explainable AI competition, HDSR 1.2, Fall, https://hdsr.mitpress.mit.edu/pub/f9kuryi8/release/6. Accessed November, 2021.

Rus, K. K., Vojko, K., & David, K. (2018). Resilience assessment of complex urban systems to natural disasters: A new literature review. *International Journal of Disaster Risk Reduction, 31*, 311–330. https://doi.org/10.1016/j.ijdrr.2018.05.015

Ryan, M., & Gregory, A. (2019). Ethics of using smart city AI and big data: The case of four large European cities. *ORBIT Journal, 2*(2). https://doi.org/10.29297/orbit.v2i2.110

Rybczynski, W. (2002). The Bilbao effect. *Atlantic Monthly, 290*(2), 138–142.

Rybczynski, W. (2008). When buildings try too hard. *Wall Street Journal*. http://www.wsj.com/articles/SB122731149503149341.

Safeguard Global (2024) Top 10 manufacturing countries in 2023, https://www.safeguardglobal.com/resources/top-10-manufacturing-countries-in-the-world-2023/.

Sagalyn, L., & Frieden, B. (1989). *Downtown Inc. how America rebuilds cities*. The MIT Press.

Sagalyn, L. (2003). *Times Square roulette. Remaking the City icon*. The MIT Press.

Sagalyn, L. (2007). Public/private development. Lessons from history, research and practice. *Journal of the American Planning Association, 73*(1), 7–22.

Sanyal, B. (Ed.). (2005). *Comparative planning cultures*. Routledge.

Sarkheyli, A., & Sarkheyli, E. (2022). Dynamic intelligent risk management in megaprojects case study: Transportation megaprojects in Sweden. *Journal of Mega Infrastructure and Sustainable Development*. Special Issue on *Artificially Intelligent Megaprojects, Benefits, Risks and Sustainable Development*, in Journal of Mega Infrastructure and Sustainable Development, forthcoming.

Sassen, S. (2014). *Expulsions. Brutality and complexity in the global economy*. Harvard University Press.

Sassen, S. (2018). *Cities in a world economy* (5th ed.). SAGE.

Saxenian, A. L. (1994). *Regional advantage: Culture and competition in Silicon Valley and route 128*. Harvard University Press.

Schaffers, H., Komninos, N., Pallot, M., Trousse, B., & Nilsson, M. (2011). Smart cities and the future internet: Toward cooperation frameworks for open innovation, lecture notes in computer science. *The Future Internet., 6656*, 431–446.

Scheffer, M. (2009). *Critical transitions in nature and society*. Princeton University Press.

Schelmetic, T. (2012) Green facts about New York's New One World Trade Center. *Industry News*, January 3. Retrieved July 6, 2015, from http://news.thomasnet.com/imt/2012/01/03/green-facts-about-new-yorks-new-one-world-trade-center.

Schiavone, F., Appio, F. P., Mora, L., & Risitano, M. (2020). The strategic, organizational and entrepreneurial evolution of smart cities. *International Entrepreneurship and Management Journal, 16*, 1155–1165.

Schindler, S., Fadaee, S., & Brockington, D. (2021). *Contemporary megaprojects: Organization, vision and resistance in the 21st century*. Berghahn Books.

Schrödinger, E. (1967). *What is life? Mind and matter*. Cambridge University Press.

Schwab, K. (2018). *Shaping the future of the 4th industrial revolution*. Penguin.

Schwärzel, K. (2017). China's fight against desertification should not be done at the cost of water security. *The Conversation*. https://theconversation.com/chinas-fight-against-desertification-should-not-be-done-at-the-cost-of-water-security-83678

Scott, W. R., Levitt, R. E., & Orr, R. J. (2011). *Global projects. Institutional and political challenges*. Cambridge University Press.

Sekhniashvili, A. (2017) Authoritarian infrastructure complex: The Turkish Tale *Center for Strategic and International Studies*, December 19th. https://www.csis.org/npfp/authoritarian-infrastructure-complex-turkish-tale.

Sengupta, R. (2020). Entropy law, social sustainability and development policy, ch. 1. In *Entropy law, sustainability and third industrial revolution*. Oxford University Press.

Sennett, R. (1992). *The conscience of the eye. The design and social life of cities*. W. W. Norton.

Sennett, R. (2012). No one likes a city that's too smart. *The Guardian*. Archived from the original on 18 March 2017. Accessed 17 March 2017.

Sennett, R. (2018). *Building and dwelling: Ethics for the city*. Farrar, Straus and Giroux.

Serebrisky, T., et al. (2015). *Financing infrastructure in Latin America and the Caribbean how, how much and by whom*. Inter-American Development Bank.

Seto, K. C., & Satterthwaite, D. (2010). Interactions between urbanization and global environmental change. *Current Opinion in Environmental Sustainability, 2*, 127–128.

Seto, K. C., Güneralp, B., & Hutyra, L. R. (2012). Global forecasts of urban expansion to 2030 and direct impacts on biodiversity and carbon pools. *Proceedings of the National Academy of Sciences of the United States of America., 109*, 16083.

Shadid, A. (2012) An ambitious Arab capital reaffirms its grand cultural vision. *The New York Times*, January 24. Retrieved May 27, 2015, from https://www.nytimes.com/2012/01/25/world/middleeast/abu-dhabi-reaffirms-its-grand-plan-for-museums.html.

Shane, D. G. (2005). *Recombinant urbanism—Conceptual modeling in architecture, Urban Design and City theory*. John Wiley & Sons Ltd.

Siemiatycki, M. (2013). Riding the wave: Explaining cycles in urban megaproject development. *Journal of Economic Policy Reform, 16*(2), 160–178. https://doi.org/10.1080/17487870.2013.797904

Silk Road Briefing. (2021). *Turkish Government Approves Canal Istanbul Project*, May 4th. https://www.silkroadbriefing.com/news/2021/05/04/turkish-government-approves-canal-istanbul-project/.

Siu, H., & Ku, A. (Eds.). (2008). *Hong Kong Mobile: Making a global population*. Hong Kong University Press.

Sjöblom, S., & Godenhjelm, S. (2009). Project proliferation and governance—Implications for environmental management. *Journal of Environmental Policy and Planning, 11*(3), 169–185.

Skinner, L. (2020). Artificial Intelligence for Megaprojects—Association for Project Management. https://hub.apm.org.uk/news/295212. Accessed November, 2021.

Sklair, L. (2017a). Sleepwalking through the Anthropocene. *The British Journal of Sociology, 68*(4), 775–784.

Sklair, L. (2017b). *The icon project. Architecture, cities and capitalist globalization*. Oxford University Press.

Skluzacek, C. R. (2010). Universality and its discontents: The Louvre and Guggenheim Abu Dhabi as a case study in the future of museums. *Art and Art History Honors Projects, 1*. https://digitalcommons.macalester.edu/art_honors/1

Smith, B. (2018). China: Environmental issues, policies and clean technology. *AzoCleanTech*, July 24th, p. 32, https://www.azocleantech.com/article.aspx?ArticleID=546.

Smith, M. P. (2005). Transnational urbanism revisited. *Journal of Ethnic and Migration Studies, 31*(2), 235–244. https://doi.org/10.1080/1369183042000339909

Smrckova, E. (2014). Basic characteristics of green cement of CEM V/a and V/B kind. *Advanced Materials Research, 894*, 196–199.

Smyth, H. (2015). *Marketing the City. The role of flagship developments in urban regeneration*. Routledge.

Snyder, G. (2008). Cited in Dana Goodyear, Zen master. Gary Snyder and the art of live. *The New Yorker*. https://www.newyorker.com/magazine/2008/10/20/zen-master

Snyder, P. (2016). *Museums after Bilbao: Neoliberal public space at the Denver art Museum*. Wesleyan University.

Söderlund, J., Sankaran, S., & Biesenthal, C. (2017). The past and present of megaprojects. *Project Management Journal, 48*(6), 5–16.

Sonmez, M. (2020) Turkey's megaprojects threaten 'mega' financial havoc, *Al- monitor*, June 1st, https://www.al-monitor.com/originals/2020/06/Turkey-megaprojects-airport-bridge-threaten-financial-havoc.Html.

Sorensen, A. (2003). Building world city Tokyo: Globalization and conflict over urban space. *Annals of Regional Science, 37*, 519–531.

Sorkin, M. (1992). *Variations on a Theme Park. The new American City and the end of public space*. Hill and Wang.

Sorkin, M. (2017). Afterword: Architecture without capitalism. In P. Deamer (Ed.), *Architecture without capitalism* (pp. 217–220). Routledge.

Southerland, D. (2017) China's global infrastructure initiative raises environmental concerns, Radio Free Asia, , https://www.rfa.org/english/commentaries/china-obor-09292017160823.html.

Spiegel Online International. (2008). Krens' museum for global contemporary art: Guggenheim Abu Dhabi will be 'pharaonic'. *Spiegel Online*, March 27. Retrieved May 29, 2015, from http://www.spiegel.de/international/world/krens-museum-for-global-contemporary-art-guggenheim-abu-dhabi-will-be-pharaonic-a-543601.html.

Stephens, S. (2007). Studio Daniel Libeskind and the Davis Partnership shake up downtown with a new addition to the Denver art Museum. *Architectural Record, 195*(1).

Stevens, G. (2002). *The favored circle: The social foundations of architectural distinction*. The MIT Press.

Stickells, Lee (2010) Instant urbanism - just add water. Architecture Australia, May–June, 2010, Vol.99(3), p.47(2).

Stiegler, B. (2019). *The age of disruption: Technology and madness in computational capitalism*. Polity.

Stiftel, B. (2021). Planners and the new urban agenda: Will we lead the agenda, or will the agenda lead us? *Town Planning Review, 92*(4), 421–442.

Storper, M. (1997). *The regional world. Territorial development in a global economy*. The Guilford Press.

Sturup, S., & Low, N. (2019). Sustainable development and mega infrastructure: An overview of the issues. *Journal of Mega Infrastructure and Sustainable Development, 1*(1), 8–26.

Taelman, E. (2009). Saadiyat Island tourist development project: Dredging and reclamation works in an ecologically sensitive area. *Terra et Aqua, 116*, 3–11.

Tafuri, M. (1979) Architecture and Utopia: *Design and Capitalist Development*, Cambridge, MA: The MIT Press.

Taleb, N. N. (2014). *Antifragile: Things that gain from disorder*. Random House.

Taylor, P. J. (2010). *Unruly Complexity. Ecology, Interpretation, Engagement*. University of Chicago Press.

Tegmark, M. (2024) Manhattan project for AGI, future of life institute, https://futureoflife.org/statement/agi-manhattan-project-max-tegmark/.

The Guardian. (2014). Helsinki v Guggenheim: the backlash against the global megabrand is on. *The Guardian*. September 11. Retrieved May 2018, https://www.theguardian.com/artanddesign/architecture-design-blog/2014/sep/11/helsinki-guggenheim-outpost-battle.

The Guardian (2015) Migrants building UAE cultural hub working in prison conditions, April 4,. Retrieved May 2018.

Thees, H. (2020). Towards local sustainability of mega infrastructure: Reviewing research on the new silk road. *Sustainability, 12*, 10612. https://doi.org/10.3390/su122410612

Thurner, S., Corominas-Murtra, B., & Hanel, R. (2017). Three faces of entropy for complex systems: Information, thermodynamics, and the maximum entropy principle. *Physical Review E, 96*(3). https://doi.org/10.1103/PhysRevE.96.032124

TNH, The Next Helsinki http://www.nexthelsinki.org.

Tollefson, J., & Weiss, K. R. (2015). Nations approve historic global climate accord. *Nature, 528*(7582), 315–316. https://doi.org/10.1038/528315a

Tomer, A. (2019) Artificial intelligence in America's digital city, Brookings Institution report, https://www.brookings.edu/research/artificial-intelligence-in-americas-digital-city/, Accessed September, 2021.

Tomlinson, R. (2016) Metropolitan governance is the missing link in Australia's reform agenda, the conversation, April 4th, https://theconversation.com/metropolitan-governance-is-the-missing-link-in-australias-reform-agenda-55872.

Tréguer, F. (2021) The virus of surveillance: How the COVID-19 pandemic is fuelling technologies of control. Political anthropological research on international social sciences (PARISS), Brill, 2 (1), pp.16-46. Halshs-03141309v3.

Tsing, A. L. (2015). *The mushroom at the end of the world: On the possibility of life in capitalist ruins*. Princeton University Press.

Tuan, Y. F. (1979). *Landscapes of fear*. University of Minnesota Press.

Tuan, Y. F. (1990). *Topophilia. A study of environmental perception, attitudes and values*. Columbia University Press.

Tuan, Y. F. (1996). *Cosmos and hearth: A cosmopolite's viewpoint*. University of Minnesota Press.

UNCTAD (2021) Catching technological waves. Innovation with equity, UNCTAD 2021 innovation report, https://unctad.org/system/files/official-document/tir2020_en.pdf.

United Nations. (1987). *Bruntland report, chapter two.* United Nations World Commission on Environment and Development. https://en.wikisource.org/wiki/Brundtland_Report/Chapter_2._Towards_SustainabLe_Development

United Nations. (2016). Executive summary of China's actions on the implementation of the 2030 Agenda for Sustainable Development. *Voluntary National Review.* https://sustainabledevelopment.un.org/memberstates/china

United Nations (2022) Climate change conference, COP 27, https://www.un.org/en/climatechange/cop27.

Urry, J. (2002). *Global complexity.* Polity.

Valdivia, A. (2023). Rural Spain could end up hosting infrastructure hubs for AI—Here's what the environmental cost could be. *The Conversation.* https://theconversation.com/rural-spain-could-end-up-hosting-infrastructure-hubs-for-ai-heres-what-the-environmental-cost-could-be-205504

van der Ploeg, L., & Vanclay, F. (2017). A human rights based approach to project induced displacement and resettlement. *Impact Assessment & Project Appraisal, 35*(1), 34–52. https://doi.org/10.1080/14615517.2016.1271538

Van Dinteren, J. (2018) The university as a catalyst in innovation district development, Conference: IASP World Conference on Science Parks and Areas of Innovation, Iran, https://www.researchgate.net/publication/327395252_The_university_as_a_catalyst_in_innovation_district_development. Accessed March 2021.

Van Meerbeeck, K., Jucker, T., & Svenning, J.-C. (2021). Unifying the concepts of stability and resilience in ecology. *Journal of Ecology, 109*(9), 3114–3132.

Vanclay, F. (2002). Conceptualising social impacts. *Environmental Impact Assessment Review, 22*(3), 183–211. https://doi.org/10.1016/S0195-9255(01)00105-6

Vanclay, F. (2017a). Project induced displacement and resettlement: From impoverishment risks to an opportunity for development? *Impact Assessment & Project Appraisal, 35*(1), 3–21. https://doi.org/10.1080/14615517.2017.1278671

Vanclay, F. (2017b). The potential contribution of social impact assessment to megaproject developments. In M. Lehtonen, P. Joly, & L. Aparicio (Eds.), *Socioeconomic evaluation of megaprojects: Dealing with uncertainties* (pp. 181–198). Routledge. https://doi.org/10.4324/9781315622125-18

Vanclay, F., & Esteves, A. M. (Eds.). (2024). *Handbook of social impact assessment and management.* Edward Elgar. https://doi.org/10.4337/9781802208870

Vermesan, O., Bahr, R., Ottella, M., Serrano, M., Karlsen, T., Wahlstrøm, T., Sand, H., Ashwathnarayan, M., M. Gamba (2020). Internet of Robotics Things. Intelligent Connectivity and Platforms, Frontiers in Robotics and AI 25, September, https://www.frontiersin.org/articles/10.3389/frobt.2020.00104/full. Accessed November, 2021.

Vincent, J. R., & Walker, B. (2013). Social-ecological systems as complex adaptive systems: Modelling and policy implications. *Environment and Development Economics, 18*(2), 111–132.

Virilio, P. (2002). The Overexposed City. In *The Blackwell City reader*. Wiley Blackwell.

Vogel, C. (2014). A new art capital, finding its own voice. Inside Frank Gehry's Guggenheim Abu Dhabi. *The New York Times.* http://www.nytimes.com/2014/12/07/arts/design/inside-frank-gehrys-guggenheim-abu-dhabi.html?_r=0

Vojnovic, I. (2014). Urban sustainability: Research, politics, policy and practice. *Cities, 41*, 530–544. https://doi.org/10.1016/j.cities.2014.06.002

Volner, I. (2015) Can the Guggenheim charm Finland? The New Yorker, May 12. Retrieved May 28, 2018, from http://www.newyorker.com/culture/culture-desk/can-the-guggenheim-charm-finland.

Vygotsky, L. S. (1962). *Thought and language.* MIT Press. (Original work published 1934). https://doi.org/10.1037/11193-000

Vygotsky, L. S. (1978). *Mind in society: The development of higher psychological processes.* Harvard University Press.

Wagner, J. (2019) Innovation districts and their dilemmas with place, https://www.brookings.edu/blog/the-avenue/2019/02/21/innovation-districts-and- their-dilemmas-with-place/. Accessed February, 2020.

Wainwright, O. (2014). Helsinki v Guggenheim: The backlash against the global megabrand is on. *The Guardian*, September 11. Retrieved May 29, 2018, from http://www.theguardian.com/artanddesign/architecture-design-blog/2014/sep/11/helsinki-guggenheim-outpost-battle.

Walker, A. (2020). Sidewalk lab's 'Smart City' was destined to fail. *Curbed*. https://archive.curbed.com/2020/5/7/21250678/sidewalk-labs-toronto-smart-city-fail

Wallerstein, I. (2004). *The uncertainties of knowledge*. Temple University Press.

Wallerstein, I. (2013). The itinerary of world-systems analysis; or, how to resist becoming a theory, chapter four in Wallerstein, I. In C. Lemert & C. A. Rojas (Eds.), *Uncertain worlds. World-systems analysis in changing times*. Paradigm Publishers.

Waltner-Toews, D., et al. (2008). *The ecosystem approach. Complexity, uncertainty and managing for sustainability*. Columbia University Press.

Wang, X., Sun, X., Liu, S., Mu, C. (2021) A preliminary exploration of factors affecting a university entrepreneurial ecosystem, Frontiers in Psychology 30 September, https://www.frontiersin.org/articles/10.3389/fpsyg.2021.732388/full. Accessed August, 2021.

Waters, M. (2022). Energy-hungry data centers are quietly moving into cities. *MIT Technology Review*. https://www.technologyreview.com/2022/06/22/1053889/city-server-farms-energy/

Watkins, D., Rebecca Lai, K. K., & Bradsher, K. (2018). The world, built by China. *The New York Times*, November 18th, https://www.nytimes.com/interactive/2018/11/18/world/asia/world-built-by-china.html.

Watson, V. (2013). African urban fantasies: Dreams or nightmares? *Environment and Urbanization, 26*(1), 215–231.

WBCSD (2021) Net-zero buildings: Where do we stand? https://www.wbcsd.org/Programs/Cities-and-Mobility/Sustainable-Cities/Transforming-the-Built-Environment/Decarbonization/Resources/Net-zero-buildings-Where-do-we-stand.

Webb, W. (2001). Quoted in Wilson Lloyd, A. Architecture for art's sake. *Atlantic Monthly, 287*, 85–88.

Weber, R. (2015). *From boom to bubble: How finance built the new Chicago*. University of Chicago Press.

WEF (2018) Harnessing artificial intelligence for the earth. Report. https://www3.weforum.org/docs/Harnessing_Artificial_Intelligence_for_the_Earth_report_2018.pdf. Accessed November, 2021.

Weichselgartner, J., & Kelman, I. (2015). Geographies of resilience: Challenges and opportunities of a descriptive concept. *Progress in Human Geography, 39*(3), 2–4. https://doi.org/10.1177/0309132513518834

Wells, M. (2010). London 2012: A rubbish Olympics… *Games Monitor*, September 26. Retrieved June 19, 2015, from http://www.gamesmonitor.org.uk/node/1108.

West, J. (2014). The freedom tower was supposed to be the greenest building in America. So what went wrong? *Mother Jones*, December 8. Retrieved July 3, 2015, from http://www.motherjones.com/environment/2014/12/freedom-tower-world-trade-center-green-sustainable-plan-sandy.

Westerman H., & Creagh, S. (2012). Barangaroo: Development interests counter the public interest. *The Conversation*, November 22nd, https://theconversation.com/barangaroo-development-interests-counter-the-public-interest-10837.

While, A., Jonas, A. E. G., & Gibbs, D. (2004). The environment and the entrepreneurial city: Searching for the urban ' sustainability fix' in Manchester and Leeds. *International Journal of Urban and Regional Research, 28*(3), 549–556.

Whitehead, A. N. (1920). *The concept of nature*. Cambridge University Press. cited in Latour, B. (2016) *Facing Gaia. Eight Lectures on the New Climatic Regime*, London: Polity, p. 85.

Wilks, A. (2021). For Erdoğan's Istanbul Canal project, critics see few winners. *Al Jazeera*. https://www.aljazeera.com/economy/2021/4/16/for-erodgans-istanbul-canal-project-critics-see-few-winners

Willsher, K. (2014) France pledges €200m to restore La Défense's crumbling cube. *The Guardian*, August. Retrieved July 5, 2015, from http://www.theguardian.com/world/2014/aug/04/france-pledges-200m-euroes-restore-la-defense-la-grande-arche.

Wilson Lloyd, A. (2001). Architecture for art's sake. *Atlantic Monthly, 287*, 85–88.

Winch. (2017). Megaproject stakeholder management, chapter 15. In B. Flyvbjerg (Ed.), *The Oxford handbook of megaproject management*. Oxford University Press.

WKCD Authority. (2013). *WKCD. Environmental impact assessment*. West Kowloon Cultural District Authority. https://ema.westkowloon.hk/eiareport/Exec%20Summary%20RevC_html.htm

WKCDA, 2013, Annual report, http://www.westkowloon.hk/en/the-authority/annual-report-20122013,. Retrieved June 2015.

WKCDA. (2017). *Annual report 2016/17*. West Kowloon Cultural District Authority. https://ema.westkowloon.hk/eiareport/Exec%20Summary%20RevC_html.htm

Wong, W. S. (2004). *Design identity of Hong Kong: Colonization, de-colonization, and re- colonization. In European academy of design*. University of the Arts.

Wood, A. (2004). Domesticating urban theory? US concepts, British cities and the limits of cross-national applications. *Urban Studies, 41*(11). https://doi.org/10.1080/0042098042000268366

World Bank. (2018). *China—systematic country diagnostic: Towards a more inclusive and sustainable development (English)*. World Bank Group. http://documents.worldbank.org/curated/en/147231519162198351/China-Systematic-Country-Diagnostic-towards-a-more-inclusive-and-sustainable-development

World Bank (2022) Urban development overview, https://www.worldbank.org/en/topic/urbandevelopment/overview.

World Trade Center (2024), Building facts, https://www.wtc.com/about/buildings/1-world-trade-center.

Woronkowicz, J., et al. (2015). *Building better arts facilities: Lessons from a U.S. National Study. (Routledge research in creative and cultural industries management)*. Routledge.

Worth, R. F. (2021) The dark reality behind Saudi Arabia's Utopian dreams, *The New York Times*. January 28, https://www.nytimes.com/2021/01/28/magazine/saudi-arabia-neom-the-line.html. Accessed June 2021.

Wu, F. (2017). China's Emergent City-region governance. A new form of state-spatial selectivity through state-orchestrated rescaling. *International Journal of Urban and Regional Research, 40*(6), 1134–1151.

Xie, L., Han, T., & Skitmore, M. (2019a). Governance of relationship risks in megaprojects: A social network analysis. *Advances in Civil Engineering, 2019*. Article ID 1426139. https://doi.org/10.1155/2019/1426139

Xie, L., Han, T., & Skitmore, M. (2019b). Governance of relationship risks in megaprojects: A social network analysis. *Advances in Civil Engineering, 2019*, Article ID 1426139. https://doi.org/10.1155/2019/1426139

XinhuaNet (2018) China approves three demonstration zones on sustainable development, February. http://www.xinhuanet.com/english/2018- 02/24/c_136997243.htm.

Yadong, Y. (2013). Eco-efficiency trends in China, 1978-2010: Decoupling environmental pressure from economic growth. *Ecological Indicators, 24*(177–184), 181. https://doi.org/10.1016/j.ecolind.2012.06.007

Yiu, M. H. L. (2018) Everyday (open) cultural space—An approach to public space for West Kowloon Cultural District.

YLE. (2012). Minister says no culture budget for Guggenheim. *YLE*, December 1st. Retrieved May 28, 2018, from https://yle.fi/uutiset/osasto/news/minister_says_no_culture_budget_for_guggenheim.

Yovanof, G., G. N. Hazapis (2009) An Architectural framework and enabling wireless technologies for digital cities & intelligent urban environments Wireless Personal Communications. 49 (3): 445–463. doi:https://doi.org/10.1007/s11277-009-9693-4.. ISSN 0929-6212. S2CID 207258536.

Zhang, J., et al. (2015). The City brain: Practice of large-scale artificial intelligence in the real world. *IET Smart Cities, 1*(1), 28–37. Accessed 4 June 2021.

Zhang, J., Bates, J., & Abbott, P. (2022). State-steered smartmentality in Chinese smart urbanism. *Urban Studies, 59*(14), 1–15. https://doi.org/10.1177/00420980211062888

Zhang, M. (2024) Cities and regions with the highest concentration of data centers, https://dgtlinfra.com/united-states-data-centers/.

Zhang, X., & Li, H. (2018a). Urban resilience and urban sustainability: What we know and what do not know? *Cities, 72* (August 2017), 141–148. http://www.xinhuanet.com/english/2018-02/24/c_136997243.htm

Zhang, X., & Li, H. (2018b). Urban resilience and urban sustainability: What we know and what do not know? *Cities, 72*, 141–148.

Zhao, K., & Ge, L. (2013). A survey on the Internet of Things security. In *2013 Ninth International Conference on Computational Intelligence and Security* (pp. 663–667). https://doi.org/10.1109/CIS.2013.145

Ziabari, K., & Urpilainen, J. (2014). Finland's economy is in the middle of a "lost decade". *Fair Observer*. October 31. Retrieved May 29, 2018, from http://www.fairobserver.com/region/europe/finlands-economy-middle-lost-decade- 01757/

Zimmermann, J., & Eber, W. (2014). Consideration of risk in PPP-projects. *Business, Management and Education, 12*(1), 30–46. https://doi.org/10.3846/bme.2014.03

Zuboff, S. (2019). *The age of surveillance capitalism: The fight for a human future at the new frontier of power*. Public Affairs.

Zukin, S. (2008). Consuming authenticity. *Cultural Studies, 22*(5), 724–748.

Zukin, S., & Papadantonakis, M. (2018). Hackathons as cooptation ritual: Socializing workers and institutionalizing innovation in the 'new' economy. *Research in the Sociology of Work, 31*, 157–181.

Zukin, S. (2020a). *The innovation complex. Cities, tech and the new economy*. Oxford University Press.

Zukin, S. (2020b). Seeing like a city: How tech became urban. *Theory and Society, 49*(5–6), 941–964.

Zukin, S. (2021). Planetary Silicon Valley. Deconstructing New York's innovation complex, *Urban Studies, 58*(1), 3–35.

Index

G. del Cerro Santamaría, *Contesting Megaprojects*,
https://doi.org/10.1007/978-3-031-98382-5

E

H

O

T